AF358650

Advances in Ultra-Precision Machining Technology and Applications

Advances in Ultra-Precision Machining Technology and Applications

Editors

Benny C. F. Cheung
Jiang Guo

Basel • Beijing • Wuhan • Barcelona • Belgrade • Novi Sad • Cluj • Manchester

Editors

Benny C. F. Cheung
Department of Industrial and
Systems Engineering
The Hong Kong Polytechnic
University
Hong Kong
China

Jiang Guo
School of Mechanical
Engineering
Dalian University of
Technology
Dalian
China

Editorial Office
MDPI AG
Grosspeteranlage 5
4052 Basel, Switzerland

This is a reprint of articles from the Special Issue published online in the open access journal *Micromachines* (ISSN 2072-666X) (available at: www.mdpi.com/journal/micromachines/special_issues/Advances_Ultra_Precision_Machining_Technology_Applications).

For citation purposes, cite each article independently as indicated on the article page online and as indicated below:

Lastname, A.A.; Lastname, B.B. Article Title. *Journal Name* **Year**, *Volume Number*, Page Range.

ISBN 978-3-7258-1560-9 (Hbk)
ISBN 978-3-7258-1559-3 (PDF)
doi.org/10.3390/books978-3-7258-1559-3

Contents

About the Editors

Benny C. F. Cheung

Prof. Benny C. F. Cheung is the Chair Professor of Ultra-Precision Machining and Metrology at the Department of Industrial and Systems Engineering (ISE) of the Hong Kong Polytechnic University. Currently, he is also the Director of State Key Laboratory of Ultra-Precision Machining Technology. He obtained his a First-Class Bachelor of Engineering (Hons) Degree in Manufacturing Engineering from the Hong Kong Polytechnic (now The Hong Kong Polytechnic University) in 1993. After graduation, he joined ASM Assembly Automation Ltd. as a Teaching Company Associate and later as a Mechanical Engineer from 1993 to 1995. He obtained his Master of Philosophy (MPhil) degree from the Hong Kong Polytechnic University in 1996 and started his career in Precision Engineering as a Technology Development Officer in ISE from 1996 to 1998. In 2000, he obtained his Doctor of Philosophy (PhD) degree from The Hong Kong Polytechnic University and then he started his academic career as a Lecturer in ISE. He was then promoted to Assistant Professorship in 2001, Associate Professorship in 2004, and Professorship in 2012. Prof. Cheung is a Chartered Engineer (CEng), a Fellow of the International Academy of Engineering and Technology (AET Fellow), a senior member of Chinese Mechanical Engineering Society (CMES), a member of the Hong Kong Institution of Engineers (MHKIE), a member of the Institution of Engineering and Technology (MIET), a member of the Institute of Electrical and Electronics Engineers (MIEEE), and a member of the American Society of Mechanical Engineers (MASME). He is also an associate member of the International Academy for Production Engineering (CIRP).

Jiang Guo

Jiang Guo is a professor at the Dalian University of Technology, a doctor of engineering at the University of Tokyo, a recipient of the National Program for the Introduction of Overseas High-Level Talents-Youth Program, a young and outstanding talent under the Liaoning Xingliao Talent Program, and a member of Elsevier's Top 2% of Top Scientists in the World in 2022. Before returning to China, he worked as a researcher at RIKEN in Japan and Singapore Science and Technology Agency (STA). His research interests include precision/ultra-precision machining, optical manufacturing, laser processing, and metal substrate construction and molding. His research achievements have been recognized as part of 2018 A*STAR Research Highlights, and he has published more than 100 academic papers in renowned international journals in the fields of manufacturing, friction, and optics. He has also authorized more than 80 international and national invention patents. He has chaired more than 20 projects, including the Japan Society for the Promotion of Science and Technology (JSPS) Youth Fund, the National Natural Science Foundation of China (NSFC), and the National Key Research and Development Program of China (NKRDP) "Key Scientific Issues of Transformative Technologies". He serves as an Editorial Board Member for the *International Journal of Extreme Manufacturing*, *Advances in Manufacturing*, the *International Journal of Hydromechatronics*, etc., as well as a reviewer for more than 80 international journals. He is also a reviewer for more than 80 international journals.

micromachines

Editorial

Editorial for the Special Issue on Advances in Ultra-Precision Machining Technology and Applications

Benny C. F. Cheung [1,*] and Jiang Guo [2]

[1] State Key Laboratory of Ultra-Precision Machining Technology, Department of Industrial and Systems Engineering, The Hong Kong Polytechnic University, Kowloon, Hong Kong
[2] School of Mechanical Engineering, Dalian University of Technology, Dalian 116024, China
* Correspondence: benny.cheung@polyu.edu.hk

Citation: Cheung, B.C.F.; Guo, J. Editorial for the Special Issue on Advances in Ultra-Precision Machining Technology and Applications. *Micromachines* **2022**, *13*, 2093.
https://doi.org/10.3390/mi13122093

Received: 21 November 2022
Accepted: 22 November 2022
Published: 28 November 2022

Publisher's Note: MDPI stays neutral with regard to jurisdictional claims in published maps and institutional affiliations.

Ultra-precision machining technology has been widely used in the manufacture of many mission-critical components for various industrial areas, such as the advanced optics, photonics aerospace, automotive, telecommunications, biomedical and energy and environmental sectors, among others. The increasing degree of geometrical complexity, requirement of high precision and the evolution of the materials used for machined workpieces have led to many research challenges in different fields, including ultra-precision machining technologies, novel machining processes, cutting mechanics, surface generation mechanisms, novel machine design, advanced sensing, machine metrology, accurate control of the machining process through modeling and simulation of ultra-precision machining processes, error compensation, materials sciences, measurement and on-machine metrology, as well as advanced applications for functional uses.

This Special Issue offers a high-quality collection of 17 papers detailing the latest research results and findings in the field of ultra-precision machining technology and related applications. These papers cover various aspects of this topic, including multi-physical coupling simulation of electrochemical machining and aerostatic spindle, error measurement and compensation, femtosecond-laser-assisted etching, numerical modeling of cutting force and stress, prediction of milling stability, influence of tribological characteristics, tribochemical mechanical polishing, low-pressure lapping, thermal effect, micromachining innovation design, single-particle erosion mechanism, dynamic performance of aerostatic thrust bearing, fly-cutting process and ultrasonic-vibration-assisted cutting.

Highlights of this collection of papers are as follows. Li et al. [1] present a study focused on the forming mechanism of cooling hole electrolytic machining using multi-physical field coupled simulation and experimental observation. An investigation of the main error sources was conducted by Xiang et al. [2] for the error motion measurement of a precision shafting based on a T-type capacitive sensor. Wang et al. [3] report a femtosecond-laser-assisted dry-etching technology that can be utilized to realize the fabrication of silicon microlenses. A new cutting force coefficient model is established Li et al. [4], revealing the influence of the cutting-edge radius on the cutting process. An updated full-discretization method is presented by Ma et al. [5] for milling stability prediction based on cubic spline interpolation. Nagît et al. [6] elaborate the tribological behavior of test piece surfaces, analyzing the changes in the values of the coefficient of friction and loss of mass that appear over time. Qi et al. [7] demonstrate the mechanism underlying oxygen production and the tribochemical reaction mechanisms of SiC during fixed abrasive tribochemical mechanical polishing. Yu et al. [8] describe the modeling and simulation of the surface generation mechanism of a novel low-pressure lapping method using the finite element method, indicating that rotational speed plays a major role in this process. A three-degrees-of-freedom (3-DOF) quasi-static kinematics model is established by Lei et al. [9] for motion errors containing the thermal effect for the hydrostatic guideway. Experimental results show that the model has a certain effect on thermal error prediction. Wang et al. [10]

present a knowledge-based holistic framework that enables process planners to achieve micromachining innovation design by analyzing innovation design procedures and available knowledge sources. Cao et al. [11] combine smoothed particle hydrodynamics (SPH) simulation and an experiment to investigate the single-particle erosion mechanism of optical glass and verify the effect of impact velocity and particle size on material removal rate. Wang et al. [12] utilize the FLUENT software to simulate and analyze the impact of throttling characteristics of small orifices on the stiffness and stability optimization of aerostatic thrust bearings. Dai et al. [13] analyze the electrochemical machining (ECM) process in a film-cooling hole by conducting a multi-physics coupling simulation on the basis of Faraday's law and a fluid heat transfer mathematical model. An et al. [14] utilize simulation to explore the causes of potassium dihydrogen phosphate (KDP) chip formation in the single-point diamond fly-cutting process, and micro-cracks on the machined surface are analyzed based on thermo-mechanical coupling and chip morphology. Guan et al. [15] propose a high-precision machining method for weak-stiffness mirrors based on the fast tool servo system and realize a clamping error with a peak-to-valley (PV) value of 5.2 μm and a cutting error with a PV value of 1.6 μm. Zhang et al. [16] demonstrate a 104 kHz ultrasonic-vibration-assisted cutting system could achieve a constant surface roughness of about 3 nm to 4 nm in machining steel optical modules with 0–15° slope degrees. Finally, Chen et al. [17] establish a multi-field coupling 5-DOF dynamics model for the aerostatic spindle considering the interaction between the air film, spindle shaft and the motor.

Funding: This research received no external funding.

Acknowledgments: We would like to take this opportunity to thank all the authors for submitting their papers to this Special Issue, the reviewers for their time in helping to improve the quality of the submitted papers for the production of this Special Issue.

Conflicts of Interest: The authors declare no conflict of interest.

References

1. Li, Z.; Li, W.; Dai, Y. Experimental Research and Multi-Physical Field Coupling Simulation of Electrochemical Machining Based on Gas–Liquid Two-Phase Flow. *Micromachines* **2022**, *13*, 246. [CrossRef] [PubMed]
2. Xiang, K.; Wang, W.; Chen, Z. Analysis of Main Error Sources for the Error Motion Measurement of a Precision Shafting Using a T-Type Capacitive Sensor. *Micromachines* **2022**, *13*, 221. [CrossRef] [PubMed]
3. Wang, B.-X.; Zheng, J.-X.; Qi, J.-Y.; Guo, M.-R.; Gao, B.-R.; Liu, X.-Q. Integration of Multifocal Microlens Array on Silicon Microcantilever via Femtosecond-Laser-Assisted Etching Technology. *Micromachines* **2022**, *13*, 218. [CrossRef] [PubMed]
4. Li, P.; Chang, Z. Numerical Modeling of the Effect of Cutting-Edge Radius on Cutting Force and Stress Concentration during Machining. *Micromachines* **2022**, *13*, 211. [CrossRef] [PubMed]
5. Ma, J.; Li, Y.; Zhang, D.; Zhao, B.; Wang, G.; Pang, X. A Novel Updated Full-Discretization Method for Prediction of Milling Stability. *Micromachines* **2022**, *13*, 160. [CrossRef] [PubMed]
6. Nagîţ, G.; Slătineanu, L.; Dodun, O.; Mihalache, A.M.; Rîpanu, M.I.; Hriţuc, A. Influence of Some Microchanges Generated by Different Processing Methods on Selected Tribological Characteristics. *Micromachines* **2022**, *13*, 29. [CrossRef] [PubMed]
7. Qi, W.; Cao, X.; Xiao, W.; Wang, Z.; Su, J. Study on the Mechanism of Solid-Phase Oxidant Action in Tribochemical Mechanical Polishing of SiC Single Crystal Substrate. *Micromachines* **2021**, *12*, 1547. [CrossRef] [PubMed]
8. Yu, N.; Li, L.; Kee, C.-s. Modeling and Simulation of the Surface Generation Mechanism of a Novel Low-Pressure Lapping Technology. *Micromachines* **2021**, *12*, 1510. [CrossRef]
9. Lei, P.; Wang, Z.; Shi, C.; Peng, Y.; Lu, F. Simulation, Modeling and Experimental Research on the Thermal Effect of the Motion Error of Hydrostatic Guideways. *Micromachines* **2021**, *12*, 1445. [CrossRef] [PubMed]
10. Zhang, D.; Wang, G.; Xin, Y.; Shi, X.; Evans, R.; Guo, B.; Huang, P. Knowledge-Driven Manufacturing Process Innovation: A Case Study on Problem Solving in Micro-Turbine Machining. *Micromachines* **2021**, *12*, 1357. [CrossRef]
11. Cao, Z.; Yan, S.; Li, S.; Zhang, Y. Theoretical Modeling and Experimental Analysis of Single-Particle Erosion Mechanism of Optical Glass. *Micromachines* **2021**, *12*, 1221. [CrossRef] [PubMed]
12. Sahto, M.P.; Wang, W.; Sanjrani, A.N.; Hao, C.; Shah, S.A. Dynamic Performance of Partially Orifice Porous Aerostatic Thrust Bearing. *Micromachines* **2021**, *12*, 989. [CrossRef] [PubMed]
13. Li, Z.; Cao, B.; Dai, Y. Research on Multi-Physics Coupling Simulation for the Pulse Electrochemical Machining of Holes with Tube Electrodes. *Micromachines* **2021**, *12*, 950. [CrossRef] [PubMed]
14. An, C.; Feng, K.; Wang, W.; Xu, Q.; Lei, X.; Zhang, J.; Yao, X.; Li, H. Interaction Mechanism of Thermal and Mechanical Field in KDP Fly-Cutting Process. *Micromachines* **2021**, *12*, 855. [CrossRef] [PubMed]

15. Li, Z.; Dai, Y.; Guan, C.; Yong, J.; Sun, Z.; Du, C. High-Precision Machining Method of Weak-Stiffness Mirror Based on Fast Tool Servo Error Compensation Strategy. *Micromachines* **2021**, *12*, 607. [CrossRef] [PubMed]
16. Zhang, C.; Cheung, C.; Bulla, B.; Zhao, C. An Investigation of the High-Frequency Ultrasonic Vibration-Assisted Cutting of Steel Optical Moulds. *Micromachines* **2021**, *12*, 460. [CrossRef] [PubMed]
17. Chen, G.; Chen, Y. Multi-Field Coupling Dynamics Modeling of Aerostatic Spindle. *Micromachines* **2021**, *12*, 251. [CrossRef] [PubMed]

 *micromachines*

MDPI

Article

Experimental Research and Multi-Physical Field Coupling Simulation of Electrochemical Machining Based on Gas–Liquid Two-Phase Flow

Zhaolong Li [1,2], Wangwang Li [2] and Ye Dai [1,2,*]

1 Key Laboratory of Advanced Manufacturing Intelligent Technology of Ministry of Education, Harbin University of Science and Technology, Harbin 150080, China; lizhaolong@hrbust.edu.cn
2 School of Mechanical and Power Engineering, Harbin University of Science and Technology, Harbin 150080, China; liwangwang19970214@163.com
* Correspondence: daiye312@163.com; Tel.: +86-0451-8639-0588

Abstract: In this paper, the forming mechanism of cooling hole electrolytic machining is studied using multi-physical field coupled simulation and experimental observation. A multi-physical field coupled simulation model was established to obtain the gas–liquid two-phase distribution law inside the machining gap, and a mathematical model of gas–liquid two-phase flow was established to analyze the change law of the size and morphology of cooling hole electrolytic machining under different process parameter conditions. The simulation and experimental results show that the size of the inlet of the cooling hole is larger, the size of the outlet is smaller, and the middle section is more stable; machining voltage and electrode feed speed have a significant influence on the size and shape of heat dissipation holes. Compared with the experimental data, simulation accuracy is good.

Keywords: electrochemical machining; gas–liquid two-phase flow; multi-physical field coupling simulation; processing voltage; feed rate

Citation: Li, Z.; Li, W.; Dai, Y. Experimental Research and Multi-Physical Field Coupling Simulation of Electrochemical Machining Based on Gas–Liquid Two-Phase Flow. *Micromachines* **2022**, *13*, 246. https://doi.org/10.3390/mi13020246

Academic Editor: Kwang-Yong Kim

Received: 21 December 2021
Accepted: 29 January 2022
Published: 1 February 2022

Publisher's Note: MDPI stays neutral with regard to jurisdictional claims in published maps and institutional affiliations.

1. Introduction

GH4169 alloy is widely used in the aerospace industry, for example, in working blades, turbine disks, and the combustion chambers of aerospace engines [1–3]. The structure and machining technology of turbine blades directly affect the performance of the engine. In order to improve the power of the turbine engine, and ensure that the turbine works in high-pressure gas above 1 MPa and a high-temperature environment of 1000 °C, it is particularly important to machine the heat dissipation holes of the turbine blades. Because of the small size of the cooling hole structure, traditional machining has higher requirements in terms of bit hardness, and it is easy to break and damage the side wall of the machining hole in the feeding process. However, EDM and laser machining are both hot machining and will inevitably form hot recast layers and microcracks on the metal surface. These issues will affect the machining accuracy, stability and working performance of the heat dissipation holes. Electrochemical machining utilizes the principle of electrochemical dissolution of electrodes in an electrolyte, so that the machined heat dissipation holes have good surface quality, no stress concentration, and no surface hardening layer. Therefore, electrochemical machining technology has a broad application prospect in the field of high-precision manufacturing, such as of thin film heat dissipation holes [4]. Electrolytic machining does not depend on the physical properties of the material, such as hardness, toughness, and mechanical strength, so it is suitable for machining deep holes of small diameter and adding such holes to difficult-to-machine cemented carbides. Electrochemical machining involves many factors, such as electrochemistry, heat transfer, hydrodynamics, etc., and its dissolution and formation state is complex and cannot be directly measured [5,6]. Many scholars have conducted numerous theoretical and experimental studies on the electrochemical dissolution mechanism of materials during electrolytic processing, mainly including simulation analysis of the steady-state process of electrolytic processing of pore structures and experimental and process optimization studies, with good research results being achieved [7–11].

A large number of scholars have studied machining performance and process parameter control and undertaken multi-physical field simulation analysis of electric, flow and temperature fields in the machining gap. They have also undertaken simulation analysis of the combination of convection heat transfer, mathematical modeling and gas–liquid two-phase flow theory, with actual hole structure electrochemical machining, shaping, and machining accuracy [12–17]. Through process testing and simulation analysis, the influence of the flow field on the machining gap in machining performance was analyzed, and servo control was optimized, thus improving machining performance [18–20].

In this paper, a simulation and an experimental study of the multi-physical field coupling forming process of hole structure electrochemical machining were carried out; a simulation model of cooling hole electrochemical machining was established; and the influence of different process parameters on gas–liquid two-phase flow field and electric field in the machining gap was analyzed. Through comparison of simulation and experiment, the influence of process parameters on the average radius, taper and other response indexes of heat dissipation holes was analyzed, and the accuracy of the multi-physical field coupling simulation model was verified, which is of great significance to multi-physical field coupling research into the process of hole structure electrochemical machining.

2. Methods and Experiments

A schematic view of the machine and processing system for cooling hole electrochemical machining is shown in Figure 1. In the cooling hole electrolytic machining experiment, the tool electrode was connected to the spindle and servo-fed in the Z direction along a motor-driven linear guide. The electrolyte was pumped to the processing area by the pressure in the flow channel, and the tool electrode was clamped by the specially designed fixture in the figure.

Figure 1. Diagram of ECM equipment processing small hole: 1. filter; 2. pump; 3. control system; 4. motor; 5. electrode; 6. working table.

The workpiece was GH4169 and the tool electrode was a tubular titanium alloy electrode with an inner diameter of 0.6 mm and an outer diameter of 1.4 mm. Titanium alloy has the advantages of good conductivity, strong corrosion resistance, high strength, good rigidity, etc., and is a good choice for electrochemical machining electrodes (covered with PTFE insulating film). The electrolyte was a sodium nitrate aqueous solution (because the workpiece material was GH4169, a strong acid electrolyte was selected as it has high current density efficiency). Central composite design was adopted and 20 sets of experiments were conducted, with a workpiece thickness of 6 mm, as shown in the process parameters (Table 1). The electrolyte flow rate was 6, 8, 10, and 12 mm/min; the voltage was 12, 16, 24, and 24 V; and the electrolyte concentration was 16. When the machining current was 0, the machining ended. The specific experimental conditions are shown in Table 1.

Table 1. Experimental conditions of electrochemical machining of heat dissipation holes.

Experimental Project	Condition
Tool electrode	Titanium alloy tube electrode (covered with insulating film)
Workpiece	GH4169 nickel-based superalloy
Processing voltage	12–24 V DC voltage [12,13]
Electrolyte concentrations	16% sodium nitrate solution
Feed rate	6–12 mm/min [12,13]

The main element composition and content of the high-temperature-resistant nickel-based alloy GH4169 are shown in Table 2.

Table 2. The main elements and content of GH4169.

Element	Ni	Cr	Nb	Mo	Ti	AI	Fe
Atomic mass	59	52	93	96	47	27	56
Percentage	50–55	17–21	4.75–5.5	2.8–3.3	0.65–1.15	0.2–0.8	Margin

The entrance morphology of the small hole on the machined GH4169 sample and the sectional morphology obtained by cutting the small hole along the axis by EDM wire cutting were characterized by SEM, as shown in Figure 2.

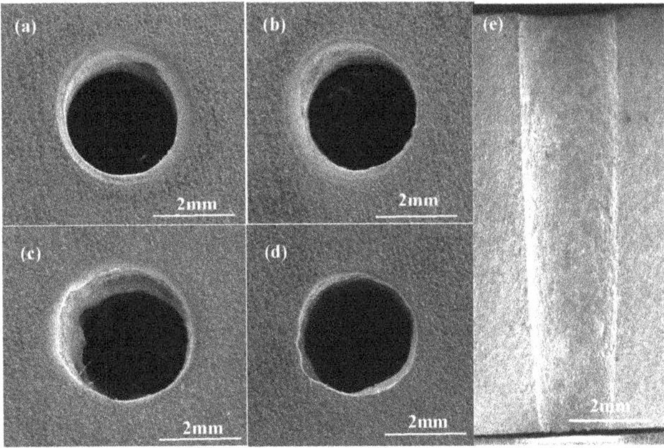

Figure 2. Morphology of cooling hole inlet. The machining speed in (**a–d**) is 6–12 mm/min, and (**e**) is the cross-sectional diagram of (**a**).

As shown in Figure 2a–e, different machining parameters were used for the electrolytic machining of the heat dissipation holes. The inlet size is slightly different; the inlet shape is better; and the inlet shape is influenced by the electric field in the machining gap. The simulation of the electric field distribution in the electrochemical machining of the tubular electrode is shown in Figure 3. It can be seen that the shape of the air intake is better. Figure 2e shows a sectional view of the cooling hole; it can be seen that the side wall of the cooling hole has a certain taper.

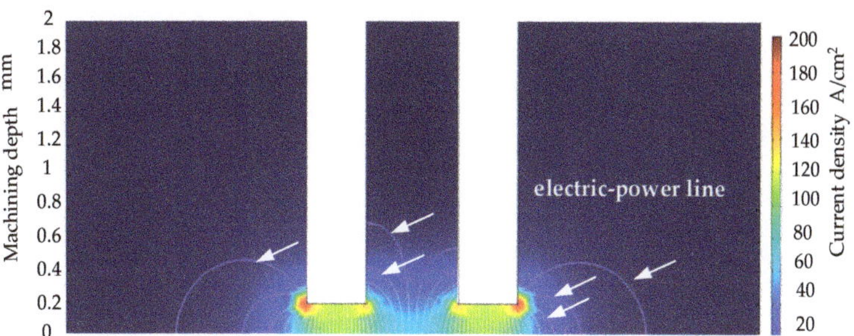

Figure 3. Schematic diagram of electric field distribution in machining gap.

3. Results and Discussion

3.1. Simulation Analysis of Multiple Physical Fields in ECM Gap of Cooling Hole

In this paper, the COMSOL built-in drawing tool was used to build a two-dimensional ax symmetric model of the cooling hole. The physical fields analyzed include electric field, flow field and temperature field. The geometric model of the machining gap of the heat dissipation holes is shown in Figure 4.

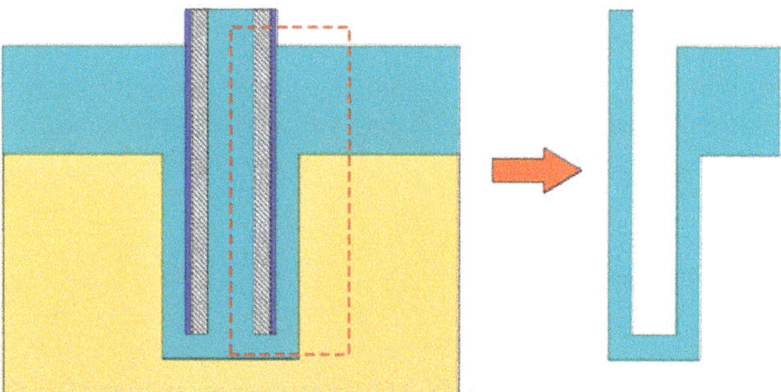

Figure 4. ECM model of cooling hole.

The specific material simulation parameters are shown in Table 3.

Table 3. Simulation material parameters.

Simulation Parameters	Values and Units
Liquid specific heat capacity	4200 J/(kg·K)
Hydrogen specific heat capacity	800 J/(kg·K)
Liquid density	1200 kg/m^3
Hydrogen density	89.9×10^{-3} kg/m^3
Liquid heat transfer coefficient	0.64 [W/(m·K)]
Hydrogen heat transfer coefficient	0.16 [W/(m·K)]
Dynamic viscosity	1.01×10^{-3} Pa·s
Bubble diameter	3×10^{-5} m
Initial temperature	293.15 K
Temperature correlation coefficient	0.025
Gas correlation coefficient	0.15

The gap in electrochemical machining contained: (1) hydrogen evolved from the electrochemical reaction; (2) a small number of metal particles, falling off due to the dissolution of the anode; (3) electrolyte. Because the volume of metal particles was very small and had little influence on the conductivity and current density of electrolyte, the gas–liquid–solid three-phase flow in the machining gap was simplified to a gas–liquid two-phase flow.

The electric field model of the machining area was regarded as a constant current electric field and the potential $\varphi(x, y)$ at any point in the machining gap between the cathode and anode metals satisfied the Laplace equation:

$$\nabla^2 \varphi = \frac{\partial^2 \varphi}{\partial x^2} + \frac{\partial^2 \varphi}{\partial y^2} + \frac{\partial^2 \varphi}{\partial z^2} = 0 \tag{1}$$

The electric field intensity E is the negative gradient of electric potential φ. According to Ohm's law, the relationship between current density i, electric field intensity E, and potential φ in the machining area is as follows:

$$E = -\nabla \varphi$$

$$i = \sigma E = -\sigma \nabla \varphi \tag{2}$$

E—the electric field intensity (V/m); σ—electrolyte conductivity (S/m); φ—potential (V).

See Formula (3) for the normal dissolution rate of the anode workpiece. By solving the displacement of each point on the surface of the workpiece with time, we could obtain the profile of the cooling hole in electrochemical machining at different times.

$$v_n = \eta \omega i = -\eta \omega \sigma \nabla \varphi \tag{3}$$

Assuming that bubbles only occupy a small volume fraction and they always move at a free-settling speed, the pressure distribution was calculated according to the mixed average continuity equation. The continuity equation of the gas–liquid two-phase flow model was established (1)–(3):

$$\varnothing_l \rho_l \frac{\partial u_l}{\partial t} + \varnothing_l \rho_l u_l \cdot \nabla u_l = -\nabla p + \nabla \cdot \left[\varnothing_l (\mu_l + \mu_T) \left(\nabla u_l + \nabla u_l^T - \frac{2}{3} (\nabla \cdot u_l) I \right) \right] + \varnothing_l \rho_l g + F \tag{4}$$

$$\frac{\partial}{\partial t} \left(\varnothing_l \rho_l + \varnothing_g \rho_g \right) + \nabla \cdot \left(\varnothing_l \rho_l + \varnothing_g \rho_g u_g \right) = 0 \tag{5}$$

$$\frac{\partial \varnothing_g \rho_g}{\partial t} + \nabla \cdot \left(\varnothing_g \rho_g u_g \right) = -m_{gl} \tag{6}$$

where $\varnothing$—phase inclusion rate; g—gravity vector; F—volume force (N); μ_l—liquid dynamic viscosity (Pa/s); μ_T—turbulent viscosity; and m_{gl}—gas to liquid mass transfer rate.

Based on the continuity equation of the gas–liquid two-phase model (4)–(6) and actual processing parameters, the distribution law of the gas–liquid two-phase flow field was simulated and analyzed.

Based on the above-mentioned modeling analysis, boundary condition setting, process parameter setting and material parameter setting, the simulation analysis was carried out in COMSOL and the conclusion was drawn. The distribution law of the hydrogen volume fraction in the machining gap under the condition of coupling of multiple physical fields is shown in Figure 5.

It can be seen from the figure that the volume fraction of hydrogen gas gradually increased in the direction of electrolyte flow; the fraction of hydrogen gas on the surface of the anode workpiece at the exit point decreased rapidly; and the maximum value was located at the "corner" of the electrode. The maximum hydrogen volume fraction in the machining gap was 0.15. The main cause was that the electrolysis reaction in the machining gap generated hydrogen gas. Due to insufficient electrolyte flow, the volume fraction of hydrogen in the sharp corners and end faces of the tool electrode continued to increase. The volume fraction of hydrogen below the outlet decreased rapidly because of the low density of hydrogen.

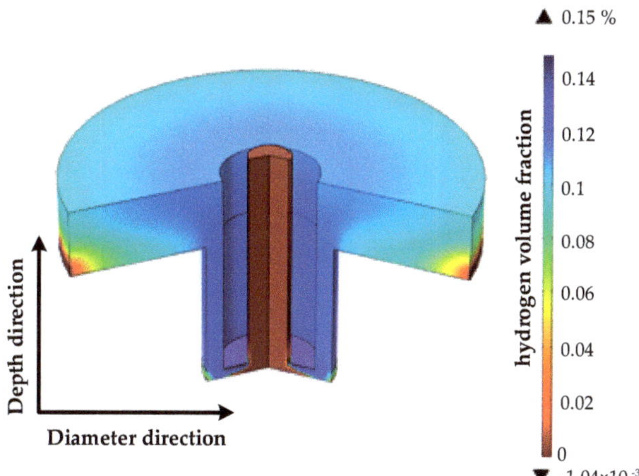

Figure 5. Distribution of hydrogen volume fraction in machining gap.

This was because the principle of ECM is that under the action of an electric field, an electric current is generated between the electrode and the workpiece, through electrolyte connection; the workpiece is corroded by electrolysis; and hydrogen and oxygen are generated near the workpiece and the electrode. The electrolyte in the machining gap formed between the electrode and the workpiece circulated continuously, in time removing the electrolytic etching products and gas from the machining gap, and thus forming the gas distribution in the electrolyte channel, which conforms to the basic principle of electrolytic machining.

The distribution of the volume fraction of hydrogen in the machining gap of heat dissipation holes under different machining voltages is shown in Figure 6. It can be seen from Figure 6a–d that the hydrogen gas precipitated during the electrochemical reaction accumulated mainly at the end faces and corners of the tool electrode, and the maximum value of the hydrogen volume fraction increased from 0.128 to 0.142 with increasing process voltage. The volume distribution of hydrogen in the end clearance varied greatly, ranging from approximately 0.12 to 0.14.

It can be seen from the figure that when the processing voltage was 12–24 V, the maximum hydrogen volume fractions on the workpiece surface were 0.019 and 0.064, respectively. It can be seen that with the increase in processing voltage, the gas volume fraction increased faster and faster. As the process voltage increased, the volume fraction of hydrogen gradually increased, but the influence range of hydrogen on the surface of the workpiece did not change significantly. When the processing voltage was 12, 16, 20 and 24 V, the maximum values of the hydrogen volume fraction on the workpiece surface were 0.019, 0.027, 0.038 and 0.064, respectively. Under different treatment voltages, the front half of the anodized surface of the workpiece was 0–0.43 mm; it was not affected by the gas and had a zero gas volume fraction. The volume fraction of gas on the surface of the workpiece measuring 0.43–0.7 mm in the middle section gradually increased. The volume fraction of gas decreased in the lower half at 0.7–0.85 mm. The increasing rate of the gas volume fraction increased with the increase in processing voltage. The main reason for this was that the hydrogen gas precipitating from the end face of the tool electrode gradually diffused within the machining gap as the electrolyte flowed under the influence of the tracing force model. Within the range of diffusion, hydrogen bubbles gradually adsorbed on the surface of the workpiece. According to Ohm's Law and Faraday's Law, the higher the processing voltage, the higher the current density, and the more hydrogen; therefore, more hydrogen adsorbed on the surface of the workpiece.

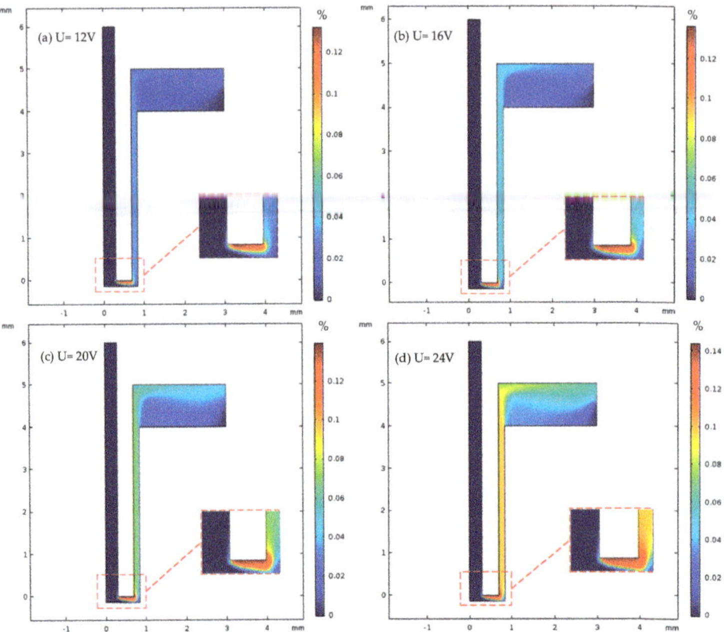

Figure 6. Distribution of gas volume fraction in machining gap of cooling hole with different machining voltages (16% sodium nitrate solution, feed rate 10 mm/min, (**a**) 12 V, (**b**) 16 V, (**c**) 20 V, (**d**) 24 V).

The volume distribution of hydrogen on the machined surface of the heat dissipation holes at different machining voltages is shown in Figure 7.

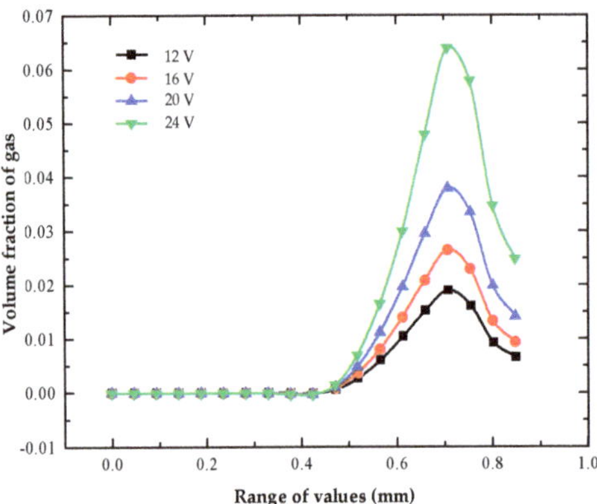

Figure 7. Distribution of gas volume fractions on anodized surface of heat dissipation holes with different processing voltages.

The magnitude and distribution of current density on the surface of the workpiece for the multi-physics field coupled cooling hole electrolytic machining simulation model without machining voltage conditions was derived, as shown in Figure 8.

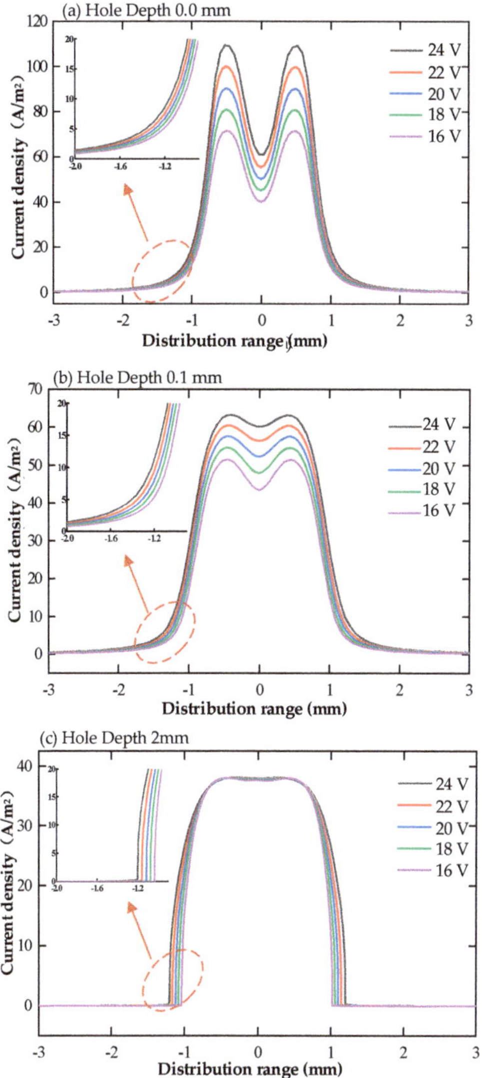

Figure 8. Current density distribution on the workpiece surface at different processing voltages (16% sodium nitrate solution, feed rate 10 mm/min, (**a**) hole depth 0 mm, (**b**) hole depth 0.1 mm, (**c**) hole depth 2 mm).

The distribution of the volume fraction of hydrogen in the machining gap of the heat dissipation holes under different inlet flow rates is shown in Figure 9. It can be seen from the figure that the volume fraction of hydrogen in the machining gap of the cooling hole gradually decreased with the increase in electrolyte inlet velocity. The hydrogen mainly gathered at the end and corner of the tool electrode, and the inlet flow rate had little effect on the maximum hydrogen volume fraction in the machining gap, which was 0.14. The hydrogen distribution in the side gaps was more uniform, while the hydrogen distribution changed more in the end gaps and at the electrolyte outlet. The hydrogen volume fraction in the end gap was also around 0.14.

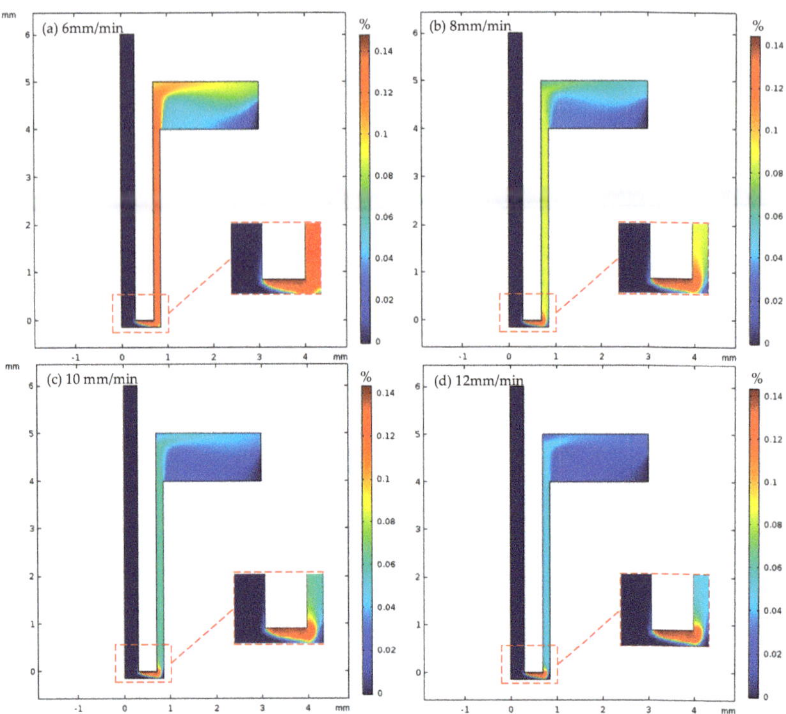

Figure 9. Hydrogen volume fraction in machining gap of cooling hole with different inlet flow rates(16% sodium nitrate solution, U 20V, (**a**) feed rate 6 mm/min, (**b**) feed rate 8 mm/min, (**c**) feed rate 10 mm/min, (**d**) feed rate 12 mm/min).

The volume fraction distribution of hydrogen on the machined surface of the heat dissipation holes under different inlet flow rates of electrolyte is shown in Figure 10.

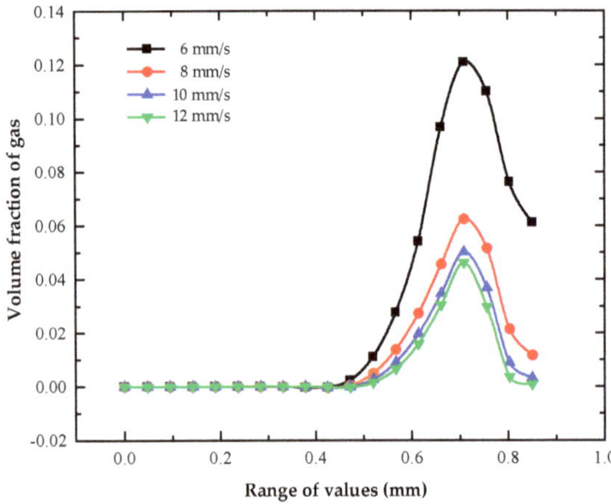

Figure 10. Volume fraction of hydrogen on the machined surface of heat dissipation holes with different inlet flow rates.

When the inlet flow rate was 6 mm/s and 12 mm/s, the maximum volume fractions of hydrogen on the machined surface were 0.121 and 0.046, respectively. At the first half of the workpiece surface (0–0.42 mm), the volume fraction of hydrogen was zero and was less influenced by the inlet flow of electrolyte. In the middle region (0.42–0.7 mm), with the increase in the electrolyte inlet velocity, the growth rate of the hydrogen volume fraction on the machined surface decreased and tended to be stable. In the second half (0.72–0.85 mm), the volume fraction of hydrogen on the machined surface decreased gradually and was less affected by the inlet flow rate. The main reason for this was the increase in the inlet flow rate of the electrolyte and the faster renewal of the electrolyte. This took more hydrogen with it, so the volume fraction of hydrogen on the machined surface was reduced. However, the fast flow rate increased the "vortex" effect and reduced the renewal rate of the hydrogen.

The magnitude and distribution of current density on the surface of the workpiece at different tool electrode feed rates for the multi-physics coupled cooling hole electrolytic machining simulation model was derived, as shown in Figure 11.

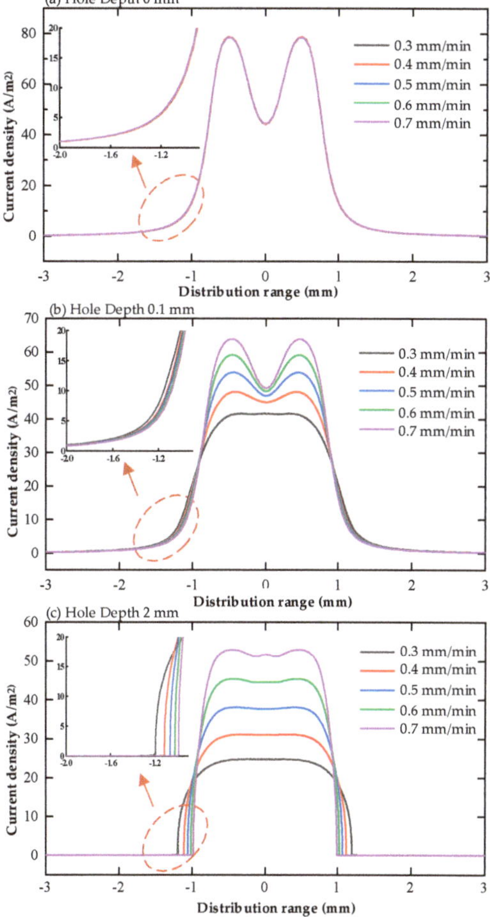

Figure 11. Distribution of current density on the workpiece surface under various electrode feed rates(16% sodium nitrate solution, U 20V, (**a**) hole depth 0mm, (**b**) hole depth 0.1mm, (**c**) hole depth 2mm).

The velocity contour diagram of the electrolyte is shown in Figure 12. The electrolyte inlet pressure was 2.4 MPa. The range of velocity contour values in the same zone was as follows: hole depth 15 mm, contour number 1.18–10.66 m/s; hole depth 35 mm, contour value 1.03–9.28 m/s. The trend of electrolyte flow rate change was basically the same when the processing depth increased; the flow rate difference at the same reference point gradually increased along the radius direction of the low-speed zone.

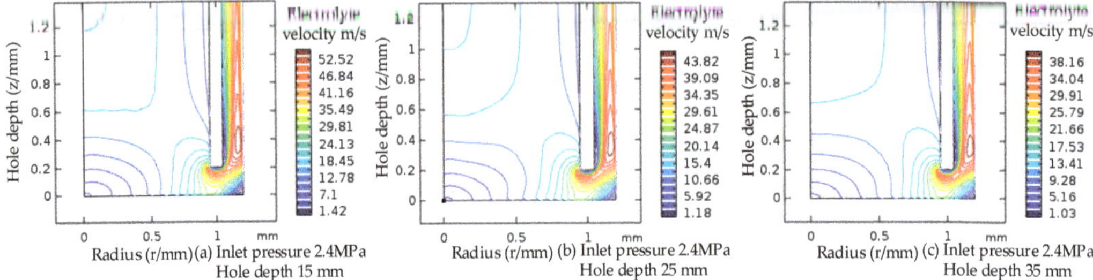

Figure 12. Velocity contour map of electrolyte in the machining gap.

To sum up, the low-speed zone in the flow field of the processing gap was greatly reduced after increasing the electrolyte pressure, and the flow speed of the electrolyte increased; this removed the electrolytic products and Joule heat in the processing gap in a timely manner and improved the stability and fixed domain of the electrolytic processing.

3.2. Analysis of Experiment and Simulation Results

In this part, the simulation model analysis is verified by experiments; the electrochemical machining process parameters and workpieces are the same as the simulation settings.

The variation trend in the cooling hole size with machining voltage obtained by comparing simulation and experiment is shown in Figure 13. The simulation results show that the inlet radius of the cooling hole was large and the outlet radius of the cooling hole was small. The mean radius and taper of the cooling hole obtained from the simulation were larger than the actual value when the machining voltage was small, and the result obtained from the experiment was larger when the voltage was larger.

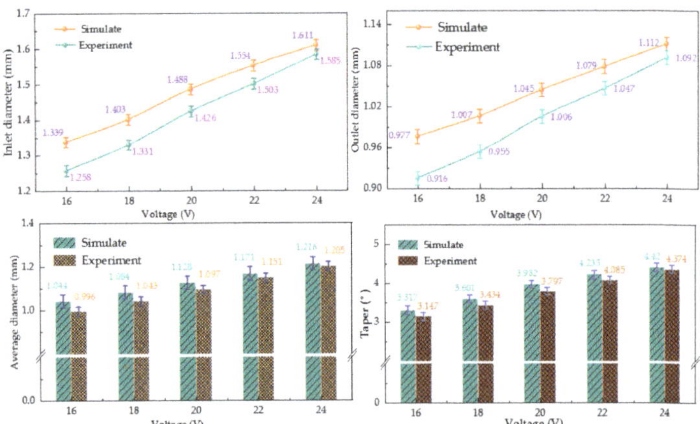

Figure 13. Effect of different machining voltage on ECM size of cooling hole.

Furthermore, the deviation between the experimental and simulated inlet diameters was 4.2%, the deviation between the experimental and simulated outlet diameters was

4.1%, the deviation between the experimental and simulated average diameters was 2.8%, and the deviation between the experimental and simulated tapers was 4%.

The larger entrance radius was due to the large electric field line scattering area and large electrolytic etching range in the initial stage of electrochemical processing, as shown in Figure 3. Therefore, in the incident stage, the range of electrolytic etching was large and the incident radius was relatively large. However, when the electrode continuously entered the workpiece, the machining was at a stable stage, and the electrolytic etching speed and the electrode feeding speed were relatively stable. However, at the exit stage, the distance between the electrode and the workpiece was decreasing, the gap electric field was shrinking, and the electrolytic etching effect became smaller.

The simulated and actual values of the electrochemical machining sizes of heat dissipation holes under different electrode feed speeds are shown in Figure 14.

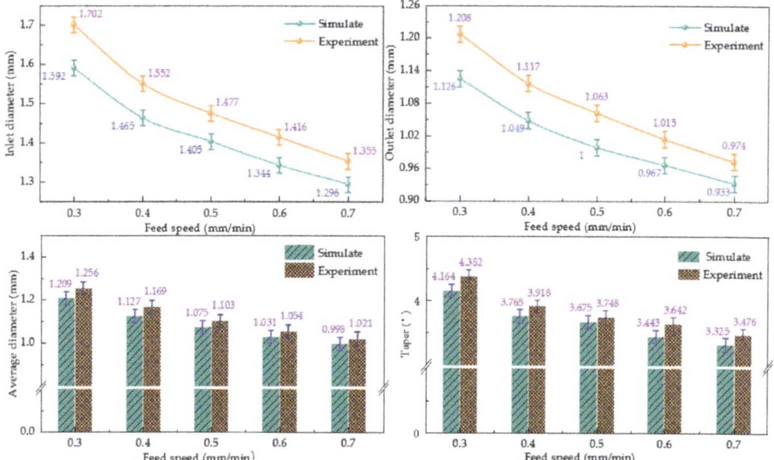

Figure 14. Effect of different electrode feed speed on ECM size of cooling hole.

It can be seen from the figure that the size of the inlet radius and outlet radius gradually decreased with the increase in feed speed. The experimental size was slightly larger than the simulation results, and the experimental size increased with the increase in feed speed. The experimental results show that the average radius of the cooling hole was larger and the taper was smaller. The reasons for this phenomenon were the stray corrosion at the inlet and the overfeed at the outlet.

Furthermore, the deviation between the experimental and simulated inlet diameters was 5.6%, the deviation between the experimental and simulated outlet diameters was 5.9%, the deviation between the experimental and simulated average diameters was 2.9%, and the deviation between the experimental and simulated tapers was 4.3%.

4. Conclusions

In this paper, the structure of the heat dissipation holes of an air turbine engine was machined by electrochemical machining with tubular electrodes. Compared with other machining methods, electrochemical machining has the advantages of good surface quality of heat dissipation holes, no stress concentration, and no surface hardening layer. In this paper, the theory of gas–liquid two-phase flow in the electrochemical machining of heat dissipation holes was mainly studied, and a simulation model of multi-physical field coupling was established. The dynamic formation of heat dissipation holes in electrochemical machining and the evolution of their size and shape were studied. The variation law of the heat dissipation aperture and taper under different process parameters was verified by experimental simulation. The main research achievements of this paper are summarized as follows.

(1) A simulation model of electrolytic machining of heat dissipation holes was established, and the influence laws of different process parameters on the gas–liquid two-phase flow field and electric field in the machining gap were analyzed. The hydrogen gas precipitated during the electrochemical reaction accumulated mainly at the end faces and corners of the tool electrode, and the maximum value of the hydrogen volume fraction increased from 0.128 to 0.142 with increasing process voltage. The volume distribution of hydrogen in end clearance varied greatly, ranging from approximately 0.12 to 0.14.

(2) The simulation analyzed the influence law of the cooling hole dimensional shape and concluded that the entrance radius and taper of the cooling hole will decrease with the increase in the tool electrode feeding speed and also with the increase in the machining point. When the voltage increased from 16 V to 24 V, the inlet diameter increased from 1.258 mm to 1.585 mm, the sidewall accuracy increased from 3.147 to 4.374, the electrode feeding speed increased from 0.3 mm/min to 0.7 mm/min, the inlet diameter decreased from 1.592 mm to 1.296 mm, and the sidewall taper decreased from 4.382 to 3.476.

(3) Simulation and experimental comparisons were conducted to analyze the influence law of machining voltage and electrode feed rate on the response indexes, such as the mean radius and taper of cooling hole electrolytic machining, and to verify the accuracy of the multi-physics field coupled simulation model. Under different voltages, the deviation between the experimental and simulated inlet diameters was 4.2%, the deviation between the experimental and simulated outlet diameters was 4.1%, the deviation between the experimental and simulated average diameter was 2.8%, and the deviation between experimental and simulated taper was 4%. Under the action of different electrode feeding speeds, the deviation between the experimental and simulated inlet diameters was 5.6%, the deviation between the experimental and simulated outlet diameters was 5.9%, the deviation between the experimental and simulated average diameters was 2.9%, and the deviation between the experimental and simulated tapers was 4.3%.

Author Contributions: Conceptualization, Z.L. and Y.D.; methodology, Z.L., Y.D. and W.L.; validation, Z.L., Y.D. and W.L.; formal analysis, Z.L., Y.D. and W.L.; investigation, Z.L. and Y.D.; resources, Z.L. and Y.D.; data curation, Z.L.; writing—original draft preparation, Z.L.; writing—review and editing, Z.L.; visualization, Z.L.; supervision, Z.L.; project administration, Z.L. and Y.D.; funding acquisition, Z.L. and Y.D. All authors have read and agreed to the published version of the manuscript.

Funding: This work was supported by the National Natural Science Foundation of China, grant number 52075134.

Institutional Review Board Statement: Not applicable.

Informed Consent Statement: Not applicable.

Data Availability Statement: Data are contained within the article.

Acknowledgments: The authors would like to thank the Key Laboratory of Advanced Manufacturing Intelligent Technology, Ministry of Education, for helpful discussions on topics related to this work. The authors are grateful to Bingren Cao for his help with the preparation of figures in this paper.

Conflicts of Interest: The authors declare no conflict of interest.

References

1. Tang, Z.G.; Guo, T.M.; Fu, Y.; Hui, Z.; Han, C.S. Research present situation and the development prospect of nickel-based super alloy. *Met. World* **2014**, *1*, 36–40.
2. Wang, F.; Xiao, J.; Yue, L.; Wu, X.Y.; Zhao, J.S. Common Key Technology of Precise Electrochemical Machining and Its Application in Aviation Manufacturing. *Electromach. Mould.* **2020**, *1*, 1–6.
3. Du, J.H.; Deng, Q.; Qu, J.L.; Lu, X.D.; Wang, M.Q.; Bi, Z.N.; Xu, T.H. Development Trend of Manufacturing Technology of Alloy GH4169 Disk Forging. *J. Iron Steel Res.* **2011**, *23*, 130–333.
4. Han, J.C. Recent developments in turbine blade internal cooling. *N. Y. Acad. Sci.* **2001**, *5*, 162–178. [CrossRef] [PubMed]

5. Zhang, C.X.; Xu, Z.Y.; Hang, Y.S.; Xing, J. Effect of solution conductivity on tool electrode wear in electrochemical discharge drilling of nickel-based alloy. *Int. J. Adv. Manuf. Technol.* **2019**, *103*, 743–756. [CrossRef]
6. Wang, G.Q.; Zhang, Y.; Li, H.S.; Tang, J. Ultrasound-assisted through-mask electrochemical machining of hole arrays in ODS super alloy. *Materials* **2020**, *13*, 5780. [CrossRef] [PubMed]
7. Niu, S.; Qu, N.S.; Yue, X.K.; Li, H.S. Effect of tool-sidewall outlet hole design on machining performance in electrochemical mill-grinding of Inconel 718. *J. Manuf. Process* **2019**, *41*, 10–22. [CrossRef]
8. Cheng, C.-P.; Wu, K.-L.; Mai, C.-C.; Yang, C.-K.; Hsu, Y.-S.; Yan, B.-H. Study of gas film quality in electrochemical discharge machining. *Int. J. Mach. Tools Manuf.* **2010**, *50*, 689–697. [CrossRef]
9. Zhang, Y.; Xu, Z.Y.; Zhu, Y.; Zhu, D. Machining of a film-cooling hole in a single-crystal super alloy by high-speed electrochemical discharge drilling. *Chin. J. Aeronaut.* **2016**, *29*, 560–570. [CrossRef]
10. Chung, D.K.; Lee, K.H.; Jeong, J.; Chu, C.N. Machining characteristics on electrochemical finish combined with micro EDM using deionized water. *Int. J. Precis. Eng. Manuf.* **2014**, *15*, 1785–1791. [CrossRef]
11. Zou, H.; Yue, X.M.; Luo, H.X.; Liu, B.H.; Zhang, S.Y. Electrochemical micromachining of micro hole using micro drill with non-conductive mask on the machined surface. *J. Manuf. Process* **2020**, *59*, 366–377. [CrossRef]
12. Li, Z.L.; Cao, B.; Di, Y. Research on Multi-Physics Coupling Simulation for the Pulse Electrochemical Machining of Holes with Tube Electrodes. *Micromachining* **2021**, *12*, 950. [CrossRef] [PubMed]
13. Li, Z.L.; Di, Y. Analysis of multi-physics coupling of small holes in gh4169 alloy by electrolytic processing of tube electrodes. *Micromachining* **2021**, *12*, 828. [CrossRef] [PubMed]
14. Wang, M.H.; Wang, J.J.; Tong, W.J.; Chen, X.; Xu, X.F.; Wang, X.D. Multi-physical Field in IEG and Micro-dimple Forming in Ultrasonic Rolling Electrochemical Micromachining. *Acta Armamentarii* **2020**, *04*, 783–791.
15. Zhou, X.C.; Cao, C.Y. Simulation of ECM temperature field based on COMSOL. *J. Qiqihar Univ.* **2018**, *34*, 41–44.
16. Jiang, W.; Zhang, Y.; Chen, Y.L.; Yu, F.D.; Ye, H.J. Multiphysics Numerical Simulation of High Frequency Short Pulse ElectrochemicalMachining for Blade. *Aeronaut. Manuf. Technol.* **2016**, *18*, 69–74.
17. Li, Z.L.; Di, S.C. Research on Accuracy Control of Deep Small Holes by Pulse Electrochemical Machining. *Acta Armamentarii* **2012**, *33*, 414–418.
18. Qu, N.S.; Liu, Y.; Zhang, J.Z.; Fang, X.L. State-of-Art and Outlook on Electrochemical Milling. *Electromach. Mould.* **2021**, *2*, 1–14.
19. Liu, W.D.; Ao, S.S.; Li, Y. Effect of anodic behavior on electrochemical machining of TB6 titanium alloy. *Electrochim. Acta* **2017**, *233*, 190–200. [CrossRef]
20. Niu, S.; Qu, U.S.; Fu, S.X. Investigation of inner-jet electrochemical milling of nickel-based alloy GH4169/Inconel 718. *Int. J. Adv. Manuf. Technol.* **2017**, *93*, 2123–2132. [CrossRef]

micromachines

Article

Integration of Multifocal Microlens Array on Silicon Microcantilever via Femtosecond-Laser-Assisted Etching Technology

Bao Xu Wang, Jia Xin Zheng, Jin Yong Qi, Ming Rui Guo, Bing Rong Gao and Xue Qing Liu *

State Key Laboratory of Integrated Optoelectronics, College of Electronic Science and Engineering, Jilin University, Changchun 130012, China; bxwang_sklio@jlu.edu.cn (B.-X.W.); zhengjx21@mails.jlu.edu.cn (J.-X.Z.); qijy20@mails.jlu.edu.cn (J.-Y.Q.); guomr21@mails.jlu.edu.cn (M.-R.G.); brgao@jlu.edu.cn (B.-R.G.)
* Correspondence: liuxueqing@jlu.edu.cn

Abstract: Micro-opto-electromechanical systems (MOEMSs) are a new class of integrated and miniaturized optical systems that have significant applications in modern optics. However, the integration of micro-optical elements with complex morphologies on existing micro-electromechanical systems is difficult. Herein, we propose a femtosecond-laser-assisted dry etching technology to realize the fabrication of silicon microlenses. The size of the microlens can be controlled by the femtosecond laser pulse energy and the number of pulses. To verify the applicability of this method, multifocal microlens arrays (focal lengths of 7–9 μm) were integrated into a silicon microcantilever using this method. The proposed technology would broaden the application scope of MOEMSs in three-dimensional imaging systems.

Keywords: MOEMS; silicon microcantilever; multifocal microlens array; femtosecond laser; dry etching

Citation: Wang, B.-X.; Zheng, J.-X.; Qi, J.-Y.; Guo, M.-R.; Gao, B.-R.; Liu, X.-Q. Integration of Multifocal Microlens Array on Silicon Microcantilever via Femtosecond-Laser-Assisted Etching Technology. *Micromachines* **2022**, *13*, 218. https://doi.org/10.3390/mi13020218

Academic Editors: Benny C. F. Cheung and Jiang Guo

Received: 31 December 2021
Accepted: 28 January 2022
Published: 29 January 2022

Publisher's Note: MDPI stays neutral with regard to jurisdictional claims in published maps and institutional affiliations.

1. Introduction

Micro-opto-electromechanical systems (MOEMSs) are a new class of micro-systems that integrate micro-optical devices and micro-electromechanical systems (MEMSs); thus, they can simultaneously realize mechanical, electrical, and optical functions [1,2]. With the characteristics of high integration, miniaturizability and accurate control, MOEMSs have great potential applications in optical communication, micro sensing and optical imaging, among others [3–5]. Owing to the limitations of fabrication technology, a micro-mirror is the most commonly used micro-optical element in MOEMSs. For example, digital micro-mirror devices are widely used as MOEMS devices in modern optics for digital optical control. To impart more functions in and expand the application scope of MOEMSs, more forms of micro-optical devices need to be integrated with MEMSs, for example, the integration of microlens on MEMS that acts as the optical scanner has wide applications in optical imaging [6,7].

As a basic component in the field of micro-optics, microlenses can realize the properties of focusing, imaging, and beam shaping. Moreover, compared with a single microlens, microlens arrays in integrated systems can obtain information on the multiple positions and angles of images [8–11]. Therefore, microlens arrays are used in optical applications, including color imaging systems [12], 3D image acquisition systems [13] and fingerprint identification systems [14]. Typically, each microlens in an array has the same size; thus, the focal length of the microlens is the same. Therefore, microlens arrays can only image objects on their common focal plane, resulting in small field-of-view angles and a low depth of field. Based on the properties of microlenses with different focal lengths, the multifocal microlens array can solve the above-mentioned problems, and hence can be applied in 3D imaging systems [15–19]. It is difficult to integrate the above-mentioned micro-optical elements with a complicated morphology and existing MEMSs, e.g., focused ion beam technology can remove materials atom by atom. Thus, high-precision micro/nanostructures can be realized using this technology. Generally, the incident ions only work in the thickness range of several atomic layers on the surface; therefore, the fabrication efficiency of this technology is very low. Photolithography is another common technique used for fabricating micro/nanostructures; however, it has a high requirement for flatness of the sample surface. Therefore, integration of complicated micro-optical elements and MEMS devices is difficult using photolithography. To overcome this, a feasible technology that realizes the integration of micro-optical elements with complicated morphologies and existing MEMSs must be developed.

Femtosecond laser processing [20–22] has been widely used to fabricate various types of micro-optical elements because of its ultra-high-precision, mask-free procedure, and in situ processing [23]. However, it has induced high surface roughness in the integration of micro-optical elements and MEMSs of hard materials owing to direct femtosecond laser ablation. In this paper, we proposed a femtosecond-laser-assisted dry etching technology to integrate a multifocal microlens array on the surface of silicon microcantilever, which realizes the integration of micro-optical devices and MEMS devices. A point on the surface of the microcantilever was modified by femtosecond laser, and the modified region was etched and expanded to a microlens via inductively coupled plasma (ICP) etching. Owing to the high degree of freedom of femtosecond laser processing and the effective reduction of surface roughness by dry etching, the fabrication efficiency of microlens arrays with high surface quality at any position on the surface of the silicon microcantilever can be realized. Moreover, a multifocal microlens array was fabricated by adjusting the modification degree of the materials, which can be realized through the adjustment of the laser parameters. Compared with the traditional microlens array, the multifocal microlens array can effectively resolve the problems of capturing 3D image depth of field and numerical aperture in a 3D integral imaging system.

2. Materials and Methods

First, undoped silicon wafers were cleaned with acetone, ethanol, and deionized water for 30 min to obtain a clean surface. The femtosecond laser (343 nm, 200 kHz, 280 fs (Light Conversion Pharos, Vilnius, Lithuania)) is tightly focused through a high numerical aperture objective lens (NA = 0.95, 40×) and matched with a three-dimensional piezoelectric platform (the strokes of the x- and y-axes are 1.5 mm, that of the z-axis is 100 µm, and accuracy is 1 nm) to realize the preparation of micro/nanostructures on the silicon surface [24–29]. In addition, the femtosecond laser can also be used to process other materials via multi-photon absorption, for example, lift off GaN [30,31]. After femtosecond laser treatment, a laser-modified region was formed on the silicon surface, changing its physical and chemical properties. Here, the femtosecond laser was used to change the etching rate in a local region with a generation of amorphous and polycrystalline phases, and composition change [23,32]. Then, the silicon sample was etched via ICP (ICP-100A, Tailong Electronics, Beijing, China) with SF_6 gas at the flow rate of 80 sccm, with the upper electrode power of 500 W, and a lower electrode power of 100 W. In the initial time of the dry etching process, the etching rate of the laser-modified area was greater than that of the unmodified area, and the modified area first etched round holes. With the further progress of etching, owing to the influence of isotropic etching, we can expand outward through the circular hole to obtain a silicon-based microlens with a smooth surface (Figure 1). The surface roughness can be decreased to about 5.56 nm (Figure 2), obtained by measuring the bottom of the microlens.

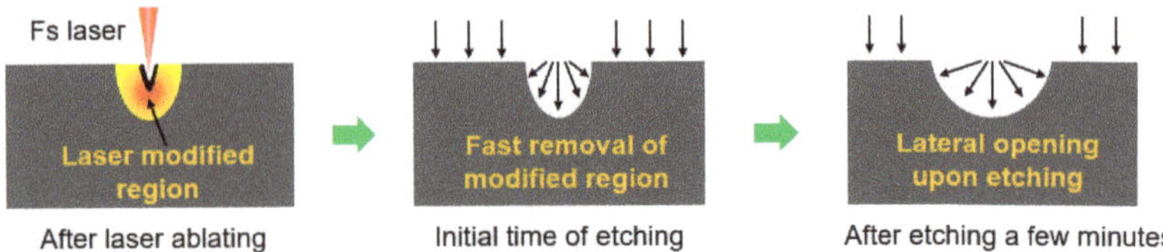

Figure 1. Schematic diagram of fabrication of silicon microlens by femtosecond laser modification with subsequent etching.

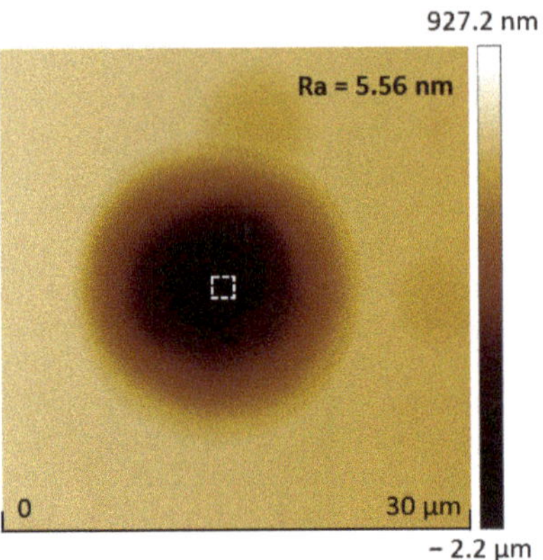

Figure 2. Atomic force microscopy (AFM) image of a silicon microlens fabricated by femtosecond laser modification with subsequent etching.

3. Results and Discussions

3.1. Preparation of Microlens

The dry etching rate of the modified region formed by the femtosecond laser is related to the degree of modification [33–35]. By adjusting parameters, such as the power of the femtosecond laser and the number of femtosecond laser pulses, regions with different degrees of modification can be realized. Based on this, we fabricated the microlens arrays with different diameters and depths on silicon wafers. According to the scanning electron microscopy (SEM) result (Figure 3a) and the cross-sectional view of the microlens array (Figure 3b,c), this method can realize the controllable preparation of microlenses with different structural parameters. The actual size of the lens changes periodically according to the experimental expectation, and the lens surface is smooth.

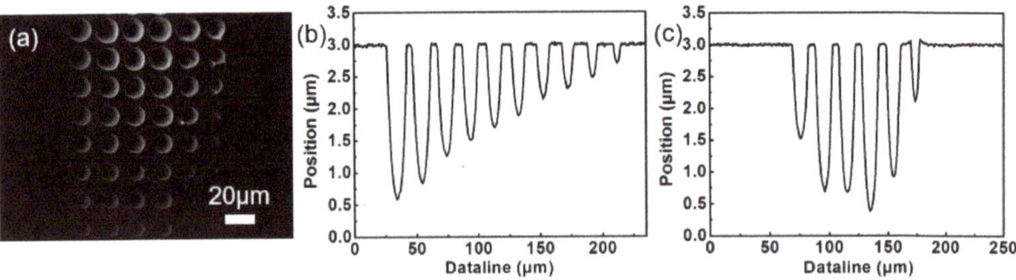

Figure 3. (**a**) Scanning electron microscopy (SEM) images of the multi-focus microlens array; (**b**) Cross-sectional profiles with the depth of lens from (**b**) deep to shallow, and (**c**) shallow to deep and then to shallow.

From further analysis of the effect of the laser on the etching morphology of the silicon-based microlens, the conclusions are as follows:

(a) Because the modified area increases with the laser power, the size of the silicon-based microlens increases after the ICP etching process. The diameter and depth of the silicon-based microlens showed an increasing trend as the power of the femtosecond laser increased, as shown in Figure 4a. By calculating the diameter and depth of the silicon-based concave microlens under different laser powers, the corresponding radius of curvature was obtained. The relationship between the radius of curvature and laser power is presented in Figure 4b.

(b) With an increase in the laser pulse number, the diameter and depth of the silicon-based concave microlens first increased and then decreased (Figure 4c). This is because the silicon surface would react with oxygen in the air with a greater number of laser pulses, generating a passivation layer on the silicon surface and preventing the etching progress. Therefore, the size of the silicon-based concave microlens decreased with an increase in the number of laser pulses. The different diameters and depths of the silicon-based concave microlens can be obtained by controlling the pulse numbers of the laser. The radius of curvature gradually decreased as the number of pulses increased, as shown in Figure 4d.

(c) The focal lengths of the microlenses with varying radii of curvature also differ. The experimental results show that the size of the silicon-based concave microlens can be flexibly adjusted by changing the femtosecond laser power and pulse number, and the controllable preparation of microlenses with different focal lengths can be realized.

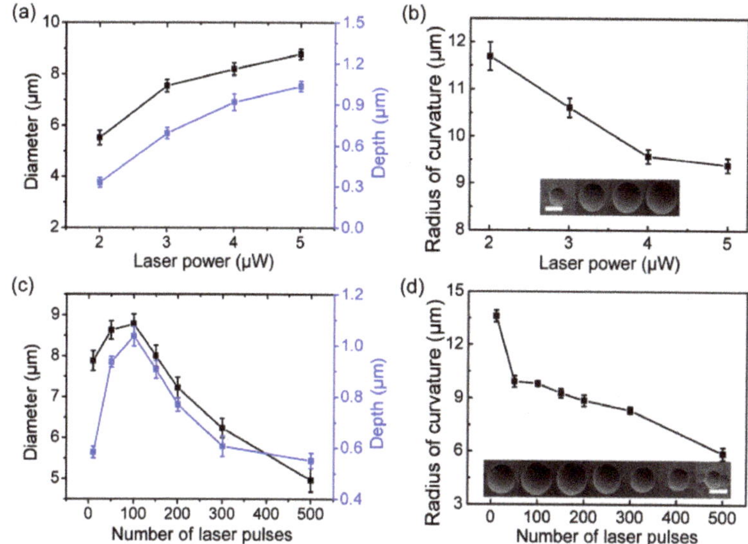

Figure 4. Relationship between the depth and the diameter of structures after etching with (**a**) laser power and (**c**) number of laser pulses. Relationship between the radius of structures after etching with (**b**) laser power and (**d**) number of laser pulses.

The energy around the laser focus is different and conforms to a gaussian distribution [36–38]. Therefore, the degree of the modification region can be affected by the different positions of the focus. The focus position of the silicon surface can be controlled by the movement of the three-dimensional piezoelectric mobile platform, which helps to study the influence of the focus position on the morphology of the microlens after etching. In the experiment, the processing of different laser focus positions was realized by controlling the depth of the silicon sample table. First, the processing was conducted at the laser focus, and gradually deviated from the focus with the processing from inside to outside, and then the circular silicon-based microlens array was obtained via ICP etching. From the SEM image displayed in Figure 5a, the diameter of the final silicon-based concave microlens decreased gradually with the shift in the focus. The atomic force microscope (AFM) image of the silicon-based microlens array in Figure 5b indicates that the depth of the prepared silicon-based concave microlens gradually decreased as the focused energy decreased. A cross-sectional profile of the silicon-based microlens array is shown in Figure 5c. The depth of the silicon-based concave microlens gradually decreased as the focused energy decreased and the processing proceeded outward. The depth of the silicon-based concave microlens was maximum when the silicon surface was at the maximum focus energy, and the depth of the concave microlens gradually decreased with the focus shift (Figure 5d).

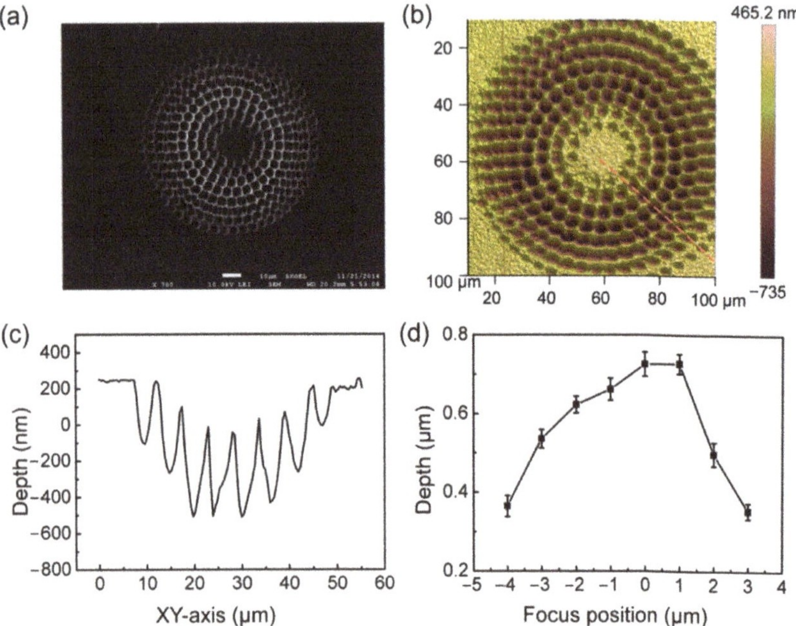

Figure 5. (**a**) SEM and (**b**) AFM images of the concave microlens arrays with different laser focus heights. (**c**) The cross-sectional profile of the concave microlens array. (**d**) Relationship between the depth of the concave microlens array with the laser focus position.

For silicon with different crystal orientations, anisotropy is often observed due to different crystal orientations in the process of wet etching. This leads to an irregular shape when etching a circular structure. To verify the effect of crystal orientation on the etching effect, we replaced silicon wafers with different crystal orientations for the experiments (Figure 6). It was found that the depths of the silicon-based concave microlenses prepared using silicon with three crystal orientations are basically the same, i.e., the morphology of the microlens is not significantly dependent on the silicon crystal orientation. This is because the principle of dry etching involves using the plasma of chemical gas to produce a chemical reaction, and accelerate physical etching and chemical etching through an electric field, so the etching is isotropic. It should be mentioned that by changing the etching gas in the cycle during the etching process, high-aspect-ratio silicon microstructures can be realized [39,40]. Therefore, different patterned silicon-based microlens arrays can be prepared using this method.

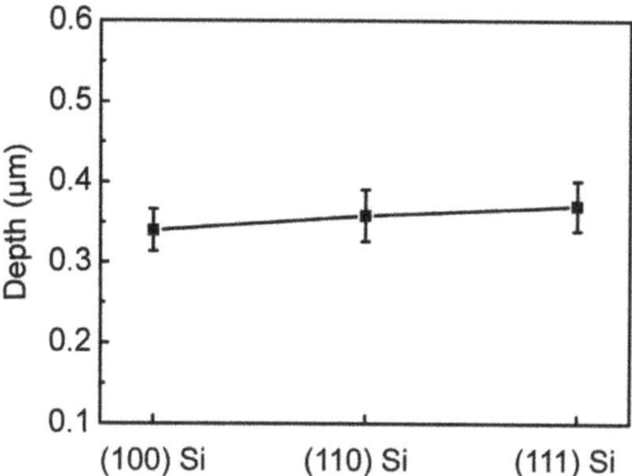

Figure 6. Relationship between the depth of the silicon microlens after etching and three types of silicon with different crystal orientations.

3.2. Microlens Array Fabrication

Femtosecond laser processing has a high degree of freedom, which can plan the trajectory of light spots and then realize the preparation of arbitrary patterned structures. In the experiment, two types of silicon-based microlens arrays were prepared by controlling the movement of a three-dimensional platform. First, the silicon wafer was ablated by a femtosecond laser to form a rectangular arrangement and a honeycomb dense arrangement lattice. Then, the silicon wafer processed using the femtosecond laser was etched via ICP. Owing to the use of the same laser parameters, the modification degree of each part was the same, thus two groups of microlens arrays with the same size and good morphology were formed after dry etching. The two silicon-based microlens arrays were characterized via SEM; the SEM images are shown in Figure 7a,b. Finally, the optical properties of the rectangular array microlens were characterized, as shown in Figure 7c,d. Thus, the silicon-based lens array has good focusing and imaging effects.

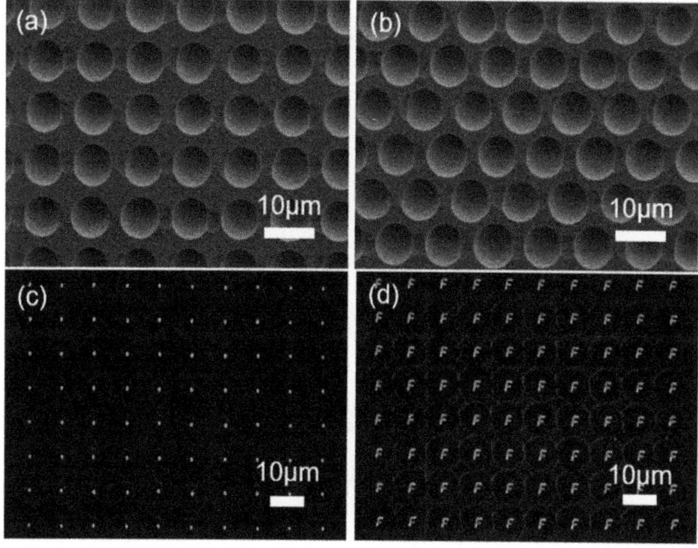

Figure 7. SEM images of (**a**) rectangular microlens array and (**b**) hexagonal microlens array. (**c**) Focal spots and (**d**) optical performance of the hexagonal microlens array.

In addition to the above regular microlens array, an arbitrary arrangement of microlenses with inconsistent structures and sizes can be realized by adjusting the laser parameters and scanning path in the experiment. Through this method, some other specific microlenses were fabricated, such as Chinese knots, petals, and hidden letters, as displayed in Figure 8a–c. The corresponding SEM images obtained after etching each pattern are shown in Figure 8d–f. It is difficult to distinguish the final designed microlens array from the laser-modified pattern. After etching, microlenses of different sizes were formed to realize a clear pattern. In particular, the letters "JLU" were difficult to observe after laser modification. Following etching, the hidden letters can be clearly seen by the distinction of the microlens diameter.

3.3. Multifocal Microlens Array Integration on Silicon Cantilever

Femtosecond lasers have the characteristics of local in situ processing and micro nanostructure preparation at any position of a nonplanar structure. Therefore, we can use this method to realize the controllable preparation of a microlens array on a silicon microcantilever. By adjusting the position of the light spot and the degree of modification, we fabricated microlens arrays of different sizes on a silicon microcantilever (Figure 9a). The size of the lens determines its focal length, and its basic relationship satisfies the following equation:

$$R = \frac{h^2 + r^2}{2h}$$

(1)

$$f = -\frac{R}{2}$$

(2)

where r and h are the radius and depth of the microlens, respectively; R is the radius of curvature of the microlens, and f is the focal length of the microlens. According to Equations (1) and (2), the focal lengths of the leftmost and rightmost microlenses were 7 and 9 μm, respectively. Figure 9b presents the cross-sectional characterization curve of the multifocal microlens array. Because of the different focal lengths of the microlenses, the focus morphology and imaging effect also differ at different positions. The focusing and imaging of the lens arrays at different focus positions are depicted in Figure 9c–f. When the large lens is at the focus, the focus and imaging of the rear small lens are unclear. Contrarily, when the small lens is in focus, the large lens will be out of focus. A multifocal microlens array was successfully fabricated on a silicon microcantilever, enabling the integration of micro-optical devices and MEMSs.

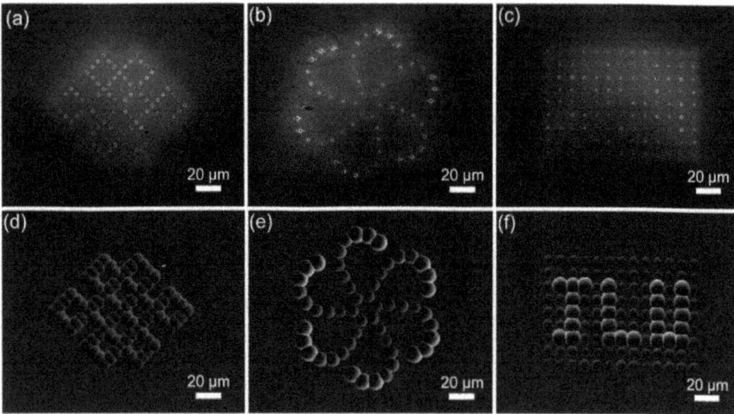

Figure 8. SEM images of (**a**) Chinese knot, (**b**) petal, and (**c**) letters "JLU" before etching; and (**d**) Chinese knot, (**e**) petal, and (**f**) letters "JLU" after etching.

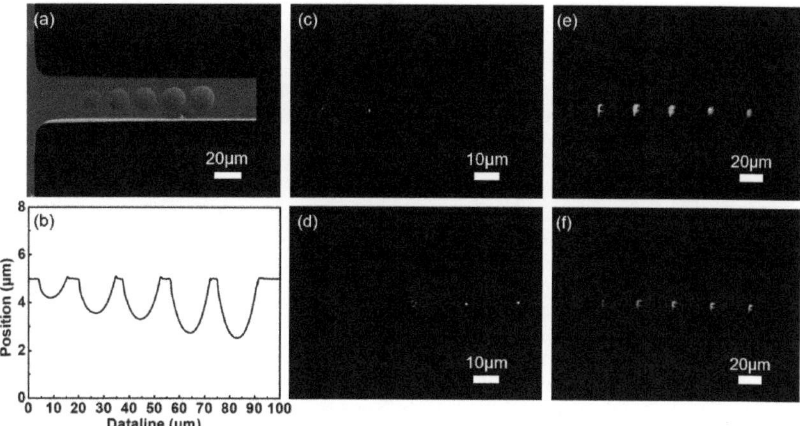

Figure 9. (**a**) SEM image of the silicon-cantilever-integrated multi-focus microlens array. (**b**) The cross-sectional profile of the multi-focus microlens array. (**c**,**d**) Focal spots at different focal locations of the multi-focus microlens array. (**e**,**f**) Optical performance at different focal locations of the multi-focus microlens array.

4. Conclusions

In summary, we investigated the effects of femtosecond laser power and pulse number on the structure of a microlens after etching. It was found that the curvature of the silicon-based concave microlens decreased with an increase in the laser power. With an increase in the number of laser pulses, the diameter and depth of the silicon-based concave microlens first increased and then decreased. The effects of focus change and silicon crystal orientation on the final effect of the experiment were studied. It was found that when the silicon surface is at the maximum focus energy, the depth of the silicon-based concave microlens is at its maximum. With the gradual shift in focus, the depth of the concave microlens gradually decreases. The degree of modification of the laser is unrelated to the crystal direction of silicon. Uniform lens arrays with different arrangements and microlens arrays with arbitrary shapes and sizes were established on the silicon surface. Finally, a multifocal microlens array was fabricated on a silicon microcantilever. Meanwhile, the focusing and imaging effects of the 7 and 9 µm microlenses at different positions were compared. The integration of micro-optical components and MEMSs was realized, providing potential applications for silicon-based MEMSs in the fields of optical communications, digital image processing and biomedicine.

Author Contributions: Conceptualization, X.-Q.L. and B.-R.G.; methodology, B.-X.W., X.-Q.L., B.-R.G. and J.-X.Z.; formal analysis, B.-X.W., J.-X.Z. and J.-Y.Q.; investigation, B.-X.W., J.-X.Z., J.-Y.Q. and M.-R.G.; data curation, X.-Q.L. and J.-X.Z.; writing—original draft preparation, B.-X.W.; writing—review and editing, B.-X.W. and J.-X.Z.; visualization, B.-X.W., J.-X.Z. and M.-R.G.; supervision, X.-Q.L. All authors have read and agreed to the published version of the manuscript.

Funding: This work was supported by the Strategic Priority Research Program of CAS (Grant No. XDC07030303) and the National Natural Science Foundation of China (NSFC, Grant Nos. 62105117, 61960206003, 61825502 and 61805100).

Institutional Review Board Statement: Not applicable.

Data Availability Statement: Not applicable.

Conflicts of Interest: The authors declare no conflict of interest.

References

1. Lu, Q.; Wang, Y.; Wang, X.; Yao, Y.; Wang, X.; Huang, W. Review of micromachined optical accelerometers: From mg to sub-mu g. *Opto-Electron. Adv.* **2021**, *4*, 200045. [CrossRef]
2. Joshi, P.K. Recent Development in Applications of Optical MEMS: A Review. *Helix* **2018**, *8*, 4345–4348. [CrossRef]
3. Tortschanoff, A.; Lenzhofer, M.; Frank, A.; Wildenhain, M.; Sandner, T.; Schenk, H.; Scherf, W.; Kenda, A. Position encoding and phase control of resonant MOEMS mirrors. *Sens. Actuators A Phys.* **2010**, *162*, 235–240. [CrossRef]
4. Shao, Y.; Dickensheets, D.L. MOEMS 3-D scan mirror for single-point control of beam deflection and focus. *J. Microlith. Microfab.* **2005**, *4*, 041502. [CrossRef]
5. Erdmann, L.; Deparnay, A.; Maschke, G.; Langle, M.L.; Brunner, R. MOEMS-Based lithography for the fabrication of micro-optical components. *J. Microlith. Microfab.* **2005**, *4*, 041601. [CrossRef]
6. Toshiyoshi, H.; Su, G.-J.; LaCosse, J.; Wu, M.C. A surface micromachined optical scanner array using photoresist lenses fabricated by a thermal reflow process. *J. Lightwave Techol.* **2003**, *21*, 1700–1708. [CrossRef]
7. Kwon, S.; Lee, L.P. Stacked Two Dimensional Microlens Scanner for Micro Confocal Imaging Array. In Proceedings of the Fifteenth IEEE International Conference on Micro Electro Mechanical Systems (Cat. No. 02CH37266), Las Vegas, NV, USA, 24 January 2002; pp. 483–486.
8. Ersumo, N.T.; Yalcin, C.; Antipa, N.; Pegard, N.; Waller, L.; Lopez, D.; Muller, R. A micromirror array with annular partitioning for high-speed random-access axial focusing. *Light Sci. Appl.* **2020**, *9*, 183. [CrossRef] [PubMed]
9. Fan, Z.-B.; Qiu, H.-Y.; Zhang, H.-L.; Pang, X.-N.; Zhou, L.-D.; Liu, L.; Ren, H.; Wang, Q.-H.; Dong, J.-W. A broadband achromatic metalens array for integral imaging in the visible. *Light Sci. Appl.* **2019**, *8*, 67. [CrossRef]
10. Zhao, R.; Huang, L.; Wang, Y. Recent advances in multi-dimensional metasurfaces holographic technologies. *PhotoniX* **2020**, *1*, 20. [CrossRef]
11. Zou, X.; Zheng, G.; Yuan, Q.; Zang, W.; Chen, R.; Li, T.; Li, L.; Wang, S.; Wang, Z.; Zhu, S. Imaging based on metalenses. *PhotoniX* **2020**, *1*, 2. [CrossRef]
12. Tanida, J.; Shogenji, R.; Kitamura, Y.; Yamada, K.; Miyamoto, M.; Miyatake, S. Color imaging with an integrated compound imaging system. *Opt. Express* **2003**, *11*, 2109–2117. [CrossRef] [PubMed]
13. Horisaki, R.; Irie, S.; Ogura, Y.; Tanida, J. Three-Dimensional information acquisition using a compound imaging system. *Opt. Rev.* **2007**, *14*, 347–350. [CrossRef]
14. Shogenji, R.; Kitamura, Y.; Yamada, K.; Miyatake, S.; Tanida, J. Bimodal fingerprint capturing system based on compound-eye imaging module. *Appl. Opt.* **2004**, *43*, 1355–1359. [CrossRef] [PubMed]
15. Zhou, X.T.; Peng, Y.Y.; Peng, R.; Zeng, X.Y.; Zhang, Y.A.; Guo, T.L. Fabrication of Large-Scale Microlens Arrays Based on Screen Printing for Integral Imaging 3D Display. *ACS Appl. Mater. Interfaces* **2016**, *8*, 24248–24255. [CrossRef] [PubMed]
16. Bae, S.I.; Kim, K.; Yang, S.; Jang, K.W.; Jeong, K.H. Multifocal microlens arrays using multilayer photolithography. *Opt. Express* **2020**, *28*, 9082–9088. [CrossRef]
17. Lee, J.H.; Chang, S.; Kim, M.S.; Kim, Y.J.; Kim, H.M.; Song, Y.M. High-Identical Numerical Aperture, Multifocal Microlens Array through Single-Step Multi-Sized Hole Patterning Photolithography. *Micromachines* **2020**, *11*, 1068. [CrossRef] [PubMed]
18. Lian, G.G.; Liu, Y.S.; Tao, K.K.; Xing, H.M.; Huang, R.X.; Chi, M.B.; Zhou, W.C.; Wu, Y.H. Fabrication and Characterization of Curved Compound Eyes Based on Multifocal Microlenses. *Micromachines* **2020**, *11*, 854. [CrossRef]
19. Park, M.K.; Lee, H.J.; Park, J.S.; Kim, M.; Bae, J.M.; Mahmud, I.; Kim, H.R. Design and Fabrication of Multi-Focusing Micro lens Array with Different Numerical Apertures by using Thermal Reflow Method. *J. Opt. Soc. Korea* **2014**, *18*, 71–77. [CrossRef]
20. Li, Z.-Z.; Wang, L.; Fan, H.; Yu, Y.-H.; Chen, Q.-D.; Juodkazis, S.; Sun, H.-B. O-FIB: Far-Field-Induced near-field breakdown for direct nanowriting in an atmospheric environment. *Light Sci. Appl.* **2020**, *9*, 41. [CrossRef]
21. Wang, H.; Zhang, Y.-L.; Han, D.-D.; Wang, W.; Sun, H.-B. Laser fabrication of modular superhydrophobic chips for reconfigurable assembly and self-propelled droplet manipulation. *PhotoniX* **2021**, *2*, 17. [CrossRef]
22. Yin, D.; Feng, J.; Ma, R.; Liu, Y.F.; Zhang, Y.L.; Zhang, X.L.; Bi, Y.G.; Chen, Q.D.; Sun, H.B. Efficient and mechanically robust stretchable organic light-emitting devices by a laser-programmable buckling process. *Nat. Commun.* **2016**, *7*, 11573. [CrossRef] [PubMed]
23. Liu, X.-Q.; Bai, B.-F.; Chen, Q.-D.; Sun, H.-B. Etching-Assisted femtosecond laser modification of hard materials. *Opto-Electron. Adv.* **2019**, *2*, 19002101–19002114. [CrossRef]
24. Fang, H.H.; Yang, J.; Ding, R.; Chen, Q.D.; Wang, L.; Xia, H.; Feng, J.; Ma, Y.G.; Sun, H.B. Polarization dependent two-photon properties in an organic crystal. *Appl. Phys. Lett.* **2010**, *97*, 101101. [CrossRef]
25. Sun, Y.L.; Dong, W.F.; Yang, R.Z.; Meng, X.; Zhang, L.; Chen, Q.D.; Sun, H.B. Dynamically Tunable Protein Microlenses. *Angew. Chem. Int. Ed.* **2012**, *51*, 1558–1562. [CrossRef]
26. Jiang, H.B.; Zhang, Y.L.; Han, D.D.; Xia, H.; Feng, J.; Chen, Q.D.; Hong, Z.R.; Sun, H.B. Bioinspired Fabrication of Superhydrophobic Graphene Films by Two-Beam Laser Interference. *Adv. Funct. Mater.* **2014**, *24*, 4595–4602. [CrossRef]
27. Feng, T.; Chen, G.; Han, H.; Qiao, J. Femtosecond-Laser-Ablation Dynamics in Silicon Revealed by Transient Reflectivity Change. *Micromachines* **2022**, *13*, 14. [CrossRef] [PubMed]
28. Phillips, K.C.; Gandhi, H.H.; Mazur, E.; Sundaram, S.K. Ultrafast laser processing of materials: A review. *Adv. Opt. Photon.* **2015**, *7*, 684–712. [CrossRef]

29. Lin, Z.; Hong, M. Femtosecond Laser Precision Engineering: From Micron, Submicron, to Nanoscale. *Ultrafast Sci.* **2021**, *2021*, 9783514. [CrossRef]

30. Yulianto, N.; Kadja, G.T.M.; Bornemann, S.; Gahlawat, S.; Majid, N.; Triyana, K.; Abdi, F.F.; Wasisto, H.S.; Waag, A. Ultrashort Pulse Laser Lift-Off Processing of InGaN/GaN Light-Emitting Diode Chips. *ACS Appl. Electron. Mater.* **2021**, *3*, 778–788. [CrossRef]

31. Yulianto, N.; Refino, A.D.; Syring, A.; Majid, N.; Mariana, S.; Schnell, P.; Wahyuono, R.A.; Triyana, K.; Meierhofer, F.; Daum, W.; et al. Wafer-Scale transfer route for top–down III-nitride nanowire LED arrays based on the femtosecond laser lift-off technique. *Microsyst. Nanoeng.* **2021**, *7*, 32. [CrossRef]

32. Juodkazis, S.; Nishimura, K.; Misawa, H.; Ebisui, T.; Waki, R.; Matsuo, S.; Okada, T. Control over the crystalline state of sapphire. *Adv. Mater.* **2006**, *18*, 1361. [CrossRef]

33. Liu, X.Q.; Chen, Q.D.; Guan, K.M.; Ma, Z.C.; Yu, Y.H.; Li, Q.K.; Tian, Z.N.; Sun, H.B. Dry-Etching-Assisted femtosecond laser machining. *Laser Photonics Rev.* **2017**, *11*, 3. [CrossRef]

34. Liu, X.Q.; Yu, L.; Chen, Q.D.; Sun, H.B. Mask-Free construction of three-dimensional silicon structures by dry etching assisted gray-scale femtosecond laser direct writing. *Appl. Phys. Lett.* **2017**, *110*, 091602. [CrossRef]

35. Liu, X.Q.; Yang, S.N.; Yu, L.; Chen, Q.D.; Zhang, Y.L.; Sun, H.B. Rapid Engraving of Artificial Compound Eyes from Curved Sapphire Substrate. *Adv. Funct. Mater.* **2019**, *29*, 1900037. [CrossRef]

36. Han, D.D.; Zhang, Y.L.; Ma, J.N.; Liu, Y.Q.; Han, B.; Sun, H.B. Light-Mediated Manufacture and Manipulation of Actuators. *Adv. Mater.* **2016**, *28*, 8328–8343. [CrossRef]

37. Zhang, Y.L.; Tian, Y.; Wang, H.; Ma, Z.C.; Han, D.D.; Niu, L.G.; Chen, Q.D.; Sun, H.B. Dual-3D Femtosecond Laser Nanofabrication Enables Dynamic Actuation. *ACS Nano* **2019**, *13*, 4041–4048. [CrossRef]

38. Wang, W.; Liu, Y.Q.; Liu, Y.; Han, B.; Wang, H.; Han, D.D.; Wang, J.N.; Zhang, Y.L.; Sun, H.B. Direct Laser Writing of Superhydrophobic PDMS Elastomers for Controllable Manipulation via Marangoni Effect. *Adv. Funct. Mater.* **2017**, *27*, 1702946. [CrossRef]

39. Huff, M. Recent Advances in Reactive Ion Etching and Applications of High-Aspect-Ratio Microfabrication. *Micromachines* **2021**, *12*, 991. [CrossRef]

40. Refino, A.D.; Yulianto, N.; Syamsu, I.; Nugroho, A.P.; Hawari, N.H.; Syring, A.; Kartini, E.; Iskandar, F.; Voss, T.; Sumboja, A.; et al. Versatilely tuned vertical silicon nanowire arrays by cryogenic reactive ion etching as a lithium-ion battery anode. *Sci. Rep.* **2021**, *11*, 19779. [CrossRef]

 *micromachines*

Article

Analysis of Main Error Sources for the Error Motion Measurement of a Precision Shafting Using a T-Type Capacitive Sensor

Kui Xiang [1], Wen Wang [2,*] and Zichen Chen [1]

[1] Key Laboratory of Advanced Manufacturing Technology of Zhejiang Province, School of Mechanical Engineering, Zhejiang University, Hangzhou 310027, China; 11125034@zju.edu.cn (K.X.); chenzc@zju.edu.cn (Z.C.)

[2] School of Mechanical Engineering, Hangzhou Dianzi University, Hangzhou 310018, China

* Correspondence: wangwn@hdu.edu.cn; Tel.: +86-571-8691-9054

Abstract: As a key indicator reflecting the working accuracy of rotary functional units, the error motions of the precision shafting are very necessary to be measured. In this paper, the main error sources for the error motion measurement of a precision shafting using a T-type capacitive sensor were investigated. The theoretical modeling error due to the approximate simplification for the output capacitance expressions was firstly analyzed. By means of the 3D-FEA method, the influence of fringe effects was subsequently investigated. Finally, the analysis of electrode installation errors was emphasized on the tilt error of the cylindrical electrode and coaxiality error of the fan-shaped electrode by establishing mathematical models and numerical simulation. Based on the theoretical analysis and simulation results, the methods of decreasing the approximate error and the nonlinear error caused by fringe effects were subsequently proposed; for the installation errors, the tilt error of cylindrical electrode only makes the solution of phase angle have a certain deviation and has almost no effect on solving the radial displacement, especially for the measurement range less than 0.1 mm; the measurement of the rotor tilt displacement was basically not affected by the coaxiality error of the fan-shaped electrode.

Keywords: precision shafting; capacitive sensor; error sources analysis

Citation: Xiang, K.; Wang, W.; Chen, Z. Analysis of Main Error Sources for the Error Motion Measurement of a Precision Shafting Using a T-Type Capacitive Sensor. *Micromachines* **2022**, *13*, 221. https://doi.org/10.3390/mi13020221

Academic Editors: Benny C. F. Cheung and Jiang Guo

Received: 7 December 2021
Accepted: 26 January 2022
Published: 29 January 2022

Publisher's Note: MDPI stays neutral with regard to jurisdictional claims in published maps and institutional affiliations.

1. Introduction

The rotary functional units with precision shafting as the core are widely used in high-end equipment such as high-precision CNC machine tools, EUV lithography equipment, axial compressors and small agile satellites [1–7]. As a key indicator reflecting the working accuracy of rotary functional units, the error motions of the precision shafting are very necessary to be measured. This measurement provides valuable insight into the error motions and performance [8], which is favorable to optimize these rotary units.

Capacitive sensors with good stability, high speed and adaptability of extreme environments are widely adopted to measure the displacement of motion targets [9–12]. Chapman et al. carried out research on the measurement of rotary axis error motions using a cylindrical capacitive sensor (CCS) with multiple electrodes of large sense area [13–15]. In order to measure the multi-degree-of-freedom (multi-DOF) error motions, several separate electrodes need to be manufactured and orderly mounted along with the guard ring, which is difficult.

Compared to the CCS used in the above research, a novel T-type capacitive sensor (T-type CS) with an integrated structure was developed to achieve the simultaneous online measurement of the 5-DOF error motions of a spindle, as shown in Figures 1 and 2 [16]. Both the radial electrode group (REG) and the end part electrode group (EPEG) were fabricated based on a flexible printed circuit board, which could produce a quite thinner electrode (0.035 mm) and a tiny gap (0.15 mm) between the electrode and guard ring.

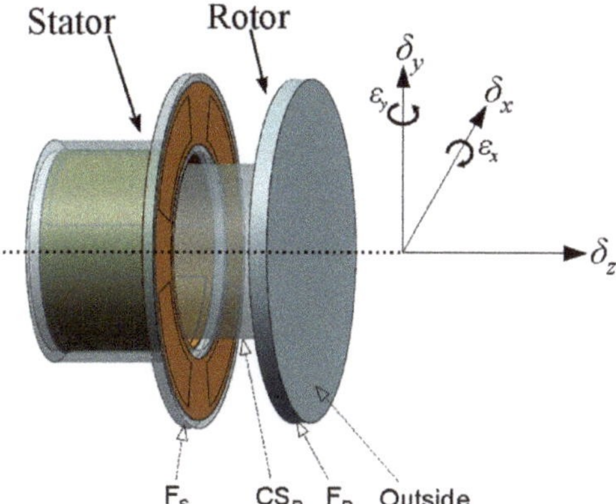

Figure 1. Schematic of a T-type CS. The error motions δ_x and δ_y are detected using the REG, which is in the annular groove of the stator cylindrical bore; δ_z and ε_x and ε_y are detected using the EPEG, which is in the annular groove of the stator end part.

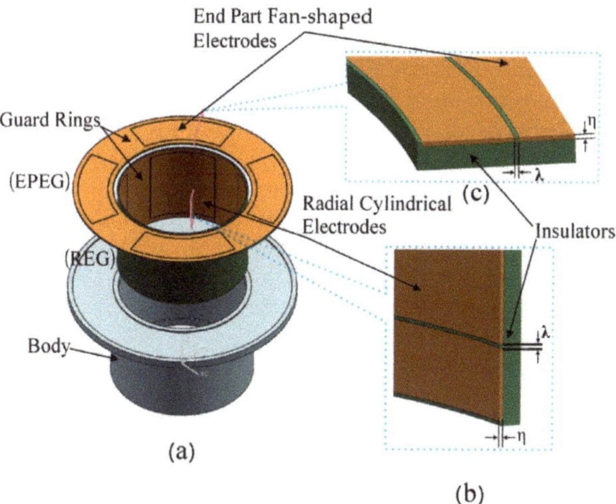

Figure 2. Schematic of the stator: (**a**) The exploded view of the structure; (**b**,**c**) partial cross-sectional view of the stator. η is the thickness of the electrode, and λ is the gap between the electrode and guard ring.

This paper focuses on the analysis of main error sources for the error motion measurement of a precision shafting using a T-type CS, as shown in Figure 3. Firstly, the theoretical modeling error due to the approximate simplification for the output capacitance expression of sensing electrodes was analyzed. Then, the influence of fringe effects was investigated using the 3D-FEA method. Finally, mathematical modeling and numerical simulation were conducted to analyze electrode installation errors, emphasizing the tilt error of the cylindrical electrode and coaxiality error of the fan-shaped electrodes.

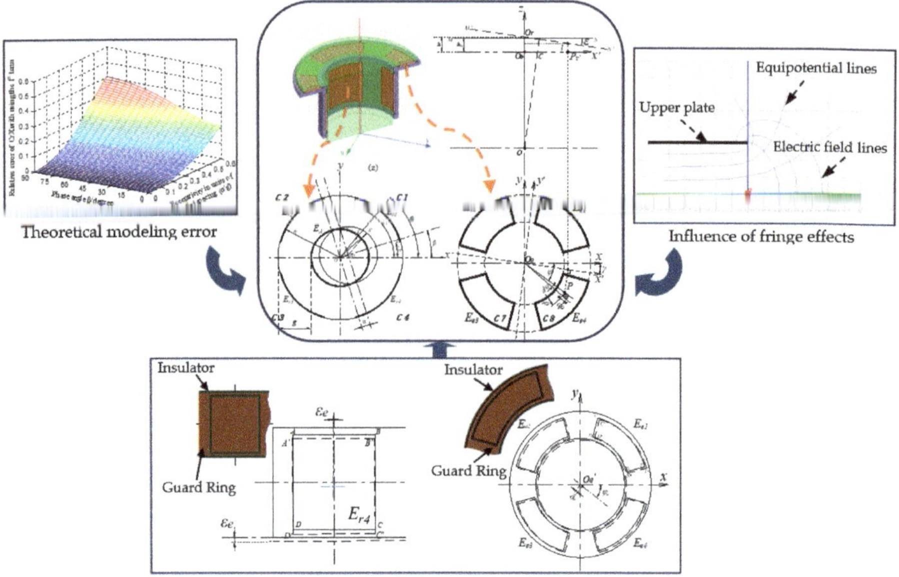

Figure 3. The general sketch of the main research content.

2. Basic Principle

2.1. Radial Error Motion Measurement

This measurement is performed by the REG, which is constituted by four cylindrical electrodes E_{r1}~E_{r4}. The four cylindrical electrodes and the radial measuring surface (CS_R) of the rotor make up four cylindrical capacitors, respectively; the corresponding capacitance are denoted by C1~C4, as shown in Figure 4 [16].

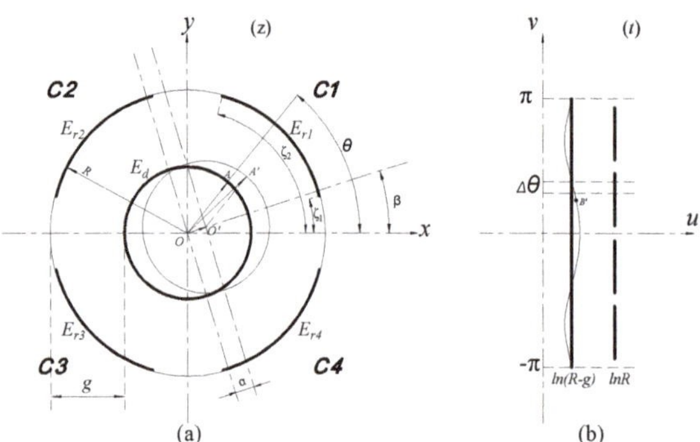

Figure 4. The calculation model of the radial error motions: (**a**) Boundary before transformation; (**b**) boundary after transformation.

Due to the arc-shaped boundary of these electrodes, the conformal mapping method was adopted to solve the Poisson Equation in electrostatic fields [17]. When the rotor has radial displacement with eccentricity α at phase angle β, caused by the generated radial run-out of a spindle, the resulting variation Δu of the equivalent spacing at any point in the complex plane t can be approximately derived as [16]:

$$\Delta u = \frac{\alpha \cos(\theta - \beta)}{R} \tag{1}$$

where R is the inside radius of the cylindrical electrode, θ is the argument of the complex number corresponding to any point on the arc-shaped boundary in the plane z.

Thereby, the differential expression of the output capacitance of a cylindrical electrode with axial width of w can be illustrated as [16]:

$$\Delta C_r = \varepsilon_0 \varepsilon_r \frac{R \Delta \theta w}{g - \alpha \cos(\theta - \beta)} \tag{2}$$

where ε_0 is the electric permittivity of vacuum, ε_r is the relative dielectric constant; g is the initial radial spacing and $\Delta \theta$ is an equivalent length of a micro-unit that is taken on the equivalent boundary in the plane t.

Then the output capacitance of the cylindrical electrode in each quadrant is approximately expressed as follows [16]:

$$C1 = \varepsilon_0 \varepsilon_r R w \int_{\zeta_1}^{\zeta_2} \frac{1}{g - \alpha \cos(\theta - \beta)} d\theta \tag{3}$$

$$C2 = \varepsilon_0 \varepsilon_r R w \int_{\pi - \zeta_2}^{\pi - \zeta_1} \frac{1}{g - \alpha \cos(\theta - \beta)} d\theta \tag{4}$$

$$C3 = \varepsilon_0 \varepsilon_r R w \int_{\pi + \zeta_1}^{\pi + \zeta_2} \frac{1}{g - \alpha \cos(\theta - \beta)} d\theta \tag{5}$$

$$C4 = \varepsilon_0 \varepsilon_r R w \int_{2\pi - \zeta_2}^{2\pi - \zeta_1} \frac{1}{g - \alpha \cos(\theta - \beta)} d\theta \tag{6}$$

where $(\zeta_2 - \zeta_1)$ denotes the angular size of the cylindrical electrode.

Based on Equations (3)–(6), the expressions of the differential output capacitance of the REG can be expressed by:

$$CrX = C1 + C4 - C2 - C3 \tag{7}$$
$$CrY = C1 + C2 - C3 - C4 \tag{8}$$

where CrX and CrY, respectively, correspond to the components δ_x and δ_y of the displacement α in the X and Y directions.

Then, the radial error motions of the spindle can be estimated by the following equations:

$$\delta_x = f_x(C1 + C4 - C2 - C3) \tag{9}$$

$$\delta_y = f_y(C1 + C2 - C3 - C4) \tag{10}$$

where f_x and f_y are denoted the transition functions between the measured capacitance of the cylindrical electrodes and the radial run-out along the x, y directions, respectively.

2.2. Axial Error Motion Measurement

As shown in Figure 2, the fan-shaped electrodes ($E_{e1} \sim E_{e4}$) of the EPEG are also distributed in the same configuration as radial electrodes. They, along with the inside annular end face of the rotor flange (F_R), make up four capacitors, respectively; the corresponding capacitance of them are denoted by C5~C8, as shown in Figure 5 [16].

When the rotor has an axial error motion along the z-axis, the variation quantity of the sum of the output capacitance of the fan-shaped electrode is given by [16]:

$$\begin{aligned} \Delta CeZ &= CeZ' - CeZ \\ &= 4\varepsilon_0 \varepsilon_r S_e \left(\frac{1}{g_{a0} + \Delta z} - \frac{1}{g_{a0}} \right) \end{aligned} \tag{11}$$

where S_e is the area of the fan-shaped electrode, g_{a0} is the initial value of the axial spacing, the variation of the axial spacing Δz is defined by $\Delta z = g_a - g_{a0}$, CeZ is the sum of the output capacitance of the fan-shaped electrodes under the axial spacing g_{a0} and CeZ' is that of the fan-shaped electrodes under the axial spacing g_a.

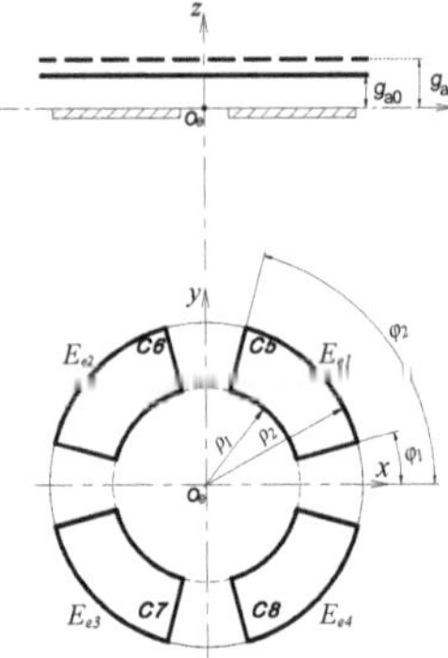

Figure 5. The calculation model of the axial error motion.

According to Equation (11), the expressions for estimating the axial error motion of the spindle can be derived as:

$$\delta_z = f_z(C5 + C6 + C7 + C8) \tag{12}$$

2.3. Tilt Error Motion Measurement

The calculation model for the measurement of tilt error motions by EPEG is shown in Figure 6 [16]. The line segment *MN* represents the axial measuring surface of the rotor (i.e., the inside annular end face of the rotor flange), and its perpendicular bisector is the medial axis of the rotor.

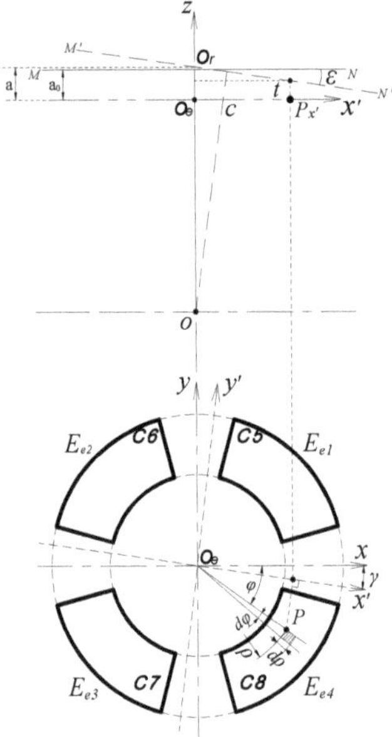

Figure 6. The calculation model of the tilt error motions.

In the boundary region of the fan-shaped electrode, a micro-plane element ΔB_P at any point P is taken as the solving unit (the area of it is denoted as ΔS_P). The micro-plane element

ΔB_P and its opposite counterpart ΔA_Q compose a plane parallel capacitor (ΔA_Q is in the axial measuring surface of the rotor), and the spacing between them is derived as [16]:

$$t = a - k \cdot \rho \cos(\varphi - \gamma) \tag{13}$$

where a is the distance from the intersection point O_r to the EPEG; $k = \tan(\varepsilon)$, ε and γ are the amplitude and yaw angle of the tilt displacement of the rotor, respectively.

Thereby, the capacitance of the plane parallel capacitor is [16]:

$$\Delta C_e = \frac{\varepsilon_0 \varepsilon_r \Delta S_P}{t} = \varepsilon_0 \varepsilon_r \frac{\rho d\varphi \cdot d\rho}{a - k \cdot \rho \cos(\varphi - \gamma)} \tag{14}$$

Then, the output capacitance of the fan-shaped electrode E_{e4} can be approximately obtained by integrating Equation (14) over the area of the fan-shaped electrode [16]:

$$C8 = \varepsilon_0 \varepsilon_r \int_{2\pi-\varphi_2}^{2\pi-\varphi_1} \int_{\rho_1}^{\rho_2} \frac{\rho d\rho}{a - k \cdot \rho \cos(\varphi + \gamma)} d\varphi \tag{15}$$

where $(\varphi_2 - \varphi_1)$ are the angular size of each fan-shaped electrode, and ρ_1 and ρ_2 are the inside and outside radius of the electrode, respectively.

By considering the minor tilt displacement of a precision spindle, the power series expansion by ignoring higher-order terms was applied to express the Equation (15) ($k \ll a$):

$$C8 = \frac{\varepsilon_0 \varepsilon_r}{a} \int_{2\pi-\varphi_2}^{2\pi-\varphi_1} \left[b + \frac{kc}{a} \cos(\varphi + \gamma) \right] d\varphi \tag{16}$$

where the factors b and c are given by $b = \frac{\rho_2^2 - \rho_1^2}{2}$, $c = \frac{\rho_2^3 - \rho_1^3}{3}$.

By analogy with the electrode E_{e4}, the output capacitance of the other electrodes can be approximately expressed as follows:

$$C7 = \frac{\varepsilon_0 \varepsilon_r}{a} \int_{\pi+\varphi_1}^{\pi+\varphi_2} \left[b + \frac{kc}{a} \cos(\varphi + \gamma) \right] d\varphi \tag{17}$$

$$C6 = \frac{\varepsilon_0 \varepsilon_r}{a} \int_{\pi-\varphi_2}^{\pi-\varphi_1} \left[b + \frac{kc}{a} \cos(\varphi + \gamma) \right] d\varphi \tag{18}$$

$$C5 = \frac{\varepsilon_0 \varepsilon_r}{a} \int_{\varphi_1}^{\varphi_2} \left[b + \frac{kc}{a} \cos(\varphi + \gamma) \right] d\varphi \tag{19}$$

By making an arrangement, Equations (16)–(19) are given as:

$$C8 = \frac{\varepsilon_0 \varepsilon_r}{a} \int_{\varphi_1}^{\varphi_2} \left[b + \frac{kc}{a} \cos(\varphi - \gamma) \right] d\varphi \tag{20}$$

$$C7 = \frac{\varepsilon_0 \varepsilon_r}{a} \int_{\varphi_1}^{\varphi_2} \left[b - \frac{kc}{a} \cos(\varphi + \gamma) \right] d\varphi \tag{21}$$

$$C6 = \frac{\varepsilon_0 \varepsilon_r}{a} \int_{\varphi_1}^{\varphi_2} \left[b - \frac{kc}{a} \cos(\varphi - \gamma) \right] d\varphi \tag{22}$$

$$C5 = \frac{\varepsilon_0 \varepsilon_r}{a} \int_{\varphi_1}^{\varphi_2} \left[b + \frac{kc}{a} \cos(\varphi + \gamma) \right] d\varphi \tag{23}$$

By combining Equations (20)–(23), the distance a can be derived as:

$$a = \frac{4\varepsilon_0 \varepsilon_r b \cdot (\varphi_2 - \varphi_1)}{C8 + C7 + C6 + C5} \tag{24}$$

Based on Equations (16)–(19), the expressions of the differential output capacitance of the EPEG can be expressed as follows:

$$\text{CeX} = \text{C5} + \text{C6} - \text{C7} - \text{C8} \tag{25}$$

$$\text{CeY} = \text{C5} + \text{C8} - \text{C6} - \text{C7} \tag{26}$$

where CeX and CeY, respectively, correspond to the components ε_x (around the x-axis), ε_y (around the y-axis) of the tilt displacement ε.

Then, the tilt error motions of the spindle can be estimated by the following equations:

$$\varepsilon_x = f_{ex}(\text{C5} + \text{C6} - \text{C7} - \text{C8}) \tag{27}$$

$$\varepsilon_y = f_{ey}(\text{C5} + \text{C8} - \text{C6} - \text{C7}) \tag{28}$$

where f_{ex} and f_{ey} are denoted as the transition functions between the measured capacitance of the fan-shaped electrodes and the tilt displacements about the x- and y-axis, respectively.

3. Main Error Sources Analysis of the T-Type CS

3.1. Theoretical Modeling Error Analysis

In order to theoretically analyze the radial error motion measurement with the T-type CS, we consider a small radial displacement of the precision spindle, i.e., $|\alpha/(R - g)| < 1$. The power series expansion with first-order approximation (Equation (1)) was applied to express the equivalent spacing changing Δu. The exact derived expression of it is:

$$\Delta u_o = \frac{1}{2}\ln\left\{\frac{(R - g)^2 + 2\alpha(R - g)\cos(\theta - \beta) + \alpha^2}{(R - g)^2}\right\} \tag{29}$$

Then, the approximate error of Equation (1) can be assessed by:

$$\Delta u_{app.error} = \left|\frac{\Delta u_o - \Delta u}{\Delta u_o}\right| \tag{30}$$

Figure 7 shows the variation situation of error $\Delta u_{app.error}$ with the changing of parameter terms $\alpha/(R - g)$ and $\cos(\theta - \beta)$. As mentioned above, the radial displacement α of a precision spindle is generally very small. In order to achieve relatively high sensitivity, the cylindrical electrode radius R is often designed to be tens of millimeters or even larger. Thus, the variation range of the term $\alpha/(R - g)$ is set to be $[0, 0.1]$. It can be observed that, when the ratio $\alpha/(R - g)$ is less than 0.01, the error $\Delta u_{app.error}$ basically keeps lower than 2% with the variation in the term $\cos(\theta - \beta)$ in the entire phase region (the value range of $\cos(\theta - \beta)$ is $[0, 1]$ correspondingly), except for a few phase angles. Therefore, a highly accurate value of the changing Δu calculated by Equation (1) can be obtained if the ratio $\alpha/(R - g)$ is less than 0.01, which can be achieved by adjusting the structural parameters of the sensor.

Since a highly accurate value of Δu can be obtained by setting the ratio $\alpha/(R - g) \leq 0.01$, the expressions (3)–(6) established based on Equation (1) could be more accurate to reflect the capacitance changing caused by the radial displacement, and is also for the expressions of the differential output capacitance of the REG derived as:

$$\text{CrX} = \frac{2\varepsilon_0\varepsilon_r Rw}{g\sqrt{1 - \left(\frac{\alpha}{g}\right)^2}}\left(\arctan\frac{2\frac{\alpha}{g}\sqrt{1 - \left(\frac{\alpha}{g}\right)^2}\sin\zeta_2\cos\beta}{1 - \left(\frac{\alpha}{g}\right)^2(\cos^2\beta + \sin^2\zeta_2)} - \arctan\frac{2\frac{\alpha}{g}\sqrt{1 - \left(\frac{\alpha}{g}\right)^2}\sin\zeta_1\cos\beta}{1 - \left(\frac{\alpha}{g}\right)^2(\cos^2\beta + \sin^2\zeta_1)}\right) \tag{31}$$

$$\text{CrY} = \frac{2\varepsilon_0\varepsilon_r Rw}{g\sqrt{1 - \left(\frac{\alpha}{g}\right)^2}}\left(\arctan\frac{2\frac{\alpha}{g}\sqrt{1 - \left(\frac{\alpha}{g}\right)^2}\cos\zeta_1\sin\beta}{1 + \left(\frac{\alpha}{g}\right)^2(\cos^2\beta + \sin^2\zeta_1 - 2)} - \arctan\frac{2\frac{\alpha}{g}\sqrt{1 - \left(\frac{\alpha}{g}\right)^2}\cos\zeta_2\sin\beta}{1 + \left(\frac{\alpha}{g}\right)^2(\cos^2\beta + \sin^2\zeta_2 - 2)}\right) \tag{32}$$

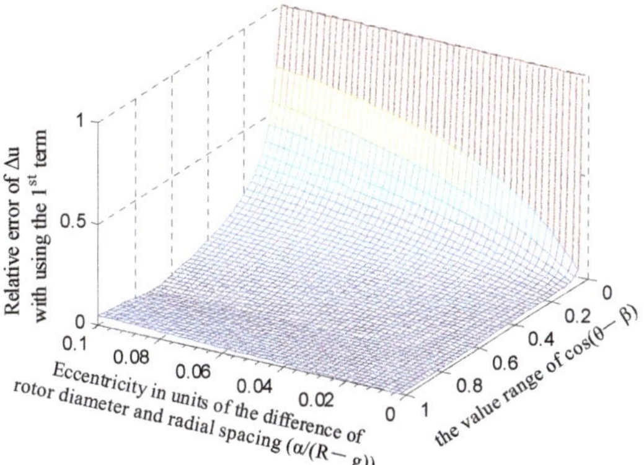

Figure 7. Error distribution of the approximate calculation of the changing Δu in the complex plane t.

In a manner similar to the simplification process on the equivalent spacing changing Δu, the first-order approximation of the Equations (31) and (32) could intuitively reflect the variation's relation of the radial displacement and the differential capacitance (CrX and CrY). Thus, the expressions of the displacement α and phase angle β can be deduced in the following form:

$$\mathrm{CrX_{tl}} = \frac{4\varepsilon_0\varepsilon_r Rw(\sin\zeta_2 - \sin\zeta_1)}{g^2} \cdot \alpha\cos\beta \tag{33}$$

$$\mathrm{CrY_{tl}} = \frac{4\varepsilon_0\varepsilon_r Rw(\cos\zeta_1 - \cos\zeta_2)}{g^2} \cdot \alpha\sin\beta \tag{34}$$

Therefore, the expression used to assess the approximate error (taking the CrX as an example) is given as:

$$\mathrm{CrX}_{app.error} = \left| \frac{\mathrm{CrX} - \mathrm{CrX_{tl}}}{\mathrm{CrX}} \right| \tag{35}$$

As shown in Figure 8, at each phase angle β, the approximate error $\mathrm{CrX}_{app.error}$ all increases with the increase in the ratio α/g. When the ratio α/g becomes larger than 0.2, the variation tendency is more significant. Figure 9 shows the error of the calculated capacitance $\mathrm{CrX_{tl}}$ relative to the simulation value CrXf as the rotor has a displacement along the direction of y = x. The dependence of the relative error on the parameter α is basically consistent with the variation tendency shown in Figure 8. Note that the value CrXf was obtained based on the simulation model of the T-type CS previously established with the design parameters in [16]; these parameters were also adopted to calculate the theoretical capacitance.

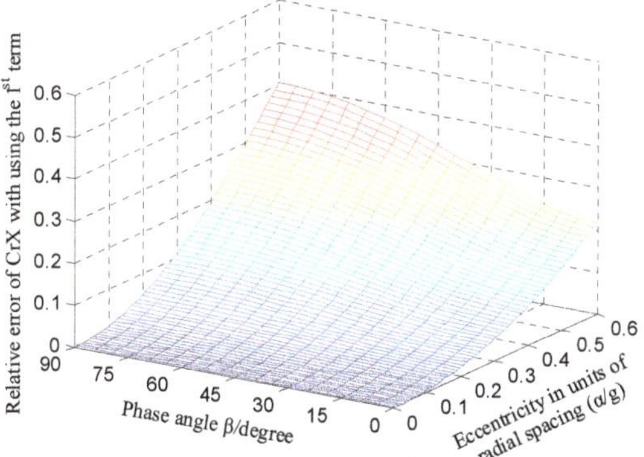

Figure 8. Calculation error distribution of CrX (expansion with first-order approximation).

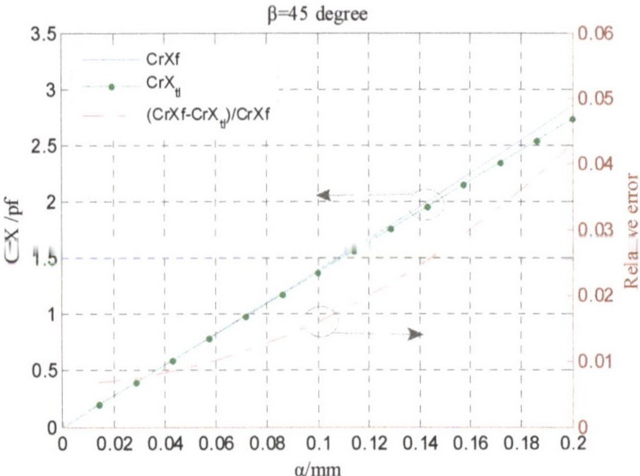

Figure 9. Simulation values and theoretical one calculated by Equation (33) comparison of radial differential output capacitance.

From the analysis in Figure 8, it can be known that diminishing the ratio α/g could make the error $\mathrm{CrX}_{app.error}$ reduce further, but the measurement range and sensitivity of the sensor will be affected to some extent. By considering the inherent nonlinearity of the capacitor output with spacing change, the power series expansion with a third-order approximation was derived to improve the calculation accuracy [18]:

$$\mathrm{CrX}_{tl3} = \frac{4\varepsilon_0\varepsilon_r Rw(\sin\zeta_2 - \sin\zeta_1)}{g} \cdot \left(\frac{\alpha}{g} + 0.95\left(\frac{\alpha}{g}\right)^3 \right) \cdot \cos\beta \tag{36}$$

$$\mathrm{CrY}_{tl3} = \frac{4\varepsilon_0\varepsilon_r Rw(\cos\zeta_1 - \cos\zeta_2)}{g} \cdot \left(\frac{\alpha}{g} + 0.95\left(\frac{\alpha}{g}\right)^3 \right) \cdot \sin\beta \tag{37}$$

Compared to the error distribution shown in Figure 8, the error $\mathrm{CrX}_{app.error}$ is significantly decreased overall for the calculation expression of the CrX with third-order approximation, as shown in Figure 10. As the phase angle β varies from $0°$ to $90°$, the error $\mathrm{CrX}_{app.error}$ remains less than 0.4% for the ratio $\alpha/g = 0.1$ and no more than 1.2% for $\alpha/g = 0.2$. Figure 11 compares the simulated differential capacitance CrXf and theoretical counterparts CrX_{tl3} calculated with Equation (36). The phase angle β equals $45°$, that is, the rotor has a displacement along the direction of y = x. It can be observed that the deviation between the value CrXf and the CrX_{tl3} is closer to the actual value.

In order to theoretically analyze the tilt error motion during the measurement, we considered tiny tilt displacement of a precision spindle and reasonable structural parameters of the sensor, that is, $|k\rho/a| < 1$ for the parameter term $k\rho/a$ in Equation (15). The power series expansion with first-order approximation (Equation (38)) was also applied to express the integrand term in Equation (15):

$$\mathrm{IGe}_{tl} = \frac{\rho}{a} \cdot \left[1 + \frac{k}{a}\rho\cos(\varphi + \gamma) \right] \tag{38}$$

Correspondingly, the expression used to assess the approximate error is given as:

$$\mathrm{IGe}_{app.error} = \left| \frac{\mathrm{IGe} - \mathrm{IGe}_{tl}}{\mathrm{IGe}} \right| \tag{39}$$

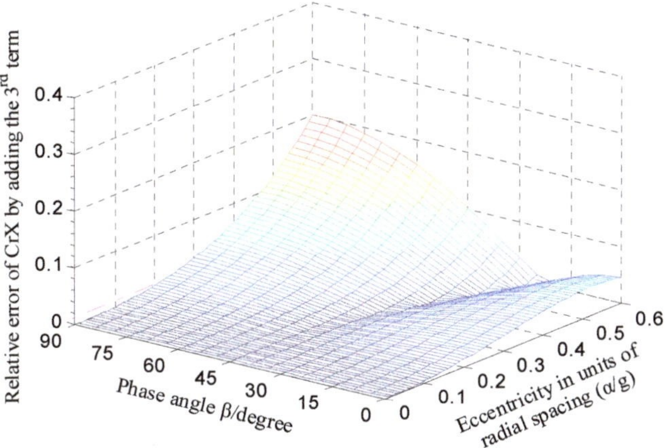

Figure 10. Calculation error distribution of CrX (expansion with third-order approximation).

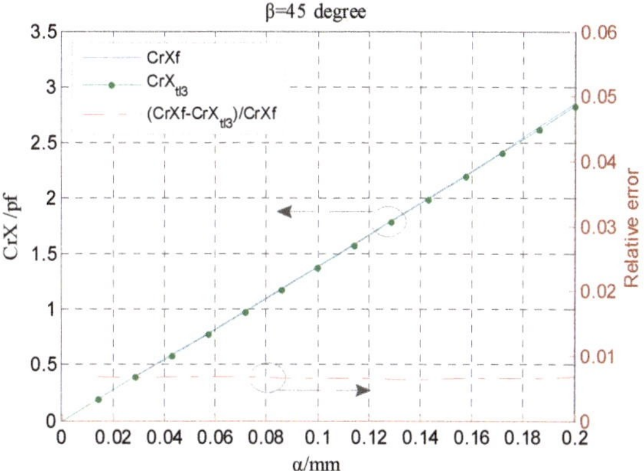

Figure 11. Simulation values and theoretical one calculated by Equation (36) comparison of radial differential output capacitance.

By considering the relative tiny tilt displacement of a precision spindle and reasonability of the sensor structural parameters, the variation range of the term $k\rho/a$ was set to be [0, 0.2]. As shown in Figure 12, the approximate error $IGe_{app.error}$ is relatively smaller over the range of the parameter terms $\cos(\varphi + \gamma)$ and $k\rho/a$. According to the simulation model parameters of the sensor [16], the ratio of the term $k\rho/a$ is about 0.04 with the tilt displacement of 200 arc-sec; thus, the error $IGe_{app.error}$ remains no more than 0.2% within the whole range of the term $\cos(\varphi + \gamma)$, i.e., the entire phase region. Figure 13a compares the theoretical and simulated capacitance of the end part electrode. The theoretical capacitance is integrated by the integrand term with a second-order approximation ($C8_{tl2}$). It can be observed that there is little difference between $C8_{tl2}$ and the counterpart using first-order approximation ($C8_{tl}$), which is also the same for the theoretical value of the CeY (the differential output capacitance of the EPEG) shown in Figure 13b. Besides, similar to the theoretical value of the CrX calculated by the expression with third-order approximation (see in Figure 11), the difference between the theoretical capacitance CeY_{tl} and the simulation one CeYf remains roughly unchanged relative to the simulation value, which indicates that the theoretical capacitance of the end part electrode solved by adopting IGe_{tl} as integrand can meet the design needs for relatively accurate results.

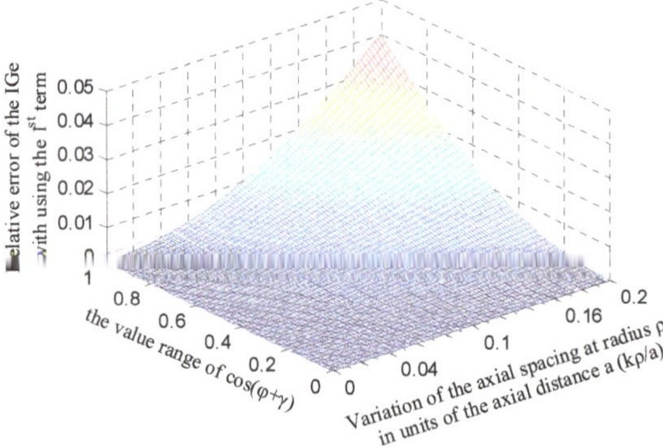

Figure 12. Error distribution of the first-order approximation of the integrand term in the capacitance expression of the end part electrode.

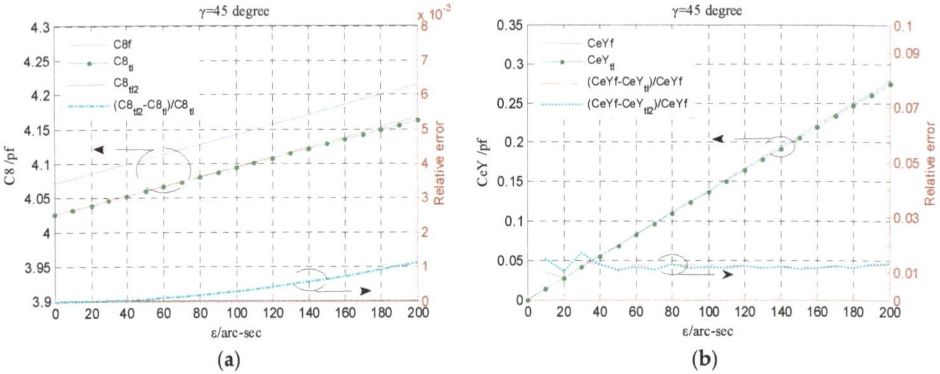

(**a**) (**b**)

Figure 13. Comparison of the simulation values of the end part electrode capacitance and the theoretical one calculated by the integrand term with first-order and second-order approximation respectively: (**a**) Single electrode capacitance; (**b**) Differential output capacitance.

3.2. Influence of Fringe Effects

Due to the existence of diverging electric fields in the plate edges in Figure 14, i.e., the "fringe effects" [19], the measured capacitance of a sensor contains some additional capacitance introduced by the fringe effects. As for the sensor electrode with finite dimensions, the variation in its output capacitance is relatively small, especially for the measurement of tiny displacement. As such, the influence of the fringe effects cannot be neglected.

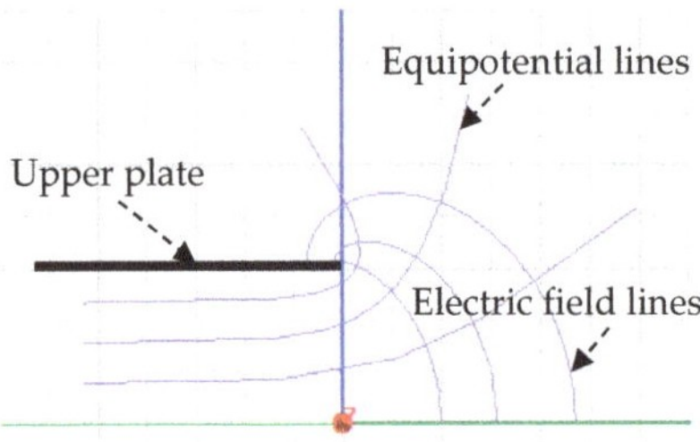

Figure 14. The fringe field distribution of a parallel plate capacitor with the gap 2π (the field lines between the two plates are symmetric, and thus only consider the distribution above the centerline).

Figure 15 shows the distribution of the electric field between the cylindrical excitation electrode (E_d) and curved sensing electrodes (E_s) for the fundamental configuration of the radial curved plate capacitor. It can be seen that the electric field concentration appears in the junction of the inner circle face of the curved plate and its side face. As shown in Figure 15a, a different intensity distribution of the electric field between the E_d and the side face of the E_s and its adjacent outer circular face formed along the radial direction. Once the equipotential guard ring (Egr) is employed outside the E_s, the electric field distribution between the E_d and the outer circular face of the E_s and between the E_d and local region of the side face near the outer circular face can be eliminated, as shown in Figure 15b. The Egr is also classically named Kelvin guard-ring.

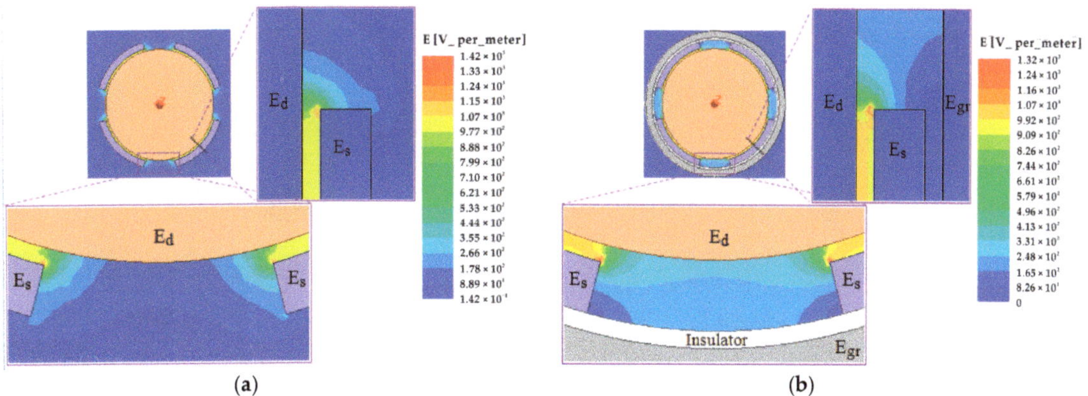

Figure 15. Electric field distribution between the electrode plates of the curved plate capacitor: (**a**) Without the Egr; (**b**) using the Egr.

The variation in the electric field distribution is also reflected in the output capacitance of the E_s, as illustrated in Figure 16. The use of the Egr can effectively reduce the additional capacitance introduced by the fringe effects (see Figure 16a,b). Moreover, the guard ring could decrease the variation in the additional capacitance with the reduction in radial spacing (see Figure 16d), which is helpful to improve the measurement accuracy and linearity. For the T-type CS, the Egr is distributed around the periphery of the sensing electrodes with a tiny gap ($\lambda = 0.1$ mm), forming a coplanar configuration between the Egr and the electrodes. Under the effect of the Egr, the amount of variation in the additional capacitance significantly decreased (about 10 times), and the corresponding nonlinear error reduced from 4.8% to 1.7%, as indicated in Figure 17.

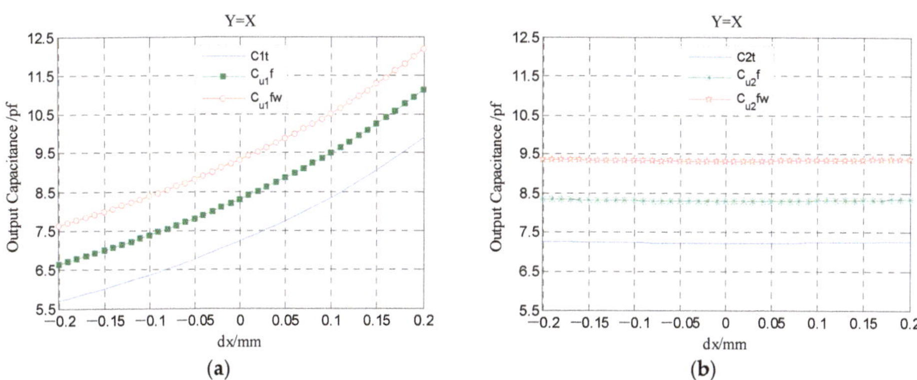

Figure 16. *Cont.*

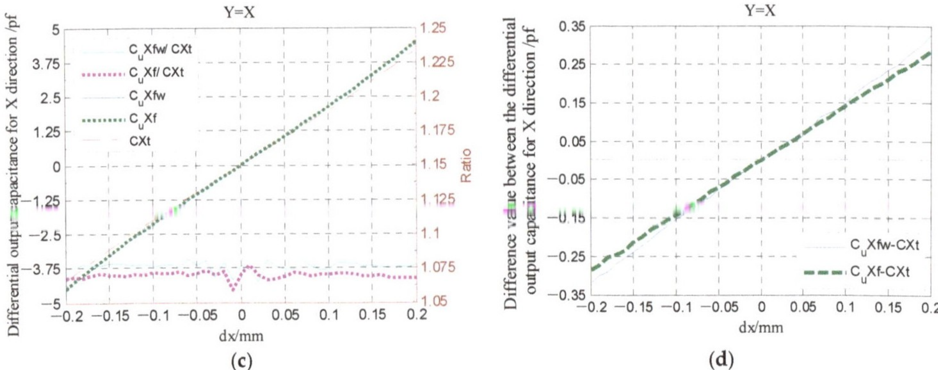

Figure 16. Simulation and theoretical value comparison for the output capacitance of the curved sensing electrodes E_s: (**a**,**b**) The 1st, 2nd quadrant; (**c**,**d**) variation tendency of the differential output capacitance of the electrode and their relative difference value. The C*fw and C*f in the curve diagram denote the simulation values under without and with the Egr, respectively; the C*t is theoretical calculated value. The dx is defined as a component of the displacement α in the x-axis direction.

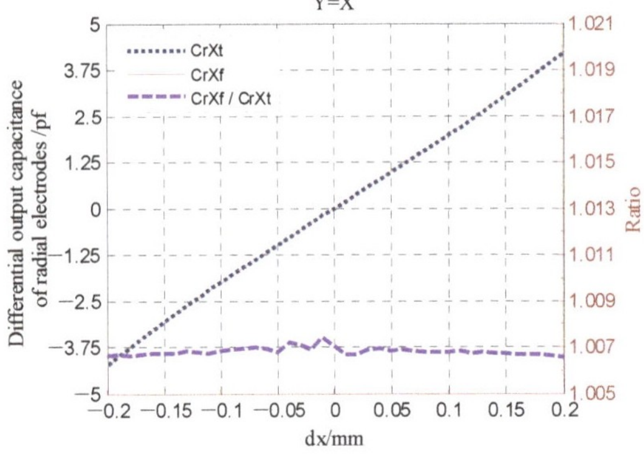

Figure 17. Simulation and theoretical values comparison of radial differential output capacitance of the T-type CS.

3.3. Installation Errors of the Sensing Electrodes

The sensing electrodes of the T-type CS include the cylindrical and fan-shaped electrodes integrated into the REG and EPEG, respectively. These electrodes are fabricated by the FPCB process in one procedure, along with the equipotential guard ring, and the sensing electrodes have a fixed position relationship with the guard ring.

The fabricated REG with planar form (as shown in Figure 18a, solid line box) is installed in the annular groove of the stator cylindrical bore by the surface mounting method. The coaxiality error, polar angle position error and tilt error (the axial boundary of the electrode relative to the axial reference of the stator) relative to the mounting reference were produced during the process. By utilizing the electrode measurement and sensor calibration method, the coaxiality and polar angle position error can be modified and removed, respectively. Thus, the analysis of the tilt error is the main focus for the cylindrical electrodes.

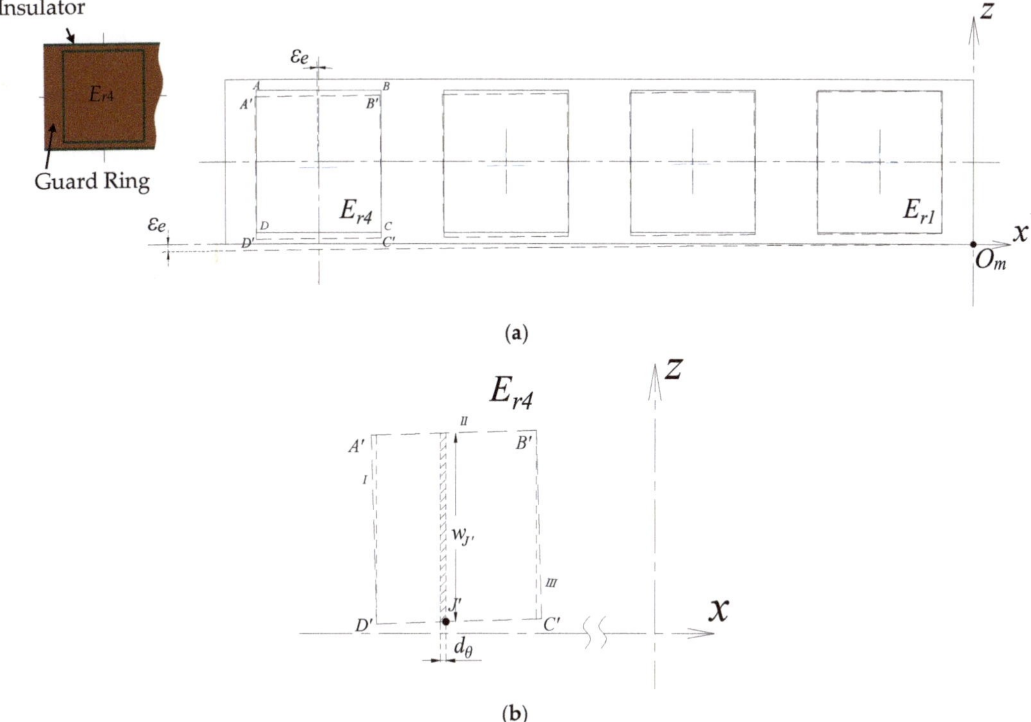

Figure 18. Tilt error analysis of the radial cylindrical electrode: (**a**) Local view of the REG structural model and schematic of the tilt error of cylindrical electrode (planar status); (**b**) geometric analysis model of single electrode. The *z*-axis is parallel to the medial axis of the stator, and the *x*-axis is parallel to axial reference of it.

By spreading out the installed REG, as shown in Figure 18a, it can be seen that a certain tilt angle ε_e is produced relative to the ideal position of the REG. The point O_m is assumed to be the tilt point. The dash line box represents the position of the cylindrical electrodes, and the dash line segment represents the axial boundary of the REG. As mentioned above, the in-plane relationship between the cylindrical electrodes is fixed during the mounting process, and thus the tilt angle ε_e is the tilt error of the electrodes.

As shown in Figure 18b, by taking a micro element with arc length $d\theta$ on the boundary $D'C'$ of the electrode E_{r4}, the axial length $w_{J'}$ of the electrode at this location can be considered as the function of the polar angle $\theta_{J'}$ at point J'. The solution region is vertically divided into three parts according to the constraint boundary of the length $w_{J'}$, and the corresponding calculation function can be expressed as follows:

$$w_I = -k_I R \cdot \theta + b_I \tag{40}$$

$$w_{II} = b_{II} \tag{41}$$

$$w_{III} = -k_{III} R \cdot \theta + b_{III} \tag{42}$$

where $k_I = k_1 + 1/k_1$, $k_{III} = -k_I$, $b_I = b_3 - b_4$, $b_{II} = b_3 - b_1$, $b_{III} = b_2 - b_1$; k_1 is the slope of the boundary $D'C'$; $b_1 \sim b_4$ are the intercepts of the lines including $D'C'$, $C'B'$, $B'A'$, $D'A'$, respectively.

Referring to Equation (6), the approximate expression of the output capacitance of the electrode E_{r4} under the tilt angle ε_e can be obtained:

$$C4' = \varepsilon_0 \varepsilon_r R \left[\int_{\theta_{D'}}^{\theta_{A'}} \frac{w_I}{g - \alpha \cos(\theta - \beta)} d\theta + \int_{\theta_{B'}}^{\theta_{D'}} \frac{w_{II}}{g - \alpha \cos(\theta - \beta)} d\theta + \int_{\theta_{C'}}^{\theta_{B'}} \frac{w_{III}}{g - \alpha \cos(\theta - \beta)} d\theta \right] \tag{43}$$

By considering the minor radial displacement of a precision spindle, at the condition $|\alpha/g| < 1$, the power series expansion with neglection of the high-order term is applied to express the integrand term in Equation (43), and thus we have:

$$C4' = \frac{\varepsilon_0 \varepsilon_r R}{g} \left\{ \begin{array}{l} \left[-k_I R \cdot \frac{\theta^2}{2} + b_I \theta - k_I R\left(\frac{\alpha}{g}\right)\left(\cos(\theta - \beta) + \left(\theta - \frac{b_I}{k_I R}\right) \cdot \sin(\theta - \beta)\right) - k_I R\left(\frac{\alpha}{g}\right)^2\left(\frac{\theta^2}{4} - \frac{b_I}{k_I R} \cdot \frac{(\theta - \beta)}{2} + \frac{\cos(2(\theta - \beta))}{8} + \left(\theta - \frac{b_I}{k_I R}\right) \cdot \frac{\sin(2(\theta - \beta))}{4}\right) \right]\Big|_{\theta_{D'}}^{\theta_{A'}} \\[4pt] + \left[\frac{2b_{II}}{g - \alpha}\sqrt{\frac{g - \alpha}{g + \alpha}} \arctan\left(\sqrt{\frac{g + \alpha}{g - \alpha}} \tan\left(\frac{\theta - \beta}{2}\right)\right) \right]\Big|_{\theta_{B'}}^{\theta_{D'}} \\[4pt] + \left[-k_{III} R \cdot \frac{\theta^2}{2} + b_{III} \theta - k_{III} R\left(\frac{\alpha}{g}\right)\left(\cos(\theta - \beta) + \left(\theta - \frac{b_{III}}{k_{III} R}\right) \cdot \sin(\theta - \beta)\right) - k_{III} R\left(\frac{\alpha}{g}\right)^2\left(\frac{\theta^2}{4} - \frac{b_{III}}{k_{III} R} \cdot \frac{(\theta - \beta)}{2} + \frac{\cos(2(\theta - \beta))}{8} + \left(\theta - \frac{b_{III}}{k_{III} R}\right) \cdot \frac{\sin(2(\theta - \beta))}{4}\right) \right]\Big|_{\theta_{C'}}^{\theta_{B'}} \end{array} \right\} \tag{44}$$

Similarly, the approximate expression of the output capacitance of the electrodes $E_{r3} \sim E_{r1}$ are derived as follows:

$$C3' = \frac{\varepsilon_0 \varepsilon_r R}{g} \left\{ \begin{array}{l} \left[-k_{2I} R \cdot \frac{\theta^2}{2} + b_{2I} \theta - k_{2I} R\left(\frac{\alpha}{g}\right)\left(\cos(\theta - \beta) + \left(\theta - \frac{b_{2I}}{k_{2I} R}\right) \cdot \sin(\theta - \beta)\right) - k_{2I} R\left(\frac{\alpha}{g}\right)^2\left(\frac{\theta^2}{4} - \frac{b_{2I}}{k_{2I} R} \cdot \frac{(\theta - \beta)}{2} + \frac{\cos(2(\theta - \beta))}{8} + \left(\theta - \frac{b_{2I}}{k_{2I} R}\right) \cdot \frac{\sin(2(\theta - \beta))}{4}\right) \right]\Big|_{\theta_{D2'}}^{\theta_{A2'}} \\[4pt] + \left[\frac{2b_{2II}}{g - \alpha}\sqrt{\frac{g - \alpha}{g + \alpha}} \arctan\left(\sqrt{\frac{g + \alpha}{g - \alpha}} \tan\left(\frac{\theta - \beta}{2}\right)\right) \right]\Big|_{\theta_{B2'}}^{\theta_{D2'}} \\[4pt] + \left[-k_{2III} R \cdot \frac{\theta^2}{2} + b_{2III} \theta - k_{2III} R\left(\frac{\alpha}{g}\right)\left(\cos(\theta - \beta) + \left(\theta - \frac{b_{2III}}{k_{2III} R}\right) \cdot \sin(\theta - \beta)\right) - k_{2III} R\left(\frac{\alpha}{g}\right)^2\left(\frac{\theta^2}{4} - \frac{b_{2III}}{k_{2III} R} \cdot \frac{(\theta - \beta)}{2} + \frac{\cos(2(\theta - \beta))}{8} + \left(\theta - \frac{b_{2III}}{k_{2III} R}\right) \cdot \frac{\sin(2(\theta - \beta))}{4}\right) \right]\Big|_{\theta_{C2'}}^{\theta_{B2'}} \end{array} \right\} \tag{45}$$

$$C2' = \frac{\varepsilon_0 \varepsilon_r R}{g} \left\{ \begin{array}{l} \left[-k_{3I} R \cdot \frac{\theta^2}{2} + b_{3I} \theta - k_{3I} R\left(\frac{\alpha}{g}\right)\left(\cos(\theta - \beta) + \left(\theta - \frac{b_{3I}}{k_{3I} R}\right) \cdot \sin(\theta - \beta)\right) - k_{3I} R\left(\frac{\alpha}{g}\right)^2\left(\frac{\theta^2}{4} - \frac{b_{3I}}{k_{3I} R} \cdot \frac{(\theta - \beta)}{2} + \frac{\cos(2(\theta - \beta))}{8} + \left(\theta - \frac{b_{3I}}{k_{3I} R}\right) \cdot \frac{\sin(2(\theta - \beta))}{4}\right) \right]\Big|_{\theta_{D3'}}^{\theta_{A3'}} \\[4pt] + \left[\frac{2b_{3II}}{g - \alpha}\sqrt{\frac{g - \alpha}{g + \alpha}} \arctan\left(\sqrt{\frac{g + \alpha}{g - \alpha}} \tan\left(\frac{\theta - \beta}{2}\right)\right) \right]\Big|_{\theta_{B3'}}^{\theta_{D3'}} \\[4pt] + \left[-k_{3III} R \cdot \frac{\theta^2}{2} + b_{3III} \theta - k_{3III} R\left(\frac{\alpha}{g}\right)\left(\cos(\theta - \beta) + \left(\theta - \frac{b_{3III}}{k_{3III} R}\right) \cdot \sin(\theta - \beta)\right) - k_{3III} R\left(\frac{\alpha}{g}\right)^2\left(\frac{\theta^2}{4} - \frac{b_{3III}}{k_{3III} R} \cdot \frac{(\theta - \beta)}{2} + \frac{\cos(2(\theta - \beta))}{8} + \left(\theta - \frac{b_{3III}}{k_{3III} R}\right) \cdot \frac{\sin(2(\theta - \beta))}{4}\right) \right]\Big|_{\theta_{C3'}}^{\theta_{B3'}} \end{array} \right\} \tag{46}$$

$$C1' = \frac{\varepsilon_0 \varepsilon_r R}{g} \left\{ \begin{array}{l} \left[-k_{4I} R \cdot \frac{\theta^2}{2} + b_{4I} \theta - k_{4I} R\left(\frac{\alpha}{g}\right)\left(\cos(\theta - \beta) + \left(\theta - \frac{b_{4I}}{k_{4I} R}\right) \cdot \sin(\theta - \beta)\right) - k_{4I} R\left(\frac{\alpha}{g}\right)^2\left(\frac{\theta^2}{4} - \frac{b_{4I}}{k_{4I} R} \cdot \frac{(\theta - \beta)}{2} + \frac{\cos(2(\theta - \beta))}{8} + \left(\theta - \frac{b_{4I}}{k_{4I} R}\right) \cdot \frac{\sin(2(\theta - \beta))}{4}\right) \right]\Big|_{\theta_{D4'}}^{\theta_{A4'}} \\[4pt] + \left[\frac{2b_{4II}}{g - \alpha}\sqrt{\frac{g - \alpha}{g + \alpha}} \arctan\left(\sqrt{\frac{g + \alpha}{g - \alpha}} \tan\left(\frac{\theta - \beta}{2}\right)\right) \right]\Big|_{\theta_{B4'}}^{\theta_{D4'}} \\[4pt] + \left[-k_{4III} R \cdot \frac{\theta^2}{2} + b_{4III} \theta - k_{4III} R\left(\frac{\alpha}{g}\right)\left(\cos(\theta - \beta) + \left(\theta - \frac{b_{4III}}{k_{4III} R}\right) \cdot \sin(\theta - \beta)\right) - k_{4III} R\left(\frac{\alpha}{g}\right)^2\left(\frac{\theta^2}{4} - \frac{b_{4III}}{k_{4III} R} \cdot \frac{(\theta - \beta)}{2} + \frac{\cos(2(\theta - \beta))}{8} + \left(\theta - \frac{b_{4III}}{k_{4III} R}\right) \cdot \frac{\sin(2(\theta - \beta))}{4}\right) \right]\Big|_{\theta_{C4'}}^{\theta_{B4'}} \end{array} \right\} \tag{47}$$

where, $k_{4I} = k_{3I} = k_{2I} = k_I$, $k_{4III} = k_{3III} = k_{2III} = k_{III}$, $b_{4II} = b_{3II} = b_{2II} = b_{II}$, $b_{iI} = b_3 - b_{i4}$, $b_{iIII} = b_{i2} - b_1 (i = 4, 3, 2)$; b_{i2}, b_{i4} are the intercepts of the lines, including $Ci'Bi'$, $Di'Ai'$ $(i = 4, 3, 2)$, respectively; i represents the electrodes $E_{r1} \sim E_{r3}$ in turn.

Further, the influence of the tilt error ε_e on the output capacitance of the cylindrical electrode (taking E_{r4} as an example) can be reflected by:

$$C4_{tilt.error} = \frac{C4' - C4}{C4} \tag{48}$$

By referring to the structural design of the T-type CS, the range of the tilt angle ε_e was set to be [0.006, 0.06] deg (about 200 arc-sec) in the numerical simulation using Matlab. A tiny circumferential displacement of the cylindrical electrode was produced due to the tilt angle ε_e, which produces a relative error of the output capacitance under the radial displacement along the same direction. The dependence of the relative error on the angle ε_e is shown in Figure 19. For the single electrode, the influence of the tilt angle ε_e on the output capacitance is very small, with respect to the amplitude of the relative error.

Figure 20 presents the dependence of the differential capacitance on the tilt angle ε_e. It can be known that the influence of the tilt angle ε_e on the differential output capacitance of the REG is slightly larger. By considering the rotor of run-out in the given directions ($\beta = 5°$, $45°$, $85°$), the differential output capacitance (CrXt-ε_e and CrYt-ε_e) can be calculated by the capacitance of single electrodes under the angle ε_e. The difference between the differential output capacitance under the angle ε_e and the counterpart under an ideal position (CrXt and CrYt) could be transformed to the difference of displacement parameters, utilizing Equations (33) and (34). The difference in the displacement parameters (magnitude α and phase angle β) are shown in Figure 21. It can be known that the produced tilt error ε_e of the cylindrical electrode only generates a certain deviation of solving the phase angle β, which is consistent in different displacement directions (brings about the overall phase advance or lag of the displacement trajectory); the effect of tilt error ε_e on solving the radial displacement α can be neglected, especially for the measurement range of less than 0.1 mm.

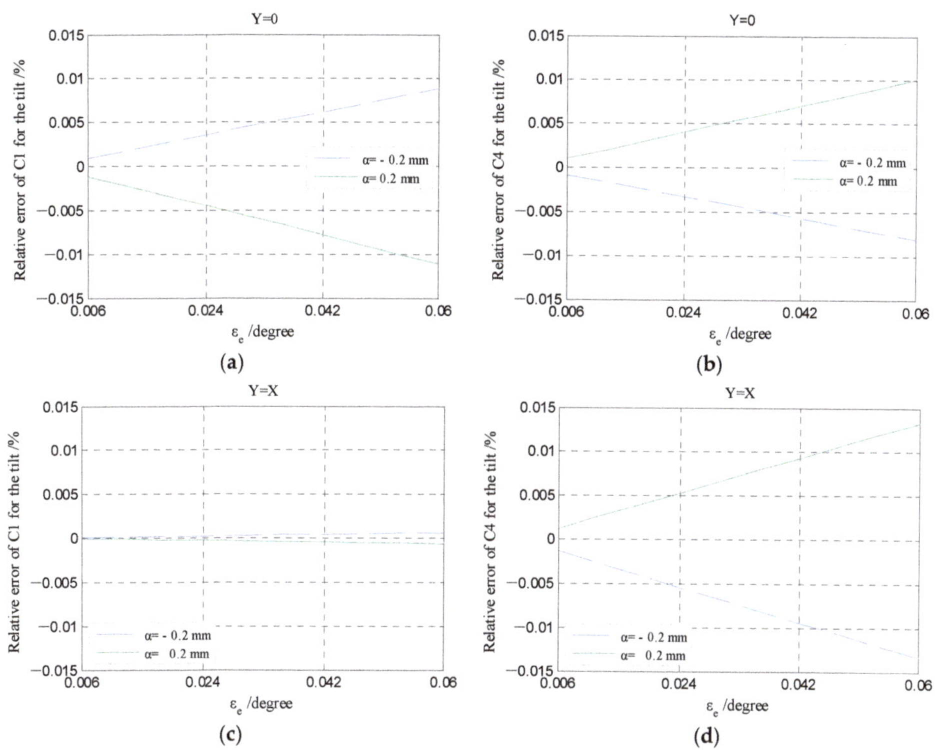

Figure 19. Relative error variations in the output capacitance of cylindrical electrodes under the different tilt angle ε_e: (**a**,**c**) The 1st quadrant; (**b**,**d**) the 4th quadrant.

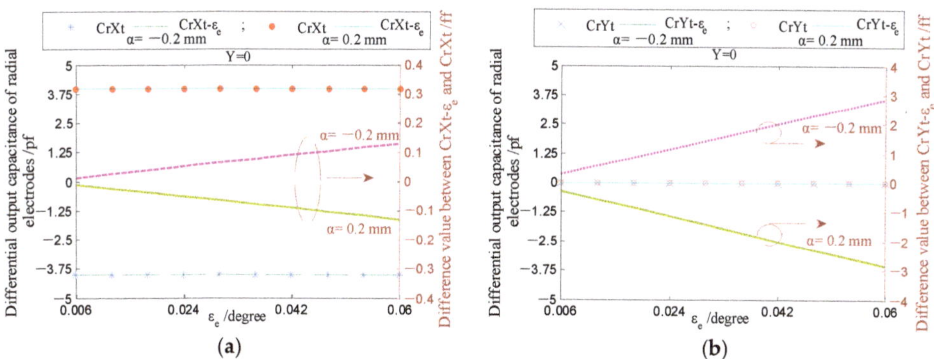

Figure 20. *Cont.*

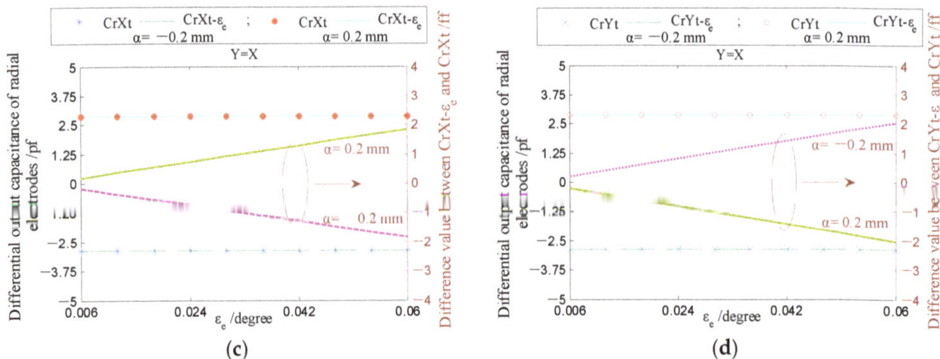

(c) (d)

Figure 20. The changing of the radial differential output capacitance under the different tilt angle ε_e: (**a**,**c**) The X-direction; (**b**,**d**) the Y-direction.

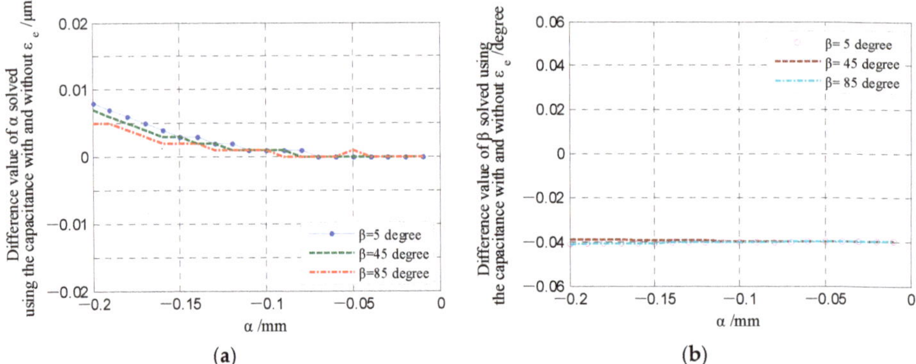

(a) (b)

Figure 21. Comparison of the solved displacement parameters of the rotor under the different electrode poses: (**a**) Displacement α; (**b**) phase angle β.

The fabricated EPEG was also in planar form (as shown in Figure 22a, solid line box) and installed in the annular groove located at the end part of the stator by the surface mounting method. The coaxiality error, polar angle position error and parallelism error (the electrode plane relative to the axial reference of the stator) relative to the mounting reference were produced during the process. For the parallelism error, it can be modified by the electrode measurement. Thus, the analysis of the fan-shaped electrodes is emphasized on the coaxiality error and polar angle position error.

Due to the position error between the shape boundary of EPEG and the sensing electrodes and the manipulation precision, there may be an eccentricity between the geometric center $O_e{}'$ of the fan-shaped electrodes in the EPEG and the medial axis of the stator (the origin position of the coordinate system). As shown in Figure 22a, δ_e denotes the magnitude of the eccentricity, and φ_e is the phase angle. δ_e is the coaxiality error of the fan-shaped electrode.

By taking the electrode E_{e1} as an example, the region confined by the points $C'D'S'T'$ (dash line) represents the position of E_{e1} under the coaxiality error δ_e. As shown in Figure 22b, the micro-plane element $\Delta B_{P'}$ (oblique line filled) at any point P' is taken as the solving unit, and the area of this element is denoted as $\Delta S_{P'}$. Then, the $\Delta S_{P'}$ is approximately given by:

$$\Delta S_{P'} = dx \cdot dy \tag{49}$$

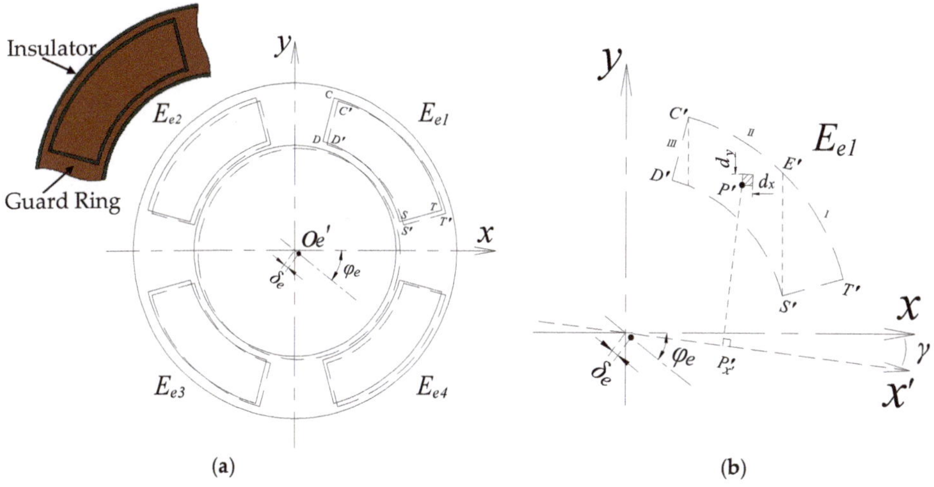

Figure 22. Coaxiality error analysis of the end part fan-shaped electrode: (a) Local view of the EPEG structural model and schematic of the coaxiality error of fan-shaped electrode (planar status); (b) geometric analysis model of single electrode.

The projection point of P' on the x'-axis is the point $P'_{x'}$. By referring to Equation (13), the expression for calculating the spacing t' at this point could be derived as:

$$t' = a - k(x \cos \gamma - y \sin \gamma) \tag{50}$$

Correspondingly, the capacitance of the plane parallel capacitor composed of the micro-planes $\Delta B_{P'}$ and $\Delta A_{Q'}$ is:

$$\Delta C_e' = \frac{\varepsilon_0 \varepsilon_r \Delta S_{P'}}{t'} = \varepsilon_0 \varepsilon_r \frac{dx \cdot dy}{a - k(x \cos \gamma - y \sin \gamma)} \tag{51}$$

Through the integration of Equation (51) over the area of the fan-shaped electrode, the output capacitance of the electrode E_{e1} is approximately expressed as:

$$C5' = \iint_D \Delta C_e' = \varepsilon_0 \varepsilon_r \iint_D \frac{dxdy}{a - k(x \cos \gamma - y \sin \gamma)} \tag{52}$$

where the integral domain D is a closed region confined by the boundary curves of the fan-shaped electrode, which can be divided into three sub-regions, i.e., D_I, D_{II} and D_{III}. These three sub-regions can be expressed as follows:

$$D_{III} = \left\{ (x,y) \,\middle|\, \sqrt{\rho_1^2 - (x - x_B)^2} + y_B \leq y \leq k_{C'}x + b_{C'}, D'_x \leq x \leq C'_x \right\},$$

$$D_{II} = \left\{ (x,y) \,\middle|\, \sqrt{\rho_1^2 - (x - x_B)^2} + y_B \leq y \leq \sqrt{\rho_2^2 - (x - x_B)^2} + y_B, C'_x \leq x \leq E'_x \right\},$$

$$D_I = \left\{ (x,y) \,\middle|\, k_{S'}x + b_{S'} \leq y \leq \sqrt{\rho_2^2 - (x - x_B)^2} + y_B, E'_x \leq x \leq T'_x \right\}.$$

where x_B and y_B are the coordinates of the geometric center O_e'; D'_x, C'_x, E'_x and T'_x represent the abscissas of the corresponding points, respectively.

Equation (52) can be rewritten as:

$$C5' = \frac{\varepsilon_0 \varepsilon_r}{a} \iint_D \frac{dxdy}{1 - \frac{k}{a}(x\cos\gamma - y\sin\gamma)} \tag{53}$$

As for multivariate function, it can be expanded to power series under certain conditions by analogy with the univariate power series expansion [20]. By considering the tiny tilt displacement of a precision spindle and reasonable structural parameters of the sensor, there is generally $|(k/a)\cdot(x\cos(\gamma) - y\sin(\gamma))| \ll 1$, the power series expansion with neglecting of a high-order term is applied to express the integrand term in Equation (53). Then, it can be expressed as:

$$C5' = \frac{\varepsilon_0 \varepsilon_r}{a} \iint_D \left[1 + \frac{k}{a}(x\cos\gamma - y\sin\gamma)\right] dxdy \tag{54}$$

By calculating Equation (54) within the integral sub-domains D_I, D_{II} and D_{III}, we have:

$$
\begin{aligned}
C5' &= \frac{\varepsilon_0 \varepsilon_r}{a}
\left\{
\begin{array}{l}
\int_{D'_x}^{C'_x} dx \int_{\sqrt{\rho_1^2 - (x-x_B)^2}+y_B}^{kc_{C'}x+b_{C'}} \left[1 + \frac{k}{a}(x\cos\gamma - y\sin\gamma)\right]dy + \int_{C'_x}^{E'_x} dx \int_{\sqrt{\rho_1^2-(x-x_B)^2}+y_B}^{\sqrt{\rho_2^2-(x-x_B)^2}+y_B} \left[1 + \frac{k}{a}(x\cos\gamma - y\sin\gamma)\right]dy \\[6pt]
+ \int_{E'_x}^{T'_x} dx \int_{k_{S'}x+b_{S'}}^{\sqrt{\rho_2^2-(x-x_B)^2}+y_B} \left[1 + \frac{k}{a}(x\cos\gamma - y\sin\gamma)\right]dy
\end{array}
\right\} \\[10pt]
&= \frac{\varepsilon_0 \varepsilon_r}{a}
\left\{
\begin{array}{l}
\left[(k_{C'} + \mu b_{C'} - \nu k_{C'} b_{C'} - \mu y_B + \nu x_B)\frac{x^2}{2} + (\mu k_{C'} - \frac{\nu}{2}k_{C'}^2 - \frac{\nu}{2})\frac{x^3}{3} - (1 - \nu y_B + \mu x_B)\cdot(\frac{x-x_B}{2}\sqrt{\rho_1^2-(x-x_B)^2} + \frac{\rho_1^2}{2}\arcsin\frac{x-x_B}{\rho_1}) \right. \\
\left. + \frac{\mu}{3}\sqrt{(\rho_1^2-(x-x_B)^2)^3} + ((b_{C'}-y_B) - \frac{\nu}{2}(b_{C'}^2-\rho_1^2) - \frac{\nu}{2}(x_B^2-y_B^2))\cdot x \right] \Big|_{D'_x}^{C'_x} \\[8pt]
+ \left[(1 - \nu y_B + \mu x_B)\cdot(\frac{x-x_B}{2}(\sqrt{\rho_2^2-(x-x_B)^2} - \sqrt{\rho_1^2-(x-x_B)^2}) + \frac{\rho_2^2}{2}\arcsin\frac{x-x_B}{\rho_2} - \frac{\rho_1^2}{2}\arcsin\frac{x-x_B}{\rho_1}) \right. \\
\left. + \frac{\mu}{3}(\sqrt{(\rho_1^2-(x-x_B)^2)^3} - \sqrt{(\rho_2^2-(x-x_B)^2)^3}) - \frac{\nu}{2}(\rho_2^2-\rho_1^2)\cdot x \right] \Big|_{C'_x}^{E'_x} \\[8pt]
+ \left[(\mu y_B - \nu x_B + \nu k_{S'} b_{S'} - \mu b_{S'} - k_{S'})\frac{x^2}{2} + (\frac{\nu}{2} + \frac{\nu}{2}k_{S'}^2 - \mu k_{S'})\frac{x^3}{3} + (1 - \nu y_B + \mu x_B)\cdot(\frac{x-x_B}{2}\sqrt{\rho_2^2-(x-x_B)^2} + \frac{\rho_2^2}{2}\arcsin\frac{x-x_B}{\rho_2}) \right. \\
\left. - \frac{\mu}{3}\sqrt{(\rho_2^2-(x-x_B)^2)^3} + ((y_B-b_{S'}) - \frac{\nu}{2}(\rho_2^2-b_{S'}^2) - \frac{\nu}{2}(y_B^2-x_B^2))\cdot x \right] \Big|_{E'_x}^{T'_x}
\end{array}
\right\}
\end{aligned}
\tag{55}
$$

Similarly, the approximate expressions of the output capacitance of the electrodes $E_{e2}\sim E_{e4}$ can be derived as follows:

$$
C6' = \frac{\varepsilon_0 \varepsilon_r}{a}
\left\{
\begin{array}{l}
\left[(k_{C'_1} + \mu b_{C'_1} - \nu k_{C'_1} b_{C'_1} - \mu y_B + \nu x_B)\frac{x^2}{2} + (\mu k_{C'_1} - \frac{\nu}{2}k_{C'_1}^2 - \frac{\nu}{2})\frac{x^3}{3} - (1 - \nu y_B + \mu x_B)\cdot(\frac{x-x_B}{2}\sqrt{\rho_1^2-(x-x_B)^2} + \frac{\rho_1^2}{2}\arcsin\frac{x-x_B}{\rho_1}) \right. \\
\left. + \frac{\mu}{3}\sqrt{(\rho_1^2-(x-x_B)^2)^3} + ((b_{C'_1}-y_B) - \frac{\nu}{2}(b_{C'_1}^2-\rho_1^2) - \frac{\nu}{2}(x_B^2-y_B^2))\cdot x \right] \Big|_{C'_{1x}}^{D'_{1x}} \\[8pt]
+ \left[(1 - \nu y_B + \mu x_B)\cdot(\frac{x-x_B}{2}(\sqrt{\rho_2^2-(x-x_B)^2} - \sqrt{\rho_1^2-(x-x_B)^2}) + \frac{\rho_2^2}{2}\arcsin\frac{x-x_B}{\rho_2} - \frac{\rho_1^2}{2}\arcsin\frac{x-x_B}{\rho_1}) \right. \\
\left. + \frac{\mu}{3}(\sqrt{(\rho_1^2-(x-x_B)^2)^3} - \sqrt{(\rho_2^2-(x-x_B)^2)^3}) - \frac{\nu}{2}(\rho_2^2-\rho_1^2)\cdot x \right] \Big|_{E'_{1x}}^{C'_{1x}} \\[8pt]
+ \left[(\mu y_B - \nu x_B + \nu k_{S'_1} b_{S'_1} - \mu b_{S'_1} - k_{S'_1})\frac{x^2}{2} + (\frac{\nu}{2} + \frac{\nu}{2}k_{S'_1}^2 - \mu k_{S'_1})\frac{x^3}{3} + (1 - \nu y_B + \mu x_B)\cdot(\frac{x-x_B}{2}\sqrt{\rho_2^2-(x-x_B)^2} + \frac{\rho_2^2}{2}\arcsin\frac{x-x_B}{\rho_2}) \right. \\
\left. - \frac{\mu}{3}\sqrt{(\rho_2^2-(x-x_B)^2)^3} + ((y_B-b_{S'_1}) - \frac{\nu}{2}(\rho_2^2-b_{S'_1}^2) - \frac{\nu}{2}(y_B^2-x_B^2))\cdot x \right] \Big|_{T'_{1x}}^{E'_{1x}}
\end{array}
\right\}
\tag{56}
$$

$$
C7' = \frac{\varepsilon_0 \varepsilon_r}{a}
\left\{
\begin{array}{l}
\left[(\mu y_B - \nu x_B + \nu k_{C'} b_{C'} - \mu b_{C'} - k_{C'})\frac{x^2}{2} + (\frac{\nu}{2} + \frac{\nu}{2}k_{C'}^2 - \mu k_{C'})\frac{x^3}{3} - (1 - \nu y_B + \mu x_B)\cdot(\frac{x-x_B}{2}\sqrt{\rho_1^2-(x-x_B)^2} + \frac{\rho_1^2}{2}\arcsin\frac{x-x_B}{\rho_1}) \right. \\
\left. + \frac{\mu}{3}\sqrt{(\rho_1^2-(x-x_B)^2)^3} + ((y_B-b_{C'}) - \frac{\nu}{2}(\rho_1^2-b_{C'}^2) - \frac{\nu}{2}(y_B^2-x_B^2))\cdot x \right] \Big|_{C'_{2x}}^{D'_{2x}} \\[8pt]
+ \left[(1 - \nu y_B + \mu x_B)\cdot(\frac{x-x_B}{2}(\sqrt{\rho_2^2-(x-x_B)^2} - \sqrt{\rho_1^2-(x-x_B)^2}) + \frac{\rho_2^2}{2}\arcsin\frac{x-x_B}{\rho_2} - \frac{\rho_1^2}{2}\arcsin\frac{x-x_B}{\rho_1}) \right. \\
\left. + \frac{\mu}{3}(\sqrt{(\rho_1^2-(x-x_B)^2)^3} - \sqrt{(\rho_2^2-(x-x_B)^2)^3}) - \frac{\nu}{2}(\rho_1^2-\rho_2^2)\cdot x \right] \Big|_{E'_{2x}}^{C'_{2x}} \\[8pt]
+ \left[(k_{S'} + \mu b_{S'} - \nu k_{S'} b_{S'} - \mu y_B + \nu x_B)\frac{x^2}{2} + (\mu k_{S'} - \frac{\nu}{2}k_{S'}^2 - \frac{\nu}{2})\frac{x^3}{3} + (1 - \nu y_B + \mu x_B)\cdot(\frac{x-x_B}{2}\sqrt{\rho_2^2-(x-x_B)^2} + \frac{\rho_2^2}{2}\arcsin\frac{x-x_B}{\rho_2}) \right. \\
\left. - \frac{\mu}{3}\sqrt{(\rho_2^2-(x-x_B)^2)^3} + ((b_{S'}-y_B) - \frac{\nu}{2}(b_{S'}^2-\rho_2^2) - \frac{\nu}{2}(x_B^2-y_B^2))\cdot x \right] \Big|_{T'_{2x}}^{E'_{2x}}
\end{array}
\right\}
\tag{57}
$$

$$
C8' = \frac{\varepsilon_0 \varepsilon_r}{a} \left\{
\begin{aligned}
&\left[\begin{aligned}
&(\mu y_B - \nu x_B + \nu k_{C'_1} b_{C'_1} - \mu b_{C'_1} - k_{C'_1})\tfrac{x^2}{2} + (\tfrac{\nu}{2} + \tfrac{\nu}{2} k^2_{C'_1} - \mu k_{C'_1})\tfrac{x^3}{3} - (1 - \nu y_B + \mu x_B) \cdot (\tfrac{x-x_B}{2}\sqrt{\rho_1^2 - (x-x_B)^2} + \tfrac{\rho_1^2}{2}\arcsin\tfrac{x-x_B}{\rho_1}) \\
&+ \tfrac{\mu}{3}\sqrt{(\rho_1^2 - (x-x_B)^2)^3} + ((y_B - b_{C'_1}) - \tfrac{\nu}{2}(\rho_1^2 - b^2_{C'_1}) - \tfrac{\nu}{2}(y_B^2 - x_B^2)) \cdot x
\end{aligned} \right]\Bigg|_{D'_{3x}}^{C'_{3x}} \\[4pt]
&+ \left[\begin{aligned}
&(1 - \nu y_B + \mu x_B) \cdot (\tfrac{x-x_B}{2}(\sqrt{\rho_2^2 - (x-x_B)^2} - \sqrt{\rho_1^2 - (x-x_B)^2}) + \tfrac{\rho_2^2}{2}\arcsin\tfrac{x-x_B}{\rho_2} - \tfrac{\rho_1^2}{2}\arcsin\tfrac{x-x_B}{\rho_1}) \\
&+ \tfrac{\mu}{3}(\sqrt{(\rho_1^2 - (x-x_B)^2)^3} - \sqrt{(\rho_2^2 - (x-x_B)^2)^3}) - \tfrac{\nu}{2}(\rho_1^2 - \rho_2^2) \cdot x
\end{aligned} \right]\Bigg|_{C'_{3x}}^{E'_{3x}} \\[4pt]
&+ \left[\begin{aligned}
&(k_{S'_1} + \mu b_{S'_1} - \nu k_{S'_1} b_{S'_1} - \mu y_B + \nu x_B)\tfrac{x^2}{2} + (\mu k_{S'_1} - \tfrac{\nu}{2}k^2_{S'_1} - \tfrac{\nu}{2})\tfrac{x^3}{3} + (1 - \nu y_B + \mu x_B) \cdot (\tfrac{x-x_B}{2}\sqrt{\rho_2^2 - (x-x_B)^2} + \tfrac{\rho_2^2}{2}\arcsin\tfrac{x-x_B}{\rho_2}) \\
&- \tfrac{\mu}{3}\sqrt{(\rho_2^2 - (x-x_B)^2)^3} + ((b_{S'_1} - y_B) - \tfrac{\nu}{2}(b^2_{S'_1} - \rho_2^2) - \tfrac{\nu}{2}(x_B^2 - y_B^2)) \cdot x
\end{aligned} \right]\Bigg|_{E'_{3x}}^{T'_{3x}}
\end{aligned}
\right\} \tag{58}
$$

where D'_{ix}, C'_{ix}, E'_{ix} and $T'_{ix(i=1,2,3)}$ represent the abscissas of boundary points of the electrodes in the 2nd to 4th quadrant, respectively; $k_{C'_1} = -k_{C'}$, $k_{S'_1} = -k_{S'}$; $\mu = k\cos(\gamma)/a$, $\nu = k\sin(\gamma)/a$.

Further, the expression used to assess the influence of the coaxiality error on the output capacitance of the fan-shaped electrode (taking E_{e1} as an example) is given as:

$$
C5_{ecc.error} = \frac{C5' - C5}{C5} \tag{59}
$$

In the numerical simulation using Matlab, the range of the coaxiality error δ_e is set to be $[-0.2, 0.2]$ mm, and the variation quantity of the phase angle φ_e is 8 degrees referring to the sensor structural design. Figure 23 shows the relative error variation in the output capacitance of fan-shaped electrodes due to the error δ_e under different phase angles φ_e. For each fan-shaped electrode, the relative error of output capacitance increases with the increase in the error δ_e. It should be pointed out that the magnitude of the relative error is quite small within the whole value range of the error δ_e. Moreover, the influence of the phase angle φ_e variation in the relative error is also very limited.

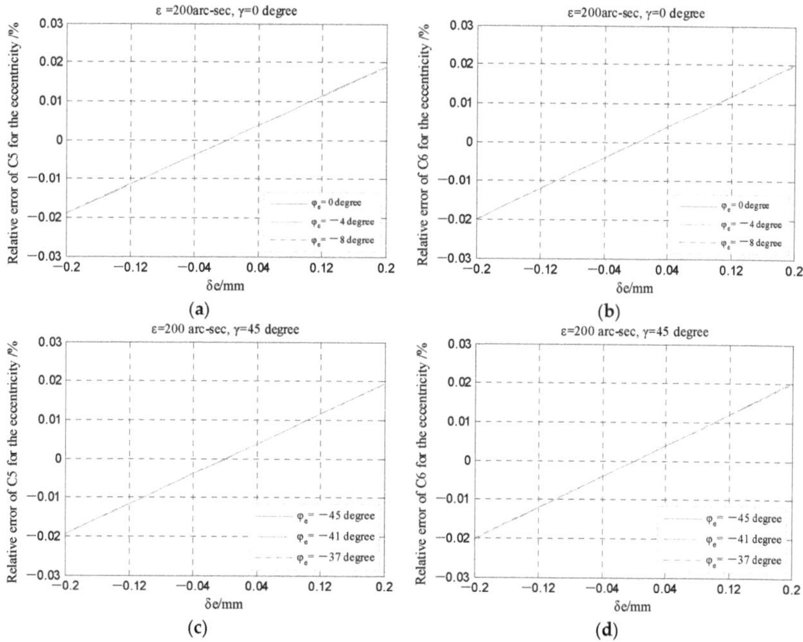

Figure 23. Relative error variations in the output capacitance of fan-shaped electrodes under the different coaxiality error δ_e: (**a**,**c**) The 1st quadrant; (**b**,**d**) the 2nd quadrant.

From Equations (27) and (28), it can be known that the tilt displacement ε and yaw angle γ are obtained based on the calculated differential output capacitance of the EPEG (CeY, CeX). As shown in Figure 24, under different yaw angle γ, the produced coaxiality error δ_e and phase angle φ_e do not cause the changing of the capacitance CeY and CeX, which indicates that the measurement of the rotor tilt displacement is basically not affected by the coaxiality error of fan-shaped electrode.

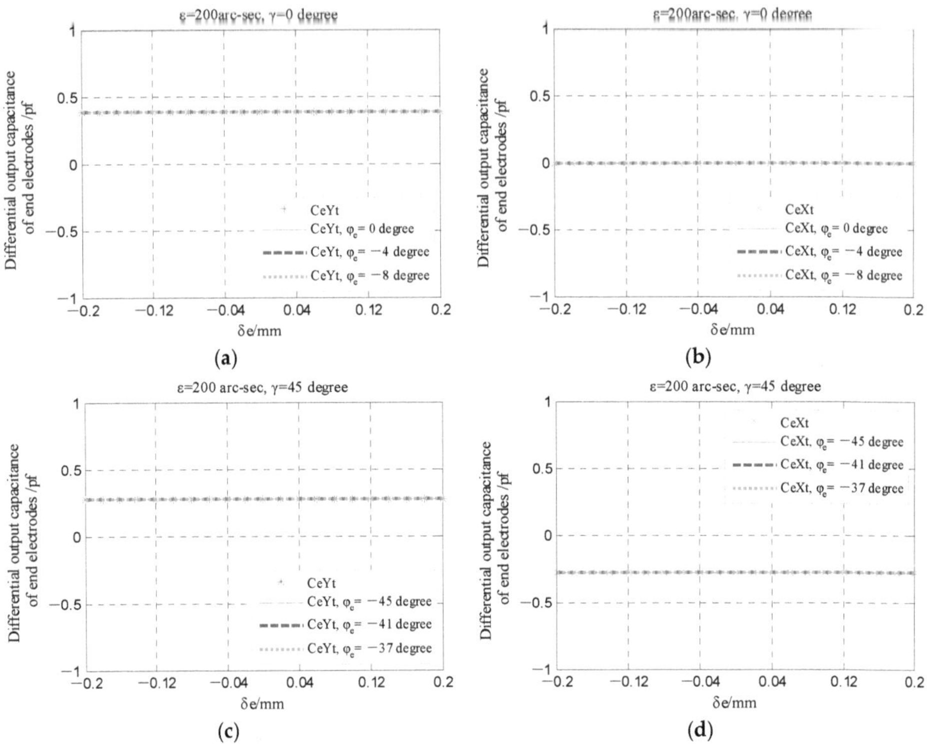

Figure 24. The changing of the end part differential output capacitance under the different coaxiality error δ_e: (**a**,**c**) The component around the y-axis; (**b**,**d**) the component around the x-axis.

The polar angle position error of the fan-shaped electrodes equals each other and does not change the increment of the electrode polar angle. Thus, the effect of the polar angle position error on the yaw angle γ measurement can be eliminated by the sensor calibration method.

4. Conclusions

The main error sources for the error motion measurement of a precision shafting using a T-type capacitive sensor (CS) were investigated in this paper. Firstly, we analyzed the theoretical modeling error due to the approximate simplification for the capacitance expressions of each sensing electrode. The approximate error of the theoretical calculations for the radial error motion measurement could be reduced under small ratios $\alpha/(R-g)$ and α/g. The introduction of the third-order term in the approximate expressions of the CrX and CrY could significantly reduce the approximate error from 5.5% to 1.2% without sacrificing the measuring range and sensitivity. For the theoretical model for the tilt error motion measurement, the theoretical capacitance of the end part electrodes was solved by adopting $\mathrm{IGe_{tl}}$ (first-order approximation) as integrand can achieve relatively accurate results with an approximate error lower than 0.2%.

Subsequently, the 3D-FEA results indicate that for the T-type CS, the coplanar configuration of the sensing electrodes and *Egr* with a tiny gap (0.1 mm) could significantly decrease the amount of variation in the additional capacitance introduced by fringe effects and reduce the corresponding nonlinear error from 4.8% to 1.7%.

Finally, among the electrode installation errors, we emphatically analyzed the tilt error of the cylindrical electrode and coaxiality error of the fan-shaped electrode. Numerical simulation was carried out based on the measurement model with the installation errors. The simulation results show that the tilt error only generates a certain deviation of solving the phase angle β (consistent in different displacement directions). The effect of the tilt error on solving the radial displacement α can be neglected, especially for the measurement range of less than 0.1 mm. Additionally, the measurement of the rotor tilt displacement is basically not affected by the coaxiality error of the fan-shaped electrode.

Author Contributions: Conceptualization, W.W. and Z.C.; methodology, W.W. and K.X.; software, K.X.; data and results analysis, W.W.and K.X.; original draft preparation, K.X.; review and editing, W.W. and Z.C.; supervision, W.W. and Z.C.; Funding Acquisition, W.W. All authors have read and agreed to the published version of the manuscript.

Funding: This research was supported by the National Natural Science Foundation of China (Grant No. U1709206, No. 51275465), Zhejiang Provincial Natural Science Foundation of China (Grant No. LZ16E050001) and the Key Laboratory of Advanced Manufacturing Technology of Zhejiang Province (Grant No. 2015KF02).

Conflicts of Interest: The authors declare no conflict of interest.

References

1. Chen, G.; Chen, Y.; Lu, Q.; Wu, Q.; Wang, M. Multi-Physics Fields Based Nonlinear Dynamic Behavior Analysis of Air Bearing Motorized Spindle. *Micromachines* **2020**, *11*, 723. [CrossRef] [PubMed]
2. IBAG. High Frequency Motor Spindles. Available online: www.ibag.ch (accessed on 20 April 2021).
3. Nanotech. Ultra-precision Machining Systems. Our-Automotive-Revolution-Rev.0717.pdf. Available online: www.nanotechsys. com (accessed on 21 April 2021).
4. Aerotech. ABRT400 Air-Bearing Direct-Drive Rotary Stage. Available online: www.aerotech.com/product/stages-actuators (accessed on 21 April 2021).
5. Gerlach, B.; Ehinger, M.; Raue, H.K.; Seiler, R. Gimballing Magnetic Bearing Reaction Wheel with Digital Controller. In Proceedings of the 11th European Space Mechanisms and Tribology Symposium, Lucerne, Switzerland, 21–23 September 2005; pp. 35–40.
6. MWI. Magnetic Bearing Momentum and Reaction Wheels with Internal Wheel Drive Electronics. Available online: www. electronicnote.com/RCG/MWI_A4.pdf (accessed on 22 April 2021).
7. Smithanik, J.; Paul, Y. Applying API 617 to Expander-Compressors with Active Magnetic Bearings. In Proceedings of the 44th Turbomachinery Symposium. Turbomachinery Laboratories, Texas A&M Engineering Experiment Station, Houston, TX, USA, 14–17 September 2015.
8. Marsh, E.; Grejda, R. Experiences with the Master Axis Method for Measuring Spindle Error Motions. *Precis. Eng.* **2000**, *24*, 50–57. [CrossRef]
9. Baxter, L.K. *Capacitive Sensors: Design and Applications*; IEEE Press: New York, NY, USA, 1997.
10. Nouira, H.; Vissiere, A.; Damak, M.; David, J.M. Investigation of the Influence of the Main Error Sources on the Capacitive Displacement Measurements with Cylindrical Artefacts. *Precis. Eng.* **2013**, *37*, 721–737. [CrossRef]
11. Lin, C.; Zheng, S.; Li, P.; Shen, Z.; Wang, S. Positioning Error Analysis and Control of a Piezo-Driven 6-DOF Micro-Positioner. *Micromachines* **2019**, *10*, 542. [CrossRef] [PubMed]
12. Bukhari, S.A.R.; Saleem, M.M.; Khan, U.S.; Hamza, A.; Iqbal, J.; Shakoor, R.I. Microfabrication Process-Driven Design, FEM Analysis and System Modeling of 3-DoF Drive Mode and 2-DoF Sense Mode Thermally Stable Non-Resonant MEMS Gyroscope. *Micromachines* **2020**, *11*, 862. [CrossRef] [PubMed]
13. Chapman, P.D. A Capacitance Based Ultra-Precision Spindle Error Analyser. *Precis. Eng.* **1985**, *7*, 129–137. [CrossRef]
14. Ahn, H.J.; Han, D.C.; Hwang, I.S. A Built-In Bearing Sensor to Measure the Shaft Motion of a Small Rotary Compressor for Air Conditioning. *Tribol. Int.* **2003**, *36*, 561–572. [CrossRef]
15. Kim, J.H.; Chang, H.K.; Han, D.C.; Jang, D.Y.; Oh, S.I. Cutting Force Estimation by Measuring Spindle Displacement in Milling Process. *CIRP Ann.-Manuf. Technol.* **2005**, *54*, 67–70. [CrossRef]
16. Xiang, K.; Wang, W.; Qiu, R.; Mei, D.; Chen, Z. A T-Type Capacitive Sensor Capable of Measuring 5-DOF Error Motions of Precision Spindles. *Sensors* **2017**, *17*, 1975. [CrossRef] [PubMed]

17. Ahn, H.J.; Jeon, S.; Han, D.C. Error Analysis of the Cylindrical Capacitive Sensor for Active Magnetic Bearing Spindles. *J. Dyn. Sys. Meas. Control.* **2000**, *122*, 102–107. [CrossRef]
18. Trusov, A.A.; Shkel, A.M. Capacitive Detection in Resonant MEMS with Arbitrary Amplitude of Motion. *J. Micromech. Microeng.* **2007**, *17*, 1583. [CrossRef]
19. Lu, Q.L. *Engineering Mathematics: Complex Function*; Higher Education Press: Beijing, China, 1996.
20. Gao, C. On Power Series of Functions in Several Variables. *Coll. Maths.* **2007**, *3*, 125–129.

micromachines

Article

Numerical Modeling of the Effect of Cutting-Edge Radius on Cutting Force and Stress Concentration during Machining

Peng Li [1] and Zhiyong Chang [1,2,*]

[1] Department of Mechanical Engineering, Northwestern Polytechnical University, Xi'an 710072, China; lip973@mail.nwpu.edu.cn
[2] Institute for Aero-Engine Smart Assembly of Shaanxi Province, Xi'an 710072, China
* Correspondence: changzy@nwpu.edu.cn

Abstract: Cutting is the primary method of material removal, and the quality of machined parts depends on the geometry of cutting tools. In this paper, a new cutting force coefficient model is established, revealing the influence of cutting-edge radius on the cutting process. The effects of cutting-edge radius on the shear angle and cutting force components are analyzed by finite element simulations. A series of simulations is conducted, and the results show that with increased cutting-edge radius, the shear angle decreases nonlinearly, and the cutting force increases gradually. Additionally, the growth rate of the feed force caused by increasing the cutting-edge radius is higher than that of the tangential force. Furthermore, the stress concentration area of the machined surface extends from the surface to the subsurface as the cutting-edge radius increases. The results of this research show that changing the cutting edge affects the cutting force component, shear angle, and stress concentration range during the cutting process. These results provide a theoretical reference for predicting the residual stress in parts.

Keywords: cutting-edge radius; shear angle; cutting force; stress concentration depth; finite element analysis

Citation: Li, P.; Chang, Z. Numerical Modeling of the Effect of Cutting-Edge Radius on Cutting Force and Stress Concentration during Machining. *Micromachines* **2022**, *13*, 211. https://doi.org/10.3390/mi13020211

Academic Editor: Benny C. F. Cheung

Received: 14 December 2021
Accepted: 27 January 2022
Published: 28 January 2022

Publisher's Note: MDPI stays neutral with regard to jurisdictional claims in published maps and institutional affiliations.

1. Introduction

Cutting machining is a common manufacturing method that is widely used in industrial applications, such as the precision machining of optical parts. Cutting-edge radius (CER) affects the cutting force [1], cutting temperature [2], and surface integrity of parts [3]. Therefore, many scholars are engaged in cutting force analysis and modeling [4–9].

Cutting force is one of the critical factors during the cutting process and is closely related to the cutting parameters. Vijayaraghavan et al. [4] used an artificial neural network to analyze the cutting force with cutting depth and feed rate. Zhuang [5] established a cutting force model of a chamfered edge and found that cutting force increased nonlinearly with the chamfering edge's length. Davis et al. [6] experimentally studied the changes in shear strain and strain rate with cutting speed and analyzed the strain rate distribution in chip flow direction. However, the above research [4–6] explored only the magnitude of cutting force and did not profoundly reveal the reasons for its change. In addition, some scholars analyzed the effect of cutting force based on elastic–plastic interactions. Chen et al. [7] established a theoretical model considering the elastic, plastic, and brittle zones. Error factors are inevitable during the cutting process, Wojciechowski et al. [8] developed a milling force model that considers the geometry errors of machining system. Wu et al. [9] found that serrated chips caused cutting force variations. Wojciechowski and Mrozek [10] minimized cutting force by optimizing the inclination angle of the tool axis.

Unfortunately, the research of [4–7] ignored the influence of CER. During the early development of cutting processes, it was assumed that the cutting edge was sharply acceptable for large-scale cutting because the CER was much smaller than the undeformed chip thickness [11]. However, in micron-scale cutting, the undeformed chip thickness is close to the CER, and the CER is a vital parameter that cannot be ignored. Fang et al. [12] proposed the relationship between the effective rake angle and the CER. With the rapid development of computer technology, Woon et al. [13,14] used the finite element method to study the relationship between chip formation and CER. The CER was constant along the tool corner [13,14]; however, Ozel [15] designed a tool in which the CER changed, and used the finite element simulation to analyze its cutting performance. Different from the finite element method, Aamir et al. [16] predicted cutting force with smooth particle hydrodynamics. In [11–13], the effects of the CER on cutting force and cutting temperature were investigated, but the studies did not reveal the evolution of involved parameters in the cutting force. Therefore, some scholars began to explore the influence of CER on parameters [17–19].

Rake angle is an important parameter of a tool, and some scholars are engaged in the effects of CER on the actual working rake angle of cutting tools. Lai et al. [17] considered that CER changes the rake angle of the tool, and the undeformed chip thickness is directly proportional to the CER. Yang et al. [18] established a two-dimensional mathematical model of the cutting force that considers the CER. Later, Dai et al. [19] proposed a method based on the CER to evaluate the cutting force coefficient. Aramcharoen and Mativenga [20] found that when the uncut chip thickness is equal to the CER, the best surface finish can be obtained; they suggest that the CER can greatly improve the burr on parts. With the development of cutting technology, ultrasonic-assisted cutting was used, and it was observed that the undeformed chip thickness increased with the CER [21].

The experimental research cost on cutting force is high, and unique fixtures need to be designed for small parts, which is cumbersome and laborious. Therefore, this paper proposes a finite element analysis model to analyze the evolution of the shear angle and explore the variational principle of cutting force components with CER; thus, it is not the magnitude of cutting force and stress of existing theories, but a new insight of evolution. In addition, the evolution of stress propagation is characterized first by the depth of stress concentration. The research results are relevant to the selection of cutting tools and optimization of tool design. Finally, this study provides a new theoretical reference for controlling the surface integrity of parts.

2. Simulation Modeling

Traditional cutting tools are assumed to be perfectly sharp, although they undergo wear quickly during the cutting process. With the upgrading of tool-manufacturing technology and the demand for energy saving and carbon emissions reduction, it is crucial to improve the lifespan of tools. For this reason, it is becoming common to use tools with a CER that can greatly increase tool strength, especially for difficult-to-machine materials. The shear angle is an important physical parameter that affects the cutting force. This section establishes a cutting force model and gives the relationship between the cutting force and shear angle.

2.1. Tool Geometry and Mechanical Model

Material undergoes large plastic deformation during cutting, and chips form when the cutting layer undergoes shear deformation. Chips move along the rake face of the tool after shear deformation, as shown in Figure 1, where F_v and F_t are the tangential force and the feed force, respectively, F_v is in the direction of velocity, and F_t is in the direction of uncut chip thickness. In the shear plane are normal force F_s and shear force F_n, respectively. The acute angle formed by F_v and F_t is η; r is the CER; φ_s is the shear angle; and h is the uncut chip thickness.

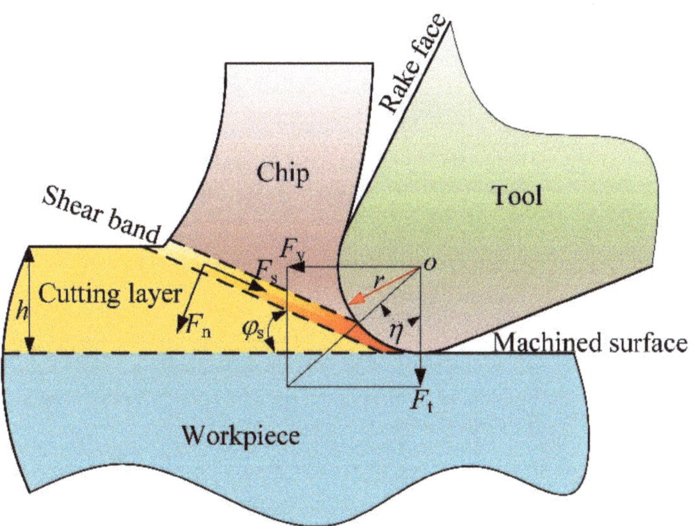

Figure 1. Geometry of the tool and workpiece simulation model.

According to [22], the shear zone can be assumed to be a thin plane, and the shear band thickness can be approximated as zero. Therefore, the normal stress σ_s and the shear stress τ_s in the shear plane can be expressed as:

$$\tau_w = \frac{F_s}{A_s}$$
$$\sigma_w = \frac{F_n}{A_s} \tag{1}$$

where A_s is the shear plane area.

According to [23], the tool–workpiece intersection forces F_v and F_t in orthogonal cutting can be expressed as

$$F_v = bh\left(\tau_w \frac{\cos(\beta_f - \beta_r)}{\sin(\varphi_s)\cos(\varphi_s + \beta_f - \beta_r)}\right)$$
$$F_t = bh\left(\tau_w \frac{\sin(\beta_f - \beta_r)}{\sin(\varphi_s)\cos(\varphi_s + \beta_f - \beta_r)}\right) \tag{2}$$

where β_f and β_r are the friction angle and rake angle, respectively. The friction angle is affected by the state of the contact between the tool and the workpiece, rake face roughness, and lubrication by the cutting fluid during the cutting process.

Based on the maximum shear stress principle, the shear angle φ_s, the friction angle β_f, and the rake angle β_r can be expressed as:

$$\varphi_s + \beta_f - \beta_r = \frac{\pi}{4} \tag{3}$$

According to [22], the average friction coefficient is assumed to be constant for a given tool–workpiece material. Substituting (3) into (2) then simplifies to

$$F_v = \sqrt{2}bh\left(\tau_w \frac{\cos(\pi/4 - \varphi_s)}{\sin(\varphi_s)}\right)$$
$$F_t = \sqrt{2}bh\left(\tau_w \frac{\sin(\pi/4 - \varphi_s)}{\sin(\varphi_s)}\right) \tag{4}$$

To characterize the effect of the shear angle on the cutting force, Equation (4) is written as

$$F_v = \sqrt{2}bh\tau_w k_v(\varphi_s)$$
$$F_t = \sqrt{2}bh\tau_w k_t(\varphi_s)$$

(5)

where $k_v(\varphi_s)$ and $k_t(\varphi_s)$ are the tangential force coefficient and the feed force coefficient, respectively. The cutting force coefficient can be written as

$$k_v(\varphi_s) = \frac{\cos(\pi/4 - \varphi_s)}{\sin(\varphi_s)}$$
$$k_t(\varphi_s) = \frac{\sin(\pi/4 - \varphi_s)}{\sin(\varphi_s)}$$

(6)

The cutting force coefficient includes the shear yield according to [22]. However, in this work, the cutting force coefficients $k_v(\varphi_s)$ and $k_t(\varphi_s)$ are different from those in [22]. Instead, $k_v(\varphi_s)$ and $k_t(\varphi_s)$ are functions of the shear angle φ_s and are nondimensional in the present model.

To understand the effect of the shear angle on cutting force coefficients $k_v(\varphi_s)$ and $k_t(\varphi_s)$, a sensitivity study was conducted by changing the φ_s, and calculating $k_v(\varphi_s)$ and $k_t(\varphi_s)$ based on Equation (6). The result is shown in Figure 2.

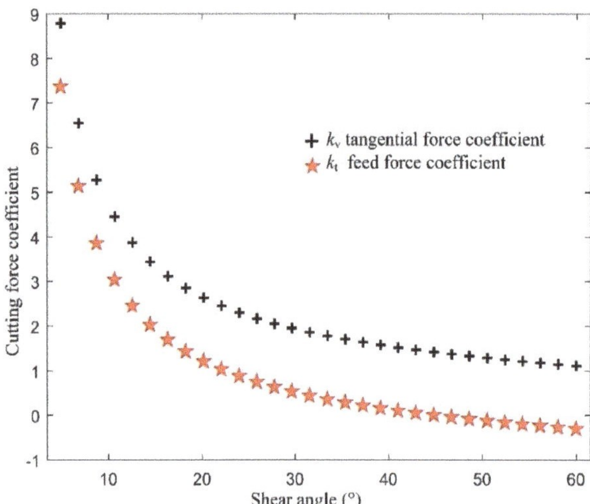

Figure 2. Cutting forces with different shear angle.

Figure 2 shows that as the shear angle increases, the cutting force coefficients $k_v(\varphi_s)$ and $k_t(\varphi_s)$ decrease nonlinearly, which means that the material deforms more easily as the shear angle increases. The tangential force coefficient $k_v(\varphi_s)$ is always larger than the feed force coefficient $k_t(\varphi_s)$ at the same shear angle.

The acute angle η between the tangential cutting force F_v and feed force F_t can be expressed as

$$\eta = \arctan\left(\frac{F_v}{F_t}\right)$$

(7)

where η reflects the direction change in the resultant cutting force.

2.2. Workpiece Property

The Johnson–Cook constitutive model is extensively utilized in metal-cutting modeling because it can adequately describe the behavior of a workpiece undergoing large strains, thermal softening, and high strain rate.

$$\sigma = \left(A + B\varepsilon^n\right)\left(1 + C\mathrm{In}\left(1 + \frac{\dot{\varepsilon}}{\varepsilon_0}\right)\right)\left(1 - \left(\frac{T - T_r}{T_m - T_r}\right)^m\right) \tag{8}$$

where $\sigma, \varepsilon^n, \dot{\varepsilon},$ and ε_0 are the flow stress, plastic strain, plastic strain rate, and equivalent plastic strain, respectively; T, T_r, and T_m are the current material temperature, melting point, and room temperature, respectively. A, B, C, m, and n are constitutive constants: A is the initial yield of materials; B is the stress hardening modulus; C is the sensitivity coefficient; m is the thermal coefficient of softening; and n refers to the hardening index. The flow behavior of GH 4169 is governed by the Johnson–Cook model. The A = 1241 Mpa, B = 622 Mpa, C = 0.0134, m = 1.3, and n = 0.652 are the parameters for GH 4169 as given by [24]. To investigate the CER on shear angle, cutting force, and stress concentrate, the values of the CER used in this simulation are shown in Table 1. The rake angle of the tool was 5°, the relief angle was 5°. The cutting speed was 30 m/min, and the feed rate was set to 0.15 mm. The tool material was cemented carbide, and the cutting tool was not coated.

Table 1. Cutting-edge radius used in this simulation.

Test No.	1	2	3	4	5
Cutting-edge radius r (μm)	5	25	45	65	85

GH 4169 is a nickel-based superalloy widely used in the aerospace industry. However, fast tool wear and high cutting temperature make cutting this material challenging. The composition of GH 4169 is presented in Table 2.

Table 2. Chemical composition of GH4169 [25].

Element	Ni	Cr	Nb	Mo	Ti	Al	Co	Mn	Cu	Si	C	Fe
Wt %	52.15	19.26	5.03	3.03	1.08	0.56	0.5	0.22	0.1	0.26	0.052	17.75

The physical properties of GH 4169 are presented in Table 3. The cutting tool is assumed to be perfectly rigid solids. The workpiece GH 4169 is a rectangular block with 2 mm length, 0.5 mm width, and 1 mm height.

Table 3. Typical mechanical properties at room temperature [25].

Tensile Strength (MPa)	Yield Strength (MPa)	Young's Modulus (GPa)	Density (g·cm^{-3})	Poisson's Ratio	Thermal Conductivity (W/m·K)
1430	1300	204	8.24	0.3	14.7

3. Results and Discussion

The cutting process is complex and involves high temperature, high pressure, and large strain. The force and stress on the tool–workpiece system are important factors that determine the energy required. The shear angle is the critical variable used to analyze the cutting force; therefore, the shear angle is investigated first in this study. This section studies the effects of CER on shear angle, cutting force, and stress concentration depth. A series of orthogonal simulations with the finite element method software AdventEdge are conducted.

3.1. Evolution of Shear Angle

The acute angle between the shear plane and the cutting speed direction is called the shear angle. The shear angle affects cutting force consumed by cutting, and a large shear angle saves energy. According to Equation (5), the cutting force is a function of the shear angle. Therefore, it is helpful to analyze the influence of the CER on the shear angle to deeply reveal the mechanism of the effect of the CER on the cutting force.

Figure 3 shows the results of the cutting simulation with different CERs, and measurement of the shear angle. Figure 3a–e shows that the cutting layer material deforms in the shear plane, so the strain rate in the shear plane is very high. As the CER increases, the position of the chip bending point shifts, resulting in a change in the shear angle.

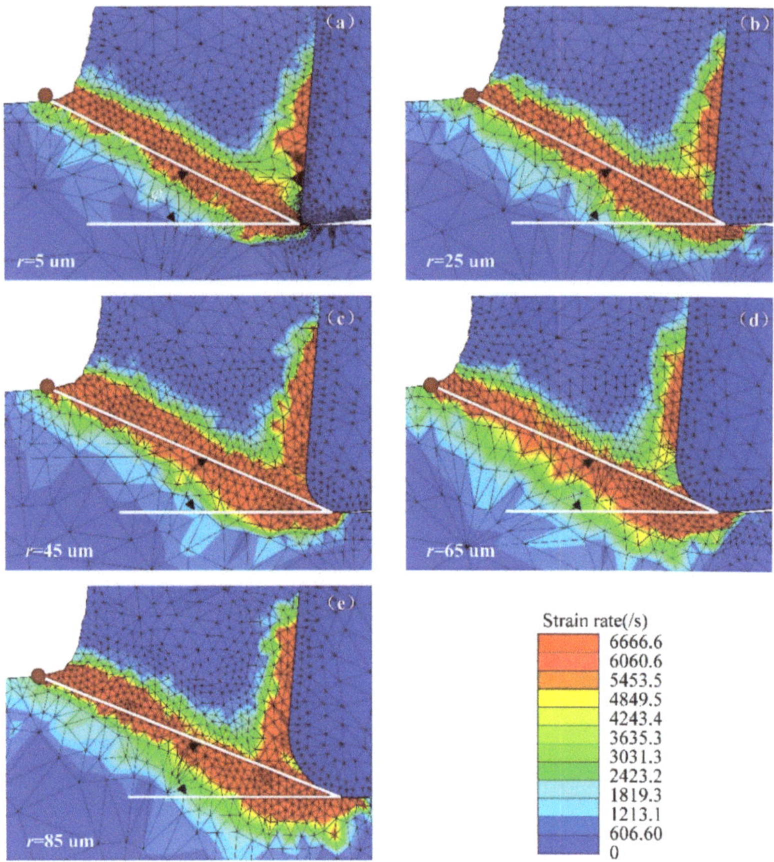

Figure 3. Bending point position versus cutting-edge radius; (**a**) r = 5 μm; (**b**) r = 25 μm; (**c**) r = 45 μm; (**d**) r = 65 μm; (**e**) r = 85 μm.

The shear angle was measured in each simulation, as shown in Figure 4. When the CER is 5 μm, the shear angle is 25°; as the CER increases to 85 μm, the shear angle decreases to 19°, a reduction of by 24%. Increasing the CER reduces the shear angle, since increasing the CER decreases the average rake angle of the tool. According to Equation (3), the shear angle decreases as the rake angle decreases.

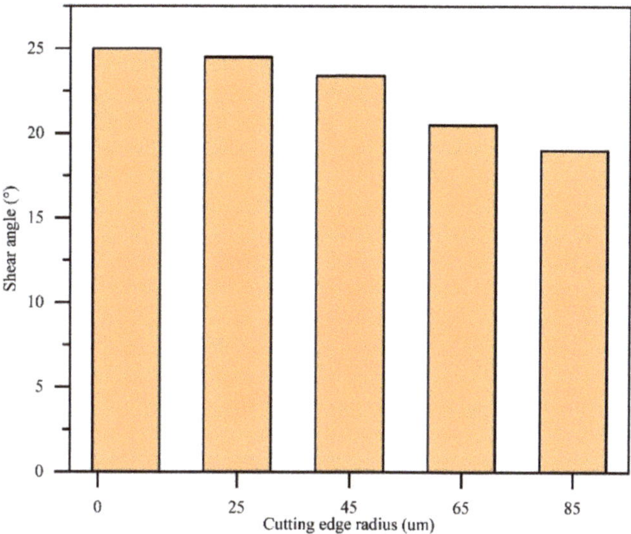

Figure 4. Shear angle for various cutting-edge radius.

3.2. Von Mises Stress of the Machined Surface

The tool–workpiece system undergoes contact stress during the cutting process, and the material forms a chip after large plastic deformation. The local area inside the workpiece where the stress is significantly higher than the surrounding region is called the stress concentration.

Figure 5a–e shows the von Mises stress distribution of the workpiece. The region of stress concentration expanded with increasing CER. In Figure 5a, the stress concentration area is mainly on both sides of the shear band due to shear deformation. However, in Figure 5e, the stress concentration area extends from the shear band to the machined surface and cutting layer, and the region of the workpiece stress concentration expands with CER. The results indicate that the CER induces stress to expand toward the cutting layer and the workpiece surface. However, the stress concentration does not extend to the chip, mainly because the cutting layer undergoes upward flow and concave bending after shear deformation, the chip surface is free, and the chip material can flow uniformly. Because the machined part is a fixed constraint during the cutting process, the stress propagates and expands to the cutting layer and the machined surface.

Although the stress expands to the machined layer, the cutting layer is removed in the next cycle. The machined surface affects the service life of the part. The depth of stress concentration is measured from the machined surface along the direction of undeformed chip thickness. The depth of stress concentration affects the fatigue life of the part and makes an important contribution to the residual stress of the surface and subsurface. The depth of stress concentration is measured to characterize the degree of stress concentration, as shown in Figure 5. The stress depth h_m is 108 µm in Figure 5a, while the stress concentration depth h_m is 348 µm in Figure 5e, and the depth of the stress concentration is 3.2 times when $r = 85$ µm than when $r = 5$ µm. As shown in Figure 5a–e, the depth of stress concentration gradually increases with increasing CER.

Figure 6 shows the maximum von Mises stress variation for different CERs. When the CER is 5 µm, the maximum stress is 1450 MPa. The maximum stress is 1680 MPa with $r = 85$ µm; the ultimate stress increases by 230 MPa compared with $r = 5$ µm. Thus, increasing the CER increases the amplitude of the maximum stress. The average rake angle of the CER is less than the design rake angle of the tool, increasing the maximum stress.

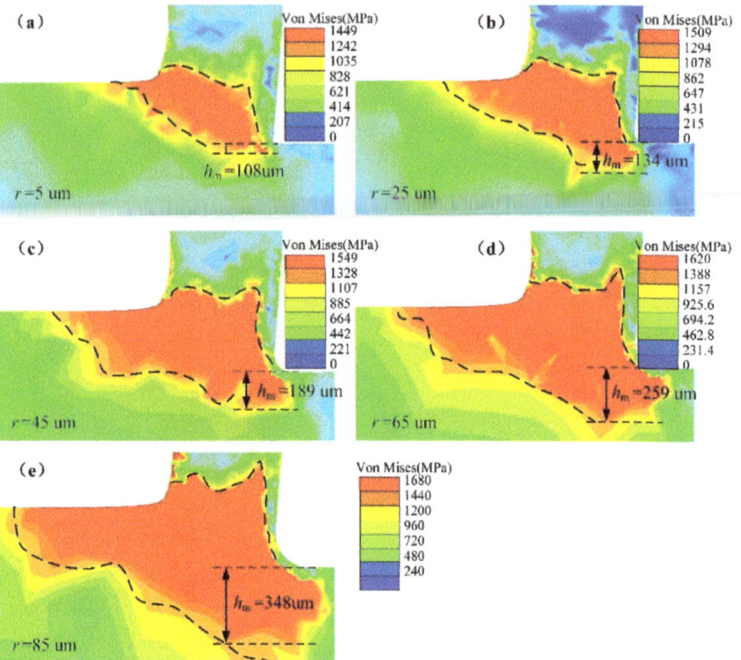

Figure 5. Von Mises stress distribution on a workpiece; (**a**) r = 5 μm; (**b**) r = 25 μm; (**c**) r = 45 μm; (**d**) r = 65 μm; (**e**) r = 85 μm.

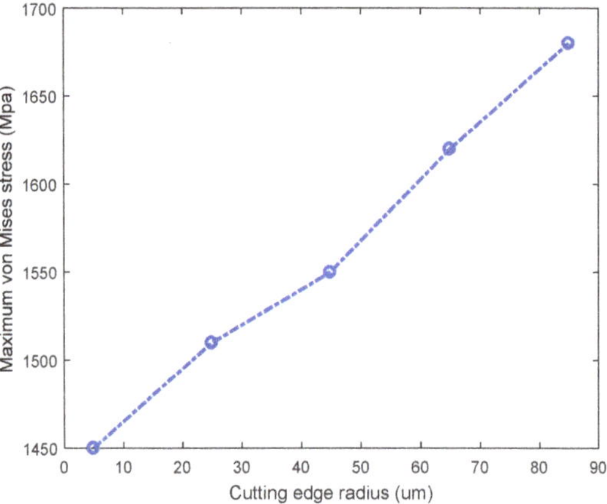

Figure 6. Maximum von Mises stress on the workpiece.

An increase in the stress concentration depth changes the residual stress of the surface and subsurface of the workpiece, increases the final residual stress of the part, and affects the fatigue of the part. Increasing the CER expands the stress concentration depth and amplitude. Additionally, the expansion of the stress concentration range increases energy consumption. These simulation results will provide a reference for efforts to control the residual stress of parts.

3.3. Effect of the CER on Cutting Forces

Cutting force is important for analyzing and controlling workpiece deformation. The influence of the CER on the cutting forces is neglected in the traditional cutting process. This section uses simulations to explore the effects of the CER on the cutting force.

As shown in Figure 7, the tangential force F_v and the feed force F_t are 600 N and 220 N in Test No.1, respectively; they increase to 730 N and 600 N in Test No.5, respectively. The tangential force and feed force rise by 22% and 173%, respectively. Therefore, the cutting force and feed force increase with the CER, mainly because the shear angle decreases as the CER increases, as shown in Section 3.1; thus, the cutting force increases. These simulation results are consistent with the theoretical analysis of Equation (5). In every case, the tangential force F_v is greater than the feed force F_t in Figure 7. The main cutting force is the tangential force during the cutting process, which determines the power required. However, the feed force F_t has a much faster growth rate than the tangential force F_v, which indicates that the CER mainly affects the feed force. Due to the vibration and deformation of the workpiece, cutting inevitably produces friction on the machined surface. The CER increases the contact area between the flank and the machined surface, so the normal pressure between the tool and the machined surface also increases.

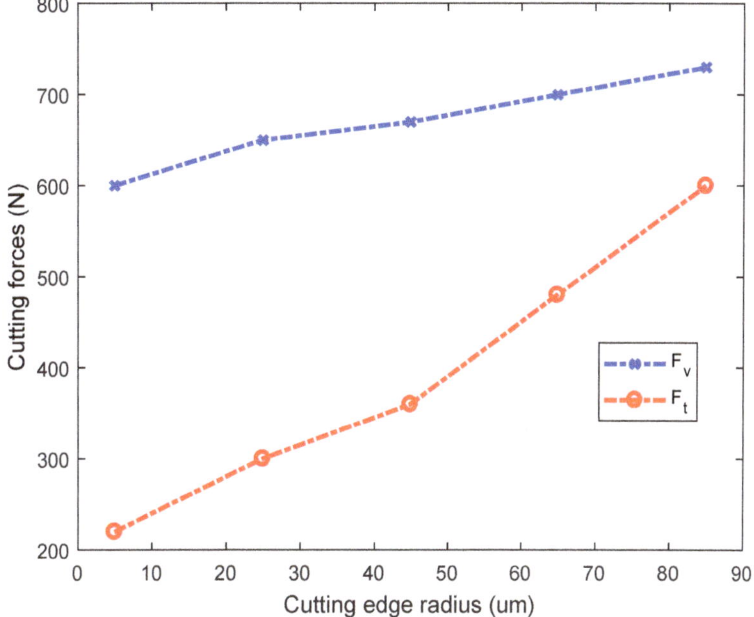

Figure 7. Cutting force for different cutting-edge radii.

The angle η between the tangential force and the feed force is shown in Figure 8. The angle η decreases with the CER mainly due to the quick increase in feed force as the CER. Additionally, increasing the CER causes the transition from the rake face to the flank face to no longer be a simple straight surface, but rather a complex one. Contact with the workpiece is also a complex surface condition. A decrease in the angle η means that the direction of the resultant force changes.

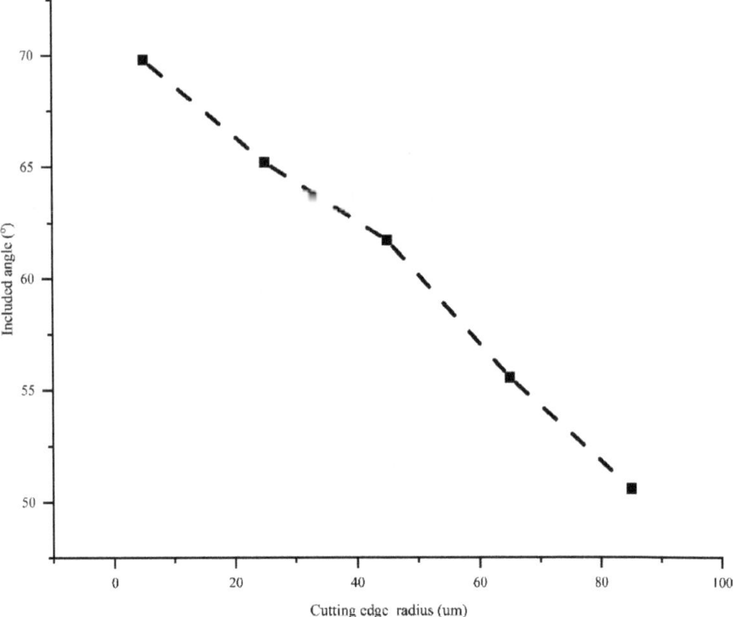

Figure 8. The included angle between the tangential force and the feed force.

4. Conclusions

Force is one of the crucial factors related to tool life and the quality of machined parts. CER is the key parameter of the geometry of the cutting tool that cannot be ignored in micromachining. This paper proposed a new method for the calculation of the cutting force coefficient based on an expression for the coefficient as a nonlinear function of the shear angle. The effect of CER on the shear angle was analyzed by simulations, and its influence on the cutting force was further studied. The main conclusions were as follows:

1. The cutting force increased with the CER. The simulation analysis showed that the cutting force coefficient decreased as the shear angle increased, which was consistent with the analytical cutting force model. As the CER increased, the shear angle decreased, which caused the cutting force to increase.
2. The increase in the feed force was quicker than the increase in the tangential force with increasing CER. With enhanced CER, the rate of feed force growth was much higher than that of the tangential force, unlike the behavior of traditional sharp cutter. The effect of the CER on the feed force must be considered in order to understand tool life and the quality of machined parts.
3. The stress concentration range of the workpiece gradually expanded with increasing CER. Simulation experiments showed that the stress concentration area extended from the shear band to the cutting layer and the subsurface of the workpiece, resulting in a substantial increase in the stress concentration depth, which affected the surface and integrity of the workpiece. Additionally, the maximum stress in the workpiece increased with the CER.

Author Contributions: P.L.: conceptualization, methodology, software, validation, formal analysis, investigation, data curation, writing—original draft preparation, writing—review and editing, and visualization; Z.C.: resources, supervision, project administration, and funding acquisition. All authors have read and agreed to the published version of the manuscript.

Funding: This work is supported by the Defense Industrial Technology Development Program (No. XXXX2018213A001).

Conflicts of Interest: The authors declare that they have no conflict of interest in this work, and no commercial or associative interests that represent a conflict of interest in connection with the submitted work.

References

1. Zhang, S.; Zhang, H.J.; Zong, W.J. Modeling and simulation on the effect of tool rake angle in diamond turning of KDP crystal. *J. Mater. Process. Technol.* **2019**, *273*, 116259. [CrossRef]
2. Liu, C.; Wan, M.; Yang, Y. Simulation of the chip morphology together with its evolution in machining of Inconel 718 by considering widely spread cutting speed. *Int. J. Adv. Manuf. Technol.* **2021**, *116*, 175–195. [CrossRef]
3. Sun, W.; Duan, C.; Yin, W. Modeling of force and temperature in cutting of particle reinforced metal matrix composites considering particle effects. *J. Mater. Process. Technol.* **2021**, *290*, 116991. [CrossRef]
4. Vijayaraghavan, V.; Garg, A.; Gao, L.; Vijayaraghavan, R.; Lu, G. A finite element based data analytics approach for modeling turning process of Inconel 718 alloys. *J. Clean. Prod.* **2016**, *137*, 1619–1627. [CrossRef]
5. Zhuang, K.; Weng, J.; Zhu, D.; Ding, H. Analytical Modeling and Experimental Validation of Cutting Forces Considering Edge Effects and Size Effects with Round Chamfered Ceramic Tools. *J. Manuf. Sci. Eng.* **2018**, *140*, 081012. [CrossRef]
6. Davis, B.; Dabrow, D.; Ifju, P.; Xiao, G.; Liang, S.Y.; Huang, Y. Study of the Shear Strain and Shear Strain Rate Progression During Titanium Machining. *J. Manuf. Sci. Eng.* **2018**, *140*, 051007. [CrossRef]
7. Li, C.; Piao, Y.; Hu, Y.; Wei, Z.; Li, L.; Zhang, F. Modelling and experimental investigation of temperature field during fly-cutting of KDP crystals. *Int. J. Mech. Sci.* **2021**, *210*, 106751. [CrossRef]
8. Wojciechowski, S.; Matuszak, M.; Powałka, B.; Madajewski, M.; Maruda, R.W.; Królczyk, G.M. Prediction of cutting forces during micro end milling considering chip thickness accumulation. *Int. J. Mach. Tools Manuf.* **2019**, *147*, 103466. [CrossRef]
9. Wu, F.; Liu, Z.; Guo, B.; Sun, Y.; Chen, J. Research on the burr-free interrupted cutting model of metals. *J. Mater. Process. Technol.* **2021**, *295*, 117190. [CrossRef]
10. Wojciechowski, S.; Mrozek, K. Mechanical and technological aspects of micro ball end milling with various tool inclinations. *Int. J. Mech. Sci.* **2017**, *134*, 424–435. [CrossRef]
11. Son, S.M.; Lim, H.S.; Ahn, J.H. Effects of the friction coefficient on the minimum cutting thickness in micro cutting. *Int. J. Mach. Tools Manuf.* **2005**, *45*, 529–535. [CrossRef]
12. Fang, F.Z.; Wu, H.; Liu, Y.C. Modelling and experimental investigation on nanometric cutting of monocrystalline silicon. *Int. J. Mach. Tools Manuf.* **2005**, *45*, 1681–1686. [CrossRef]
13. Woon, K.S.; Rahman, M.; Fang, F.Z.; Neo, K.S.; Liu, K. Investigations of tool edge radius effect in micromachining: A FEM simulation approach. *J. Mater. Process. Technol.* **2008**, *195*, 204–211. [CrossRef]
14. Woon, K.S.; Rahman, M.; Neo, K.S.; Liu, K. The effect of tool edge radius on the contact phenomenon of tool-based micromachining. *Int. J. Mach. Tools Manuf.* **2008**, *48*, 1395–1407. [CrossRef]
15. Özel, T. Computational modelling of 3D turning: Influence of edge micro-geometry on forces, stresses, friction and tool wear in PcBN tooling. *J. Mater. Process. Technol.* **2009**, *209*, 5167–5177. [CrossRef]
16. Mir, A.; Luo, X.; Siddiq, A. Smooth particle hydrodynamics study of surface defect machining for diamond turning of silicon. *Int. J. Adv. Manuf. Technol.* **2017**, *88*, 2461–2476. [CrossRef]
17. Lai, M.; Zhang, X.D.; Fang, F.Z. Study on critical rake angle in nanometric cutting. *Appl. Phys. A Mater. Sci. Process.* **2012**, *108*, 809–818. [CrossRef]
18. Yang, S.; Tong, X.; Liu, X.; Ji, W.; Zhang, Y. Investigation on the characteristic of forces of the tool edge in finish machining of titanium alloys. *Int. J. Adv. Manuf. Technol.* **2018**, *96*, 2431–2441. [CrossRef]
19. Dai, X.; Zhuang, K.; Ding, H. A systemic investigation of tool edge geometries and cutting parameters on cutting forces in turning of Inconel 718. *Int. J. Adv. Manuf. Technol.* **2019**, *105*, 531–543. [CrossRef]
20. Aramcharoen, A.; Mativenga, P.T. Size effect and tool geometry in micromilling of tool steel. *Precis. Eng.* **2009**, *33*, 402–407. [CrossRef]
21. Arefin, S.; Zhang, X.Q.; Neo, D.W.K.; Kumar, A.S. Effects of cutting edge radius in vibration assisted micro machining. *Int. J. Mech. Sci.* **2021**, *208*, 106673. [CrossRef]
22. Yusuf, A. Manufacturing Automation: Metal Cutting Mechanics, Machine Tool Vibrations, and CNC Design. *Appl. Mech. Rev.* **2001**, *54*, B84.
23. Albrecht, P. New Developments in the Theory of the Metal-Cutting Process: Part I. The Ploughing Process in Metal Cutting. *J. Eng. Ind.* **1960**, *82*, 348–357. [CrossRef]
24. Ozel, T.; Llanos, I.; Soriano, J.; Arrazola, P.-J. 3D Finite Element Modelling of Chip Formation Process for Machining Inconel 718: Comparison of FE Software Predictions. *Mach. Sci. Technol.* **2011**, *15*, 21–46. [CrossRef]
25. Li, P.; Chang, Z. Accurate modeling of working normal rake angles and working inclination angles of active cutting edges and application in cutting force prediction. *Micromachines* **2021**, *12*, 1207. [CrossRef]

Article

A Novel Updated Full-Discretization Method for Prediction of Milling Stability

Junjin Ma [1,*], Yunfei Li [1], Dinghua Zhang [2], Bo Zhao [1], Geng Wang [1] and Xiaoyan Pang [3]

[1] School of Mechanical and Power Engineering, Henan Polytechnic University, Jiaozuo 454000, China; iyfhpu369@163.com (Y.L.); zhaob@hpu.edu.cn (B.Z.); wanggeng@hpu.edu.cn (G.W.)
[2] Key Laboratory of Contemporary Design and Integrated Manufacturing Technology, Ministry of Education, Northwestern Polytechnical University, Xi'an 710072, China; dhzhang@nwpu.edu.cn
[3] College of Computer Science and Technology, Henan Polytechnic University, Jiaozuo 454000, China; pangxy@hpu.edu.cn
* Correspondence: majhpu@hpu.edu.cn

Abstract: This paper presents an updated full-discretization method for milling stability prediction based on cubic spline interpolation. First, the mathematical model of the time-delay milling system considering regenerative chatter is represented by a dynamic delay differential equation. Then, in a single tooth passing period, the time is divided into a finite time intervals, the state item and the time-delay item are approximated in each time interval by cubic spline interpolation and third-order Newton interpolation, respectively. Afterward, a transition matrix is constructed to represent the transfer relationship of the teeth in a period. Finally, based on Floquet theory, the milling stability lobes can be obtained. Meanwhile, in order to improve computational efficiency, an optimized method is proposed based on the traditional algorithm and the proposed method has high precision without losing high efficiency. Finally, several milling experiments are conducted to verify the accuracy of the proposed method, and the results show that the predicted results agree well with the experimental results.

Keywords: milling stability; delay-differential equation; computational efficiency; Floquet theory

Citation: Ma, J.; Li, Y.; Zhang, D.; Zhao, B.; Wang, G.; Pang, X. A Novel Updated Full-Discretization Method for Prediction of Milling Stability. *Micromachines* **2022**, *13*, 160. https://doi.org/10.3390/mi13020160

Academic Editors: Benny C. F. Cheung and Jiang Guo

Received: 30 December 2021
Accepted: 19 January 2022
Published: 21 January 2022

Publisher's Note: MDPI stays neutral with regard to jurisdictional claims in published maps and institutional affiliations.

1. Introduction

Chatter is a serious problem in the milling of a thin-walled workpiece such as aero-engine blades, casings, impellers, blisks etc., which not only reduces the surface quality and production efficiency but also shortens the life of machine spindles and cutters. Therefore, it is necessary to investigate chatter in the milling of thin-walled workpieces for obtaining chatter-free operations.

Up to now, chatter has been studied by considering the time-varying milling process system by several researchers, including T. Insperger [1–3], Chandiramani [4], and Eksioglu [5] et al. From these studies, we found that in the milling process, considerable valuable results on chatter are investigated by a mathematical model of a time-periodic delay differential equation (DDE), and stability lobes diagram can be used to obtain the chatter-free process parameters more accurately.

To investigate machine stability considering regeneration chatter, a considerable number of methods, including analytical methods, numerical methods, and experimental methods, have been proposed to predict the stability lobes diagram (SLD), e.g., the analytical methods [6], the temporal finite element methods [7], the semi-discretization and full-discretization methods [8,9], the time domain numerical simulation methods [10], and the experimental methods [11]. Among the methods mentioned above, the semi-discretization methods and full-discretization methods are widely used due to their efficiency and accuracy.

In recent years, Altintas [12] proposed a zero-order analytical (ZOA) method by the Fourier series average method of dynamic milling force coefficient. Then, Merdol [13]

transformed the low radial immersion milling dynamics into an eigenvalue problem by considering the tooth spacing angle and tooth passing frequencies for accurately predicting the stability lobes diagram. The semi-discrete cycloid method based on the nonlinear cutting force milling model was presented by Faassen [14] for investigating the stability of a milling system, which was proved effectively.

Furthermore, to obtain a high-precision stability lobe diagram, the numerical methods, including the numerical integration method [15,16], Runge–Kutta-based discretization method [17], and precise integration method [18] were developed. Shortly after that, a semi-discrete method (SDM) was proposed by Insperger [19] and then, a full-discretization method (FDM) was presented by Ding et al. [20]. Later on, a second-order FDM method [21], third-order FDM method [22], high-order FDM method [23], Hermite interpolation FDM [24], and the update FDM [25] are successively proposed. It is proved that the accuracy and efficiency of these methods were improved to some extent. However, these extended methods have more complex algorithm structures, which will take more calculation time and obtain the low convergence accuracy to a certain degree.

Therefore, to obtain a higher convergence rate and computational efficiency, a novel update FDM based on Spline–Newtons interpolation is proposed in this paper. The most important difference of the proposed method compared with the existing methods is that the cubic spline interpolation method was utilized to handle the state item and the third-order Newton interpolation method was used to approximate the time-delay item. The remainder of the paper is organized as follows. In Section 2, a systematic mathematical model and algorithm are described in detail. In Section 3, the rate of convergence estimates of the proposed method is calculated compared with some existing method. In Sections 4 and 5, two benchmark examples for a one degrees of freedom (DOF) milling model are given to illustrate the accuracy and efficiency of the proposed approach. Some verified experiments are conducted and analyzed in Section 6. Finally, some conclusions are presented.

2. Systematic Mathematical Model and Algorithm

For a conventional milling process, a schematic diagram of milling a thin-walled section while considering regenerative chatter is given in Figure 1. Without loss of generality, the dynamic model of the milling process system considering the regenerative effect can be expressed by a n-dimensional linear time periodic system with a single discrete time delay as follows:

$$\dot{X}(t) = A_0 X(t) + A(t)[X(t) - X(t - T)] \tag{1}$$

where A_0 is a constant matrix, $A(t)$ is a time-periodic matrix, and $A(t) = A(t + T)$, T is the time period which equals to the time delay, and $X(t)$ is the relative displacement between the cutter and workpiece. In order to solve Equation (1), time period T can be equally divided into m small-time intervals, and $T = mh$, where m is an integer and h is the range of each interval. Then, the dynamic response of Equation (1) can be obtained by a direct integration method on each time interval $kh \leq t \leq (k + 1)h$:

$$X(t) = e^{A_0(t - kh)} X(kh) + \int_{kh}^{t} \left\{ e^{A_0(t - \varepsilon)} [A(\varepsilon)X(\varepsilon) - A(\varepsilon)X(\varepsilon - T)] \right\} d\varepsilon. \tag{2}$$

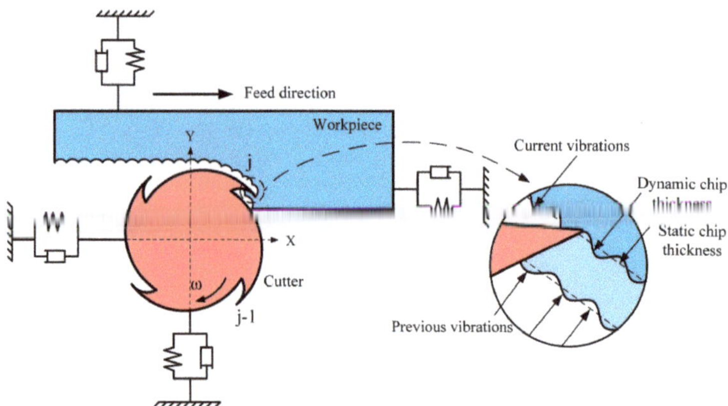

Figure 1. Dynamic model of milling a workpiece.

Let $\mathbf{X}(kh) = \mathbf{X}_k$ with $k = 0, 1, \ldots, m$, when $t = (k + 1)h$, Equation (2) can be equivalently converted to the following form:

$$\mathbf{X}_{k+1} = e^{\mathbf{A}_0 h}\mathbf{X}_k + \int_0^h e^{\mathbf{A}_0 \varepsilon}[\mathbf{A}(kh + h - \varepsilon)\mathbf{X}(kh + h - \varepsilon) - \mathbf{A}(kh + h - \varepsilon)\mathbf{X}(kh + h - \varepsilon - T)]d\varepsilon \tag{3}$$

Next, the integral term in Equation (3) should be handled. The state item $\mathbf{X}(kh + h - \varepsilon)$ can be approximately represented by cubic spline interpolation using $\mathbf{X}_{k+1}, \mathbf{X}_k, \mathbf{X}_{k-1}, \mathbf{X}_{k-2}$. In addition, two other constraints $\dot{\mathbf{X}}(kh - 2h)$ and $\dot{\mathbf{X}}(kh + h)$ are used in cubic spline interpolation. Namely, let $\mathbf{A}_k$ stands for $\mathbf{A}(kh)$:

$$\begin{cases} \dot{\mathbf{X}}(kh - 2h + 0) = \mathbf{A}_0\mathbf{X}_{k-2} + \mathbf{A}_{k-2}(\mathbf{X}_{k-2} - \mathbf{X}_{k-2-m}) \\ \dot{\mathbf{X}}(kh + h - 0) = \mathbf{A}_0\mathbf{X}_{k+1} + \mathbf{A}_{k+1}(\mathbf{X}_{k+1} - \mathbf{X}_{k+1-m}) \end{cases} . \tag{4}$$

At the time interval $[t_k, t_{k+1}]$, the state item $\mathbf{X}(kh + h - \varepsilon)$ can be approximated by cubic spline interpolation, resulting in:

$$\mathbf{X}(kh + h - \varepsilon) = \mu_1\mathbf{X}_{k+1} + \mu_2\mathbf{X}_k + \mu_3\mathbf{X}_{k-1} + \mu_4\mathbf{X}_{k-2} \tag{5}$$

where

$$\mu_1 = \frac{-11h\mathbf{A}_0 + 18\mathbf{I}}{15h^3}\varepsilon^3 + \frac{26h\mathbf{A}_0 - 33\mathbf{I}}{15h^2}\varepsilon^2 - \mathbf{A}_0\varepsilon + \mathbf{I} \tag{6}$$

$$\mu_2 = \frac{-9\mathbf{I}}{5h^3}\varepsilon^3 + \frac{14\mathbf{I}}{5h^2}\varepsilon^2 \tag{7}$$

$$\mu_3 = \frac{4\mathbf{I}}{5h^3}\varepsilon^3 - \frac{4\mathbf{I}}{5h^2}\varepsilon^2 \tag{8}$$

$$\mu_4 = \frac{-h\mathbf{A}_0 - 3\mathbf{I}}{15h^3}\varepsilon^3 + \frac{h\mathbf{A}_0 + 3\mathbf{I}}{15h^2}\varepsilon^2. \tag{9}$$

The time delay item $\mathbf{X}(kh + h - \varepsilon - T)$ in Equation (3) can be approximately expressed at four points $\mathbf{X}_{k-m+3}, \mathbf{X}_{k-m+2}, \mathbf{X}_{k-m+1}, \mathbf{X}_{k-m}$ by the third-order Newton interpolation method as follows:

$$\mathbf{X}(kh + h - \varepsilon - T) = \lambda_1\mathbf{X}_{k-m} + \lambda_2\mathbf{X}_{k-m+1} + \lambda_3\mathbf{X}_{k-m+2} + \lambda_4\mathbf{X}_{k-m+3} \tag{10}$$

where

$$\lambda_1 = \frac{\varepsilon^3}{6h^3} + \frac{\varepsilon^2}{2h^2} + \frac{\varepsilon}{3h} \tag{11}$$

$$\lambda_3 = \frac{\varepsilon^3}{2h^3} + \frac{\varepsilon^2}{2h^2} - \frac{\varepsilon}{h} \tag{12}$$

$$\lambda_3 = \frac{\varepsilon^3}{2h^3} + \frac{\varepsilon^2}{2h^2} - \frac{\varepsilon}{h} \tag{13}$$

$$\lambda_4 = \frac{-\varepsilon^3}{6h^3} + \frac{\varepsilon}{6h}. \tag{14}$$

Subsequently, the time-periodic item $\mathbf{A}(kh + h - \varepsilon)$ in Equation (3) can be expressed by linear interpolation which using points $\mathbf{A}(kh)$ and $\mathbf{A}(kh + h)$, and:

$$\mathbf{A}(kh + h - \varepsilon) = \mathbf{A}_u + \mathbf{A}_v \varepsilon \tag{15}$$

where

$$\mathbf{A}_u = \mathbf{A}_{k+1} \tag{16}$$

$$\mathbf{A}_v = (\mathbf{A}_k - \mathbf{A}_{k+1})/h. \tag{17}$$

Then, substituting Equation (5), Equation (10), and Equation (15) into Equation (3) leads to the interpolated item $\mathbf{X}(kh + h - \varepsilon)$, $\mathbf{X}(kh + h - \varepsilon - T)$ and $\mathbf{A}(kh + h - \varepsilon)$ are taken into Equation (3), and the DDE is approximated by an ordinary differential equation (ODE), which can be simplified as follows:

$$\mathbf{P}_{k+1}\mathbf{X}_{k+1} = \begin{aligned} &\mathbf{P}_k\mathbf{X}_k + \mathbf{P}_{k-1}\mathbf{X}_{k-1} + \mathbf{P}_{k-2}\mathbf{X}_{k-2} + \mathbf{P}_{k-m+3}\mathbf{X}_{k-m+3} \\ &+ \mathbf{P}_{k-m+2}\mathbf{X}_{k-m+2} + \mathbf{P}_{k-m+1}\mathbf{X}_{k-m+1} + \mathbf{P}_{k-m}\mathbf{X}_{k-m} \end{aligned} \tag{18}$$

where

$$\mathbf{P}_{k+1} = \mathbf{I} - (\mathbf{a}_1\mathbf{A}_u + \mathbf{a}_2\mathbf{A}_v) \tag{19}$$

$$\mathbf{P}_k = \mathbf{\Phi}_0 + (\mathbf{b}_1\mathbf{A}_u + \mathbf{b}_2\mathbf{A}_v) \tag{20}$$

$$\mathbf{P}_{k-1} = \mathbf{c}_1\mathbf{A}_u + \mathbf{c}_2\mathbf{A}_v \tag{21}$$

$$\mathbf{P}_{k-2} = \mathbf{d}_1\mathbf{A}_u + \mathbf{d}_2\mathbf{A}_v \tag{22}$$

$$\mathbf{P}_{k-m+3} = \mathbf{e}_1\mathbf{A}_u + \mathbf{e}_2\mathbf{A}_v \tag{23}$$

$$\mathbf{P}_{k-m+2} = \mathbf{f}_1\mathbf{A}_u + \mathbf{f}_2\mathbf{A}_v \tag{24}$$

$$\mathbf{P}_{k-m+1} = \mathbf{g}_1\mathbf{A}_u + \mathbf{g}_2\mathbf{A}_v \tag{25}$$

$$\mathbf{P}_{k-m} = \mathbf{h}_1\mathbf{A}_u + \mathbf{h}_2\mathbf{A}_v. \tag{26}$$

Define:

$$\mathbf{\Phi}_0 = e^{\mathbf{A}_0 h} \tag{27}$$

$$\mathbf{\Phi}_1 = \int_0^h e^{\mathbf{A}_0 \varepsilon} d\varepsilon = \mathbf{A}_0^{-1}(\mathbf{\Phi}_0 - \mathbf{I}) \tag{28}$$

$$\mathbf{\Phi}_2 = \int_0^h \varepsilon e^{\mathbf{A}_0 \varepsilon} d\varepsilon = \mathbf{A}_0^{-1}(h\mathbf{\Phi}_0 - \mathbf{\Phi}_1) \tag{29}$$

$$\mathbf{\Phi}_3 = \int_0^h \varepsilon^2 e^{\mathbf{A}_0 \varepsilon} d\varepsilon = \mathbf{A}_0^{-1}(h^2\mathbf{\Phi}_0 - 2\mathbf{\Phi}_2) \tag{30}$$

$$\mathbf{\Phi}_4 = \int_0^h \varepsilon^3 e^{\mathbf{A}_0 \varepsilon} d\varepsilon = \mathbf{A}_0^{-1}(h^3\mathbf{\Phi}_0 - 3\mathbf{\Phi}_3) \tag{31}$$

$$\mathbf{\Phi}_5 = \int_0^h \varepsilon^4 e^{\mathbf{A}_0 \varepsilon} d\varepsilon = \mathbf{A}_0^{-1}(h^4\mathbf{\Phi}_0 - 4\mathbf{\Phi}_4). \tag{32}$$

In addition, the coefficients in Equations (19)–(26) can be expressed as:

$$\mathbf{a}_1 = \frac{-11h\mathbf{A}_0 + 18\mathbf{I}}{15h^3}\mathbf{\Phi}_4 + \frac{26h\mathbf{A}_0 - 33\mathbf{I}}{15h^2}\mathbf{\Phi}_3 - \mathbf{A}_0\mathbf{\Phi}_2 + \mathbf{\Phi}_1 \tag{33}$$

$$\mathbf{a}_2 = \frac{-11h\mathbf{A}_0 + 18\mathbf{I}}{15h^3}\mathbf{\Phi}_5 + \frac{26h\mathbf{A}_0 - 33\mathbf{I}}{15h^2}\mathbf{\Phi}_4 - \mathbf{A}_0\mathbf{\Phi}_3 + \mathbf{\Phi}_2 \tag{34}$$

$$\mathbf{b}_1 = \frac{-9}{5h^3}\mathbf{\Phi}_4 + \frac{14}{5h^2}\mathbf{\Phi}_3 \tag{35}$$

$$\mathbf{b}_2 = \frac{-9}{5h^3}\mathbf{\Phi}_5 + \frac{14}{5h^2}\mathbf{\Phi}_4 \tag{36}$$

$$\mathbf{c}_1 = \frac{4}{5h^3}\mathbf{\Phi}_4 - \frac{4}{5h^2}\mathbf{\Phi}_3 \tag{37}$$

$$\mathbf{c}_2 = \frac{4}{5h^3}\mathbf{\Phi}_5 - \frac{4}{5h^2}\mathbf{\Phi}_4 \tag{38}$$

$$\mathbf{d}_1 = \frac{-h\mathbf{A}_0 - 3\mathbf{I}}{15h^3}\mathbf{\Phi}_4 + \frac{h\mathbf{A}_0 + 3\mathbf{I}}{15h^2}\mathbf{\Phi}_3 \tag{39}$$

$$\mathbf{d}_2 = \frac{-h\mathbf{A}_0 - 3\mathbf{I}}{15h^3}\mathbf{\Phi}_5 + \frac{h\mathbf{A}_0 + 3\mathbf{I}}{15h^2}\mathbf{\Phi}_4 \tag{40}$$

$$\mathbf{e}_1 = \frac{1}{6h^3}\mathbf{\Phi}_4 - \frac{1}{6h}\mathbf{\Phi}_2 \tag{41}$$

$$\mathbf{e}_2 = \frac{1}{6h^3}\mathbf{\Phi}_5 - \frac{1}{6h}\mathbf{\Phi}_3 \tag{42}$$

$$\mathbf{f}_1 = \frac{-1}{2h^3}\mathbf{\Phi}_4 - \frac{1}{2h^2}\mathbf{\Phi}_3 + \frac{1}{h}\mathbf{\Phi}_2 \tag{43}$$

$$\mathbf{f}_2 = \frac{-1}{2h^3}\mathbf{\Phi}_5 - \frac{1}{2h^2}\mathbf{\Phi}_4 + \frac{1}{h}\mathbf{\Phi}_3 \tag{44}$$

$$\mathbf{g}_1 = \frac{1}{2h^3}\mathbf{\Phi}_4 + \frac{1}{h^2}\mathbf{\Phi}_3 - \frac{1}{2h}\mathbf{\Phi}_2 - \mathbf{\Phi}_1 \tag{45}$$

$$\mathbf{g}_2 = \frac{1}{2h^3}\mathbf{\Phi}_5 + \frac{1}{h^2}\mathbf{\Phi}_4 - \frac{1}{2h}\mathbf{\Phi}_3 - \mathbf{\Phi}_2 \tag{46}$$

$$\mathbf{h}_1 = \frac{-1}{6h^3}\mathbf{\Phi}_4 - \frac{1}{2h^2}\mathbf{\Phi}_3 - \frac{1}{3h}\mathbf{\Phi}_2 \tag{47}$$

$$\mathbf{h}_2 = \frac{-1}{6h^3}\mathbf{\Phi}_5 - \frac{1}{2h^2}\mathbf{\Phi}_4 - \frac{1}{3h}\mathbf{\Phi}_3. \tag{48}$$

Then, if the matrix $\mathbf{P}_{k+1}$ in Equation (18) is nonsingular, Equation (18) can be given by:

$$\begin{aligned}\mathbf{X}_{k+1} = \ &\mathbf{P}_{k+1}^{-1}\mathbf{P}_k\mathbf{X}_k + \mathbf{P}_{k+1}^{-1}\mathbf{P}_{k-1}\mathbf{X}_{k-1} + \mathbf{P}_{k+1}^{-1}\mathbf{P}_{k-2}\mathbf{X}_{k-2} + \mathbf{P}_{k+1}^{-1}\mathbf{P}_{k-m+3}\mathbf{X}_{k-m+3} \\ &+\mathbf{P}_{k+1}^{-1}\mathbf{P}_{k-m+2}\mathbf{X}_{k-m+2} + \mathbf{P}_{k+1}^{-1}\mathbf{P}_{k-m+1}\mathbf{X}_{k-m+1} + \mathbf{P}_{k+1}^{-1}\mathbf{P}_{k-m}\mathbf{X}_{k-m}\end{aligned} \tag{49}$$

According to Equation (49), a discrete map could be defined as:

$$\mathbf{Q}\mathbf{N}_{k+m} = \mathbf{R}\mathbf{N}_k \tag{50}$$

where

$$\mathbf{N}_k = [\mathbf{X}_{k-m}, \mathbf{X}_{k-m+1}, \cdots, \mathbf{X}_{k-1}, \mathbf{X}_k]^T. \tag{51}$$

In addition, $\mathbf{Q}$ and $\mathbf{R}$ can be expressed as:

$$\mathbf{R} = \begin{bmatrix}
0 & 0 & 0 & 0 & 0 & 0 & \cdots & 0 & 0 & 0 & \mathbf{I} \\
\mathbf{P}_{k,m} & \mathbf{P}_{k,m-1} & \mathbf{P}_{k,m-2} & \mathbf{P}_{k,m-3} & 0 & 0 & \cdots & 0 & \mathbf{P}_{k,-2} & \mathbf{P}_{k,-1} & 0 \\
0 & \mathbf{P}_{k+1,m} & \mathbf{P}_{k+1,m-1} & \mathbf{P}_{k+1,m-2} & \mathbf{P}_{k+1,m-3} & 0 & \cdots & 0 & 0 & \mathbf{P}_{k+1,-2} & 0 \\
0 & 0 & \mathbf{P}_{k+2,m} & \mathbf{P}_{k+2,m-1} & \mathbf{P}_{k+2,m-2} & \mathbf{P}_{k+2,m-3} & \cdots & 0 & 0 & 0 & 0 \\
\vdots & \vdots & \vdots & \vdots & \vdots & \vdots & \ddots & \vdots & \vdots & \vdots & \vdots \\
0 & 0 & 0 & 0 & 0 & 0 & \cdots & \mathbf{P}_{k+m-3,-2} & \mathbf{P}_{k+m-3,-1} & \mathbf{P}_{k+m-3,0} & \mathbf{P}_{k+m-3,1} \\
0 & 0 & 0 & 0 & 0 & 0 & \cdots & 0 & \mathbf{P}_{k+m-2,-2} & \mathbf{P}_{k+m-2,-1} & \mathbf{P}_{k+m-2,0} \\
0 & 0 & 0 & 0 & 0 & 0 & \cdots & 0 & 0 & \mathbf{P}_{k+m-1,-2} & \mathbf{P}_{k+m-1,-1}
\end{bmatrix} \tag{52}$$

$$\mathbf{Q} = \begin{bmatrix}
\mathbf{I} & 0 & 0 & 0 & \cdots & 0 & 0 & 0 & 0 & 0 & 0 \\
\mathbf{P}_{k,0} & \mathbf{P}_{k,1} & 0 & 0 & \cdots & 0 & 0 & 0 & 0 & 0 & 0 \\
\mathbf{P}_{k+1,-1} & \mathbf{P}_{k+1,0} & \mathbf{P}_{k+1,1} & 0 & \cdots & 0 & 0 & 0 & 0 & 0 & 0 \\
\mathbf{P}_{k+2,-2} & \mathbf{P}_{k+2,-1} & \mathbf{P}_{k+2,0} & \mathbf{P}_{k+2,1} & \cdots & 0 & 0 & 0 & 0 & 0 & 0 \\
\vdots & \vdots & \vdots & \vdots & \ddots & \vdots & \vdots & \vdots & \vdots & \vdots & \vdots \\
0 & 0 & 0 & 0 & \cdots & \mathbf{P}_{k+m-3,-2} & \mathbf{P}_{k+m-3,-1} & \mathbf{P}_{k+m-3,0} & \mathbf{P}_{k+m-3,1} & 0 & 0 \\
0 & 0 & 0 & 0 & \cdots & 0 & \mathbf{P}_{k+m-2,-2} & \mathbf{P}_{k+m-2,-1} & \mathbf{P}_{k+m-2,0} & \mathbf{P}_{k+m-2,1} & 0 \\
0 & -\mathbf{P}_{k+m-1,m-2} & -\mathbf{P}_{k+m-1,m-3} & 0 & \cdots & 0 & 0 & \mathbf{P}_{k+m-1,-2} & \mathbf{P}_{k+m-1,-1} & \mathbf{P}_{k+m-1,0} & \mathbf{P}_{k+m-1,1}
\end{bmatrix} \tag{53}$$

It is clear that $\mathbf{Q}$ and $\mathbf{R}$ are both a $(2m + 2) \times (2m + 2)$ dimensional matrix. Therefore, the transition matrix $\mathbf{V}$ in a single tooth passing period can be defined as:

$$\mathbf{V} = \mathbf{Q}^{-1}\mathbf{R}. \tag{54}$$

Now, according to Floquet theory [26], the stability of the system can be determined by judging whether the modulus of the eigenvalues of the transition matrixes are less than 1 or not. If not, the system is unstable, otherwise, it will be stable.

Remark: If $\mathbf{P}_{k+1}$ is singular, the processing method in Ref. [20] can be utilized. From Equations (52) and (53), it can be seen that $8m$ variables need to be calculated complicatedly compared with the first-order and second-order FDM, which leads to the increase of calculating time. To enhance the calculation efficiency, the traditional algorithms compressed the $2m + 2$ dimensional matrix into a $m + 2$ dimensional matrix, and calculated the eigenvalues of the transition matrix in one period in the whole region, then, the stability boundary is drawn. However, a novel algorithm is proposed to obviously improve computational efficiency. It is well known that the machining process is stable below the boundary of the SLD and is unstable on the upper boundary of the SLD, while on the stable boundary the eigenvalue of the transition matrix is 1, represented by the spindle speed and depth of cut. According to the constraints of the spindle speed and depth of cut, the modulus of the transition matrix eigenvalues are calculated and judged with 1. If the value is more than 1, the algorithm stops, otherwise it continues, which is only the modulus of transition matrix eigenvalues that are less than 1 and calculated for the stability boundary, which can greatly improve the computational efficiency by reducing the calculation of the eigenvalues in the unstable region.

3. Rate of Convergence Estimates

To verify the fast convergence rate of the proposed method, a one DOF dynamic milling system is selected, and the proposed method is compared to the 0th SDM, 1st FDM, 2nd FDM, 3rd FDM, and Hermite FDM.

The rate of convergence can be clearly determined by the local discretization error (LDE). As stated in the literature [19], for the 0th SDM, the LDE is $O(h^2)$. The LDE of FDM with the 1st, 2nd, 3rd, and Hermite are $O(h^2)$, $O(h^3)$, $O(h^4)$, and $O(h^3)$ [24,27], respectively. For the proposed method, the LDE is $O(h^4)$.

All operations are from the same computer environment: Matlab 2018b, Inter(R) Core(TM) i5-4210H CPU @ 2.90 GHz 2.90 GHz. The milling system parameters are derived from the Ding [20]: The damping ratio ζ, model mass m_t, and natural frequency w_n are 0.011, 0.03993 kg, and 1844 Hz, respectively. The cutter has two flutes. The cutting force coefficients are $K_{tc} = 6 \times 10^8$ and $K_{rc} = 2 \times 10^8$.

The spindle speed n is selected as 5000 rpm, and the axial depths of cut a_p is selected as 0.1 mm, 0.2 mm, 0.5 mm, and 0.80 mm, respectively. The exact eigenvalues $|\mu_0|$ corresponding to different axial depths of cut are 0.7368, 0.8192, 1.0726, and 1.2880, respectively. The LDE can be known as the absolute value of difference between the current eigenvalue $|\mu|$ and exact eigenvalue $|\mu_0|$.

As shown in Figure 2, the LDE of the 0th SDM, 1st FDM, 2nd FDM, 3rd FDM, Hermite FDM, and the proposed method are analyzed. From Figure 2, it can be clearly seen that the proposed method has a faster convergence rate. It should be mentioned that the proposed method is able to converge to a sufficient accuracy when the discrete number m is small.

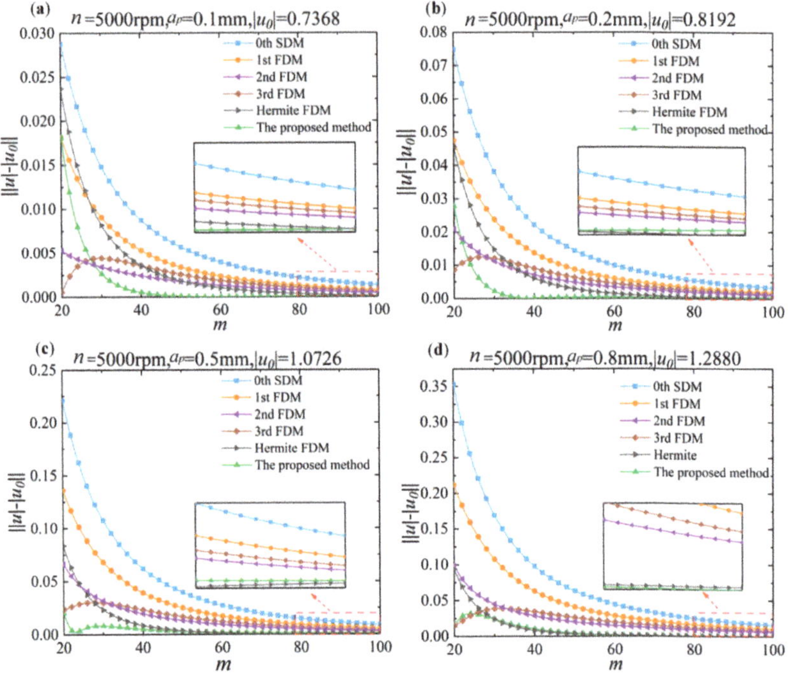

Figure 2. The convergence rate of eigenvalues for different discrete numbers m for the 0th SDM, 1st FDM, 2nd FDM, 3rd FDM, Hermite FDM, and the proposed method. (**a**) Axial depth of cut a_p is 0.1 mm, exact eigenvalue $|\mu_0|$ is 0.7368. (**b**) Axial depth of cut a_p is 0.2 mm, exact eigenvalue $|\mu_0|$ is 0.8192. (**c**) Axial depth of cut a_p is 0.5 mm, exact eigenvalue $|\mu_0|$ is 1.0726. (**d**) Axial depth of cut a_p is 0.8 mm, exact eigenvalue $|\mu_0|$ is 1.2880.

4. Computational Accuracy Analysis

Under low and high immersion ratio conditions, the effectiveness of the proposed method is verified in terms of both computational efficiency and accuracy of milling stability prediction by comparing with other methods. The modal parameter selection is consistent with Section 3 of this paper. Equation (1) is the dynamic state-space model of the milling system, where constant matrix $\mathbf{A}_0$ and time-periodic matrix $\mathbf{A}(t)$ can be express as:

$$A_0 = \begin{bmatrix} -\zeta w_n & 1/m_t \\ m_t(\zeta w_n)^2 - m_t w_n{}^2 & -\zeta w_n \end{bmatrix} \tag{55}$$

$$\mathbf{A}(t) = \begin{bmatrix} 0 & 0 \\ -a_p h(t) & 0 \end{bmatrix} \tag{56}$$

$$h(t) = \sum_{j=1}^{N} g(\varphi_j(t)) \sin(\varphi_j(t))(K_{tc} \cos(\varphi_j(t)) + K_{rc} \sin(\varphi_j(t))) \tag{57}$$

$$\varphi_j(t) = \frac{2\pi n}{60} t + (j-1)\frac{2\pi}{N} \tag{58}$$

$$g(\varphi_j(t)) = \begin{cases} 1 & \varphi_{st} < \varphi_j(t) < \varphi_{ex} \\ 0 & otherwise \end{cases} \tag{59}$$

where $\varphi_j(t)$ is the angular position of the j-th cutter tooth, and φ_{st} and φ_{ex} are the starting and exiting edge positions of the tool in contact with the workpiece, respectively. For down-milling, $\varphi_{st} = \arccos(2a/D - 1)$ and $\varphi_{ex} = \pi$; for up-milling, $\varphi_{st} = 0$ and $\varphi_{ex} = \arccos(1 - 2a/D)$, where a/D is the radial immersion ratio.

For $a/D = 1$, all methods are calculated over a 200×100-sized grid of parameters under the condition of $n = 5000$–10,000 rpm and $a_p = 0$–4 mm. For $a/D = 0.1$, all methods are calculated over a 200×100-sized grid of parameters under the condition of $n = 5000$–12,000 rpm and $a_p = 0$–5 mm. The SLDs for $a/D = 1$ and $a/D = 0.1$ are shown in Figures 3 and 4. The red curves in Figures 3 and 4 are the reference curves, calculated by the Hermite FDM at the discrete number $m = 100$. It can be seen that the proposed method has a high accuracy both at a low and high radial immersion ratio. In particular, the 0th SDM, 1st FDM, 2nd FDM, 3rd FDM, and Hermite FDM all have large errors with the reference curve when $m = 20$ at the $a/D = 1$. However, the proposed method almost coincides with the reference curve. For $a/D = 0.1$, the 1st FDM agrees best with the reference curve when $m = 20$, and the proposed method agrees with the reference curve immediately as m increases.

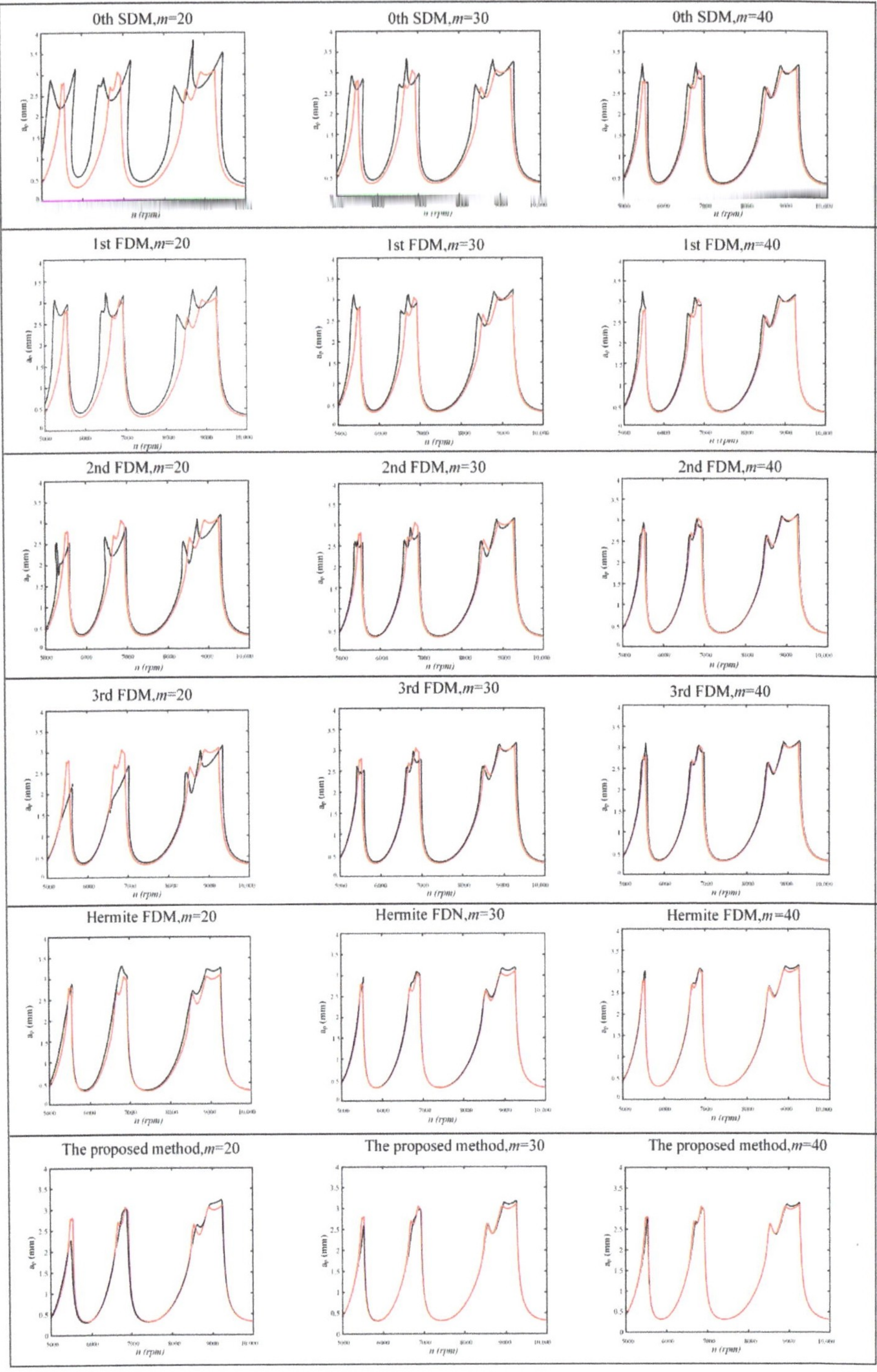

Figure 3. A comparison of computational accuracy in $a/D = 1$.

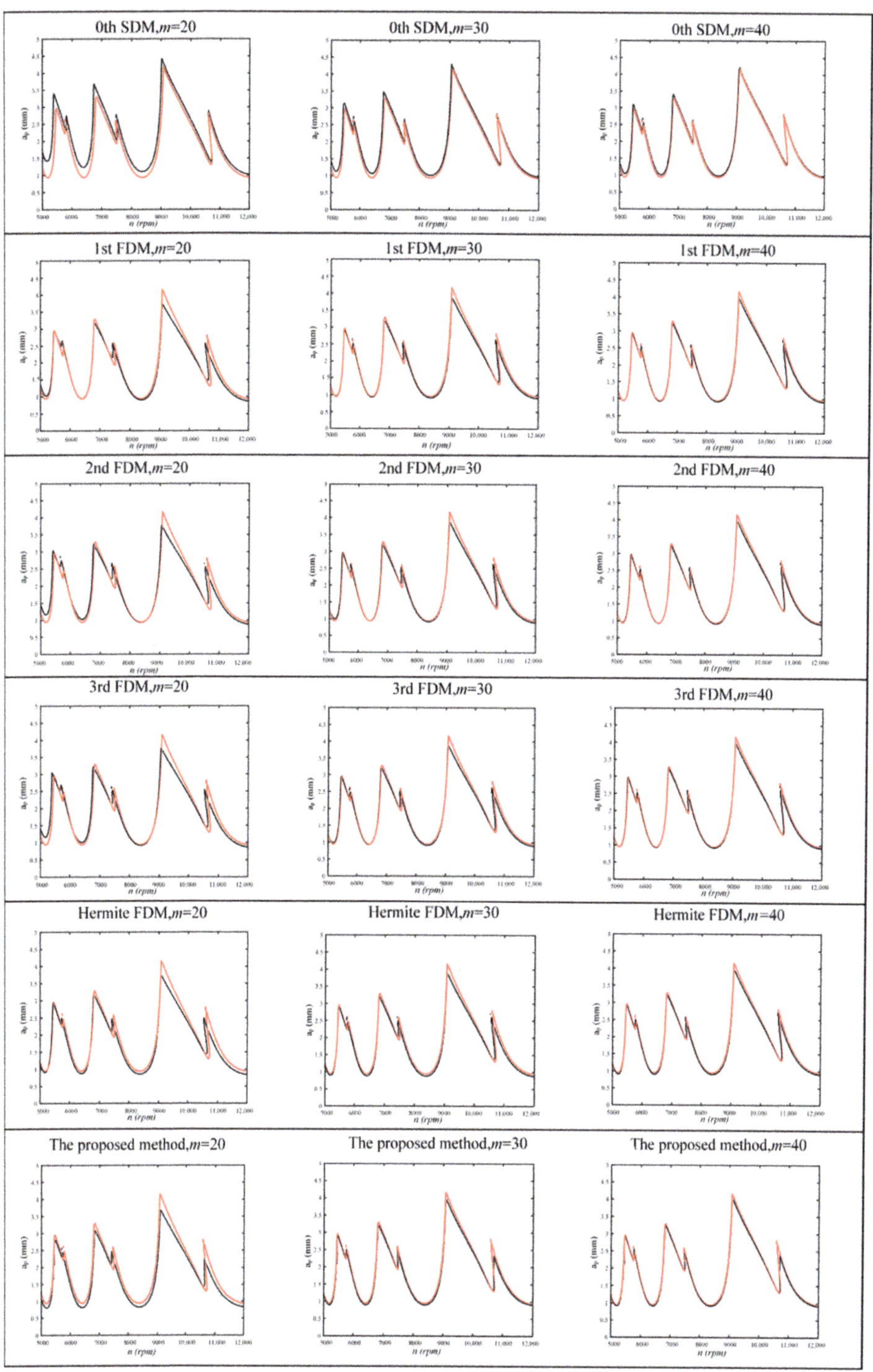

Figure 4. A comparison of computational accuracy in $a/D = 0.1$.

5. Computational Efficiency Analysis

To verify the computational efficiency of the proposed method, the time required for the computation of the FDM in Section 4 for different discrete numbers m is discussed. The time required for the calculation is shown in Figure 5. From Figure 5, it can be seen that the proposed method has a faster computational efficiency compared to other methods. When $a/D = 1$, the proposed method saves an average of 69.2%, 73.3%, 75.4%, and 66.7% of time compared to the 1st FDM, 2nd FDM, 3rd FDM, and Hermite FDM, respectively. When $a/D = 0.1$, the proposed method saves an average of 53.3%, 58.8%, 63.8%, and 47.5% of time compared to the 1st FDM, 2nd FDM, 3rd FDM, and Hermite FDM, respectively. It can be seen that the proposed method has a higher computational linear efficiency when $a/D = 1$. The main reason is that when $a/D = 1$, the contact time between the cutter and workpiece is at a maximum, while the transfer matrix Equation (54) needs to be calculated multiple times, and the proposed method saves the calculation time required for the stability region.

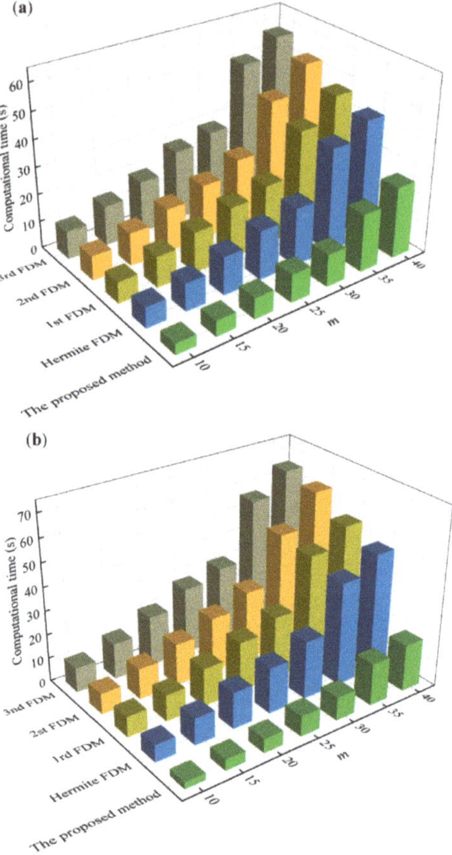

Figure 5. Comparison of calculated time among the 0th SDM, 1st FDM, 2nd FDM, 3rd FDM, Hermite FDM, and the proposed method with different radial immersion ratios. (**a**) Calculated time at $a/D = 0.1$ and (**b**) Calculated time at $a/D = 1$.

6. Verification

To verify the effectiveness of the proposed method in the milling of the thin-walled plate, some experiments are conducted in this section. The dimension of the plate used in modal test and machining experiments is $80 \times 40 \times 3$ mm, and all experiments were carried out on the three-axis milling center (VMC-850E), which is shown in Figure 6. The material properties of the workpiece and the cutter parameters are given in Table 1.

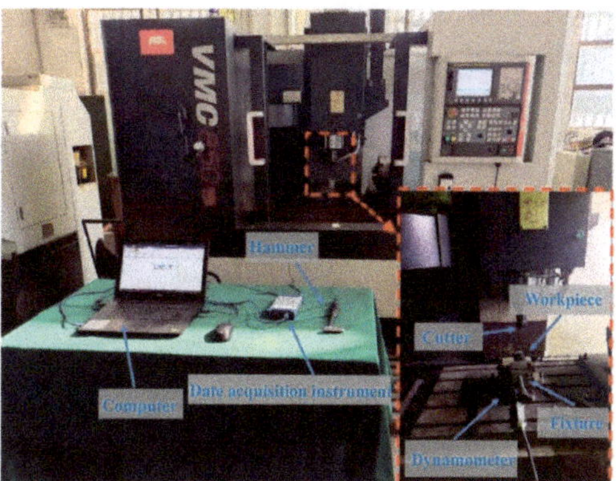

Figure 6. Configuration of the experiments.

Table 1. The properties of the workpiece and the specifications of the cutter.

Cutter	Diameter (mm)	Number of Flutes	Helix Angle (°)	Length (mm)
	12	2	30	75
Workpiece	Density (g/cm^3)	Possion's Ration	Young's Modulus (GPa)	Material
	4.6	0.34	108	Ti-6Al-4V

6.1. Cutting Force Coefficients Calibration

For cutting force coefficients calibrated, as is known to all, when full-immersion milling (slot milling) is used, the average milling forces are expressed as:

$$\begin{cases} \overline{F}_x = -\frac{Na}{4}K_{rc}f - \frac{Na}{\pi}K_{re} \\ \overline{F}_y = \frac{Na}{4}K_{tc}f + \frac{Na}{\pi}K_{te} \end{cases}. \tag{60}$$

Then, for slot milling, five groups of full-immersion milling experiments were carried out. The machining parameters are the spindle speed 1000 rpm, axial depth of cut at 0.5 mm, and feed rate at 40 mm/min, 80 mm/min, 120 mm/min, 160 mm/min, and 200 mm/min.

Therefore, the average milling forces at each feed rate are measured by Kistler9257B, and the cutting-edge components are estimated by a linear regression of the accumulated data. Next, based on the literature [28], the cutting force coefficients are evaluated as K_{tc} = 1120.8 N/mm^2, K_{rc} = 2285.6 N/mm^2, K_{te} = 9.16 N/mm, and K_{re} = 13.21 N/mm.

6.2. Modal Parameters Identification

An impact experiment is conducted for obtaining the modal parameters of the thin-walled workpiece. The modal parameters of the milling system are obtained by an acquisition instrument DH5981, acceleration sensors (Ref. sensitivity 10.25 mV/g), and modal hammer (500 N).

In tests, for a different measured position on the workpiece, the dynamic response is different. Therefore, considering the clamping constraints, the impact measured points 1, 2, and 3 distributed on the thin-walled plate are shown in Figure 7a. Point 1 and point 3 are symmetric with respect to point 2, and point 2 locates the middle of the thin-walled plate edge. Next, all the vibration responses on the different measured points are obtained, and according to the experimental results, we found that the vibration response at point 1 is the same as that at point 3. In addition, considering the unstable state in the cut-in and cut-out region, the representative point 2 are chosen for measuring responses, as shown in Figure 7b,c, and the experimental setup is shown in Figure 6. Therefore, the modal parameters in point 2 are identified and listed in Table 2.

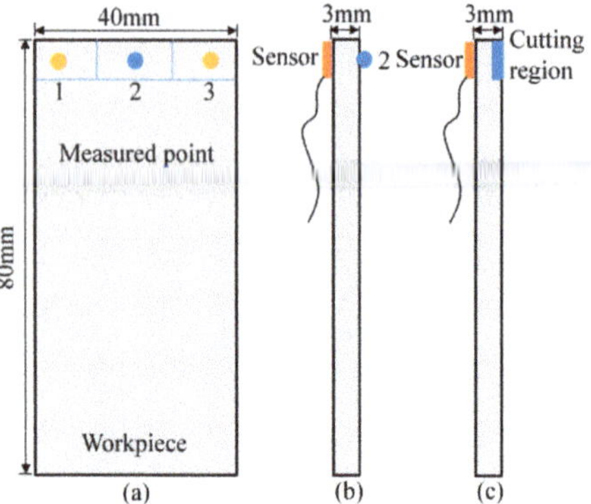

Figure 7. (**a**) Distributed points 1, 2, and 3 on the thin-walled plate for impact tests. (**b**) Position of the sensor and point 2 for measuring responses. (**c**) Position of the sensor and cutter region for machining.

Table 2. Modal parameters of the cutter and workpiece.

System	Natural Frequency ω (Hz)	Damping Ratio ζ	Stiffness k (N·m^{-1})
Workpiece (Mode no.1)	575	0.007	1.19×10^6
Workpiece (Mode no.2)	1820	0.012	1.31×10^7
Cutter in X direction	2126	0.036	1.45×10^8
Cutter in Y direction	2134	0.033	1.47×10^8

6.3. Machining Tests

In this section, in order to validate the accuracy of the proposed approach for quickly and accurately predicting the stability of the milling system, the four degree of freedom in the X and Y direction for the milling cutter and workpiece in the X and Y direction for thin-walled section are considered. According to the proposed method, the stability lobe diagram with the discrete number $m = 40$ at the $a/D = 0.1$ is calculated, and the milling parameters are determined based on the stability lobe diagram as shown in Figure 8. The milling parameters in points A($n = 1500$ rpm, $a_p = 0.2$ mm) and C($n = 2500$ rpm, $a_p = 0.4$ mm) are stable parameters, while the points B($n = 1500$ rpm, $a_p = 0.6$ mm) and D($n = 3000$ rpm, $a_p = 0.4$ mm) are located in the unstable cutting region. All the dynamic responses in different points are measured and investigated, and only the dynamic response and its spectrum in points A, B, C, and D are shown in Figure 9. From Figure 9a,c, it can be seen that there is only the tool tooth passing frequency (i.e., 200 Hz, 400 Hz, 600 Hz, 650 Hz, 666 Hz, 833 Hz, 875 Hz, and 917 Hz.). From Figure 9b,d, it can be seen that the chatter frequency (i.e., 520 Hz, 580 Hz, 620 Hz, 660 Hz, 690 Hz, 790 Hz, 860 Hz, and 910 Hz.) occurs besides at the tool tooth passing frequency (i.e., 100 Hz, 200 Hz, 300 Hz, 400 Hz, 500 Hz, 600 Hz).

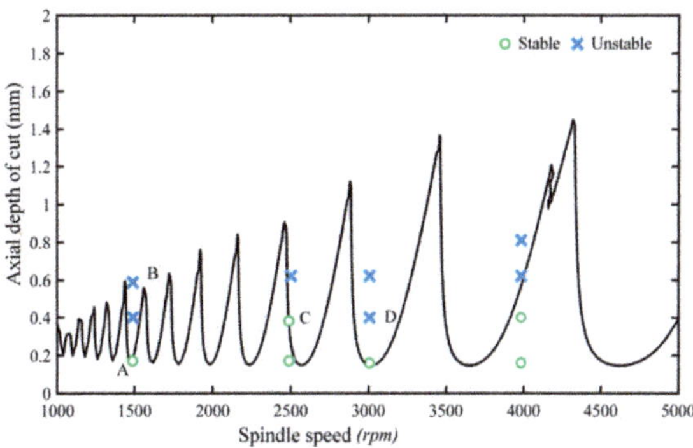

Figure 8. Stability lobe diagrams at the measured point 2.

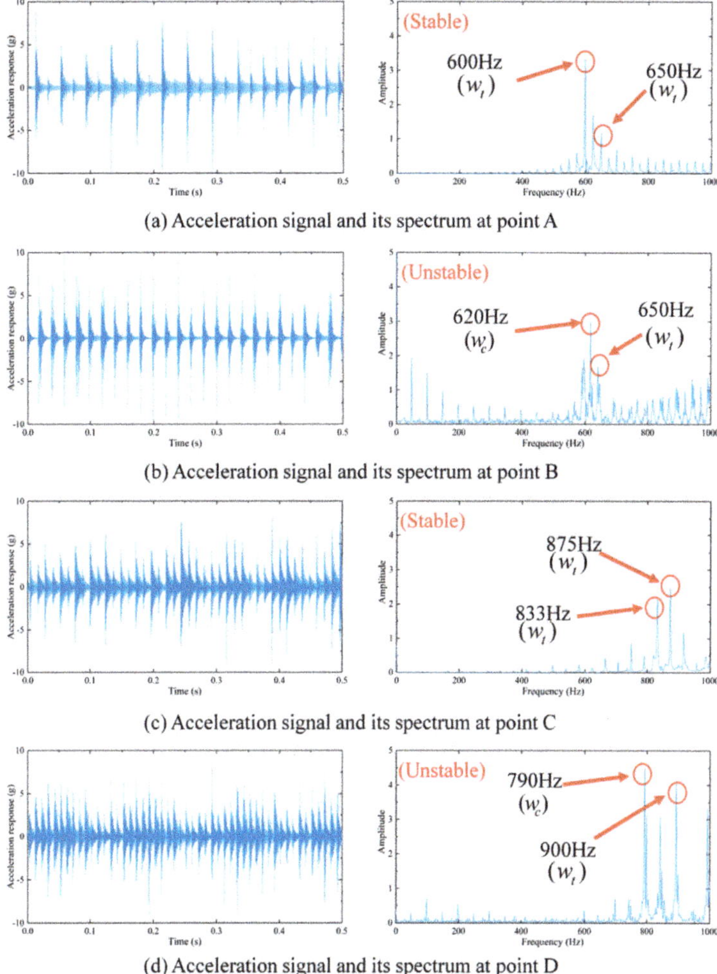

(a) Acceleration signal and its spectrum at point A

(b) Acceleration signal and its spectrum at point B

(c) Acceleration signal and its spectrum at point C

(d) Acceleration signal and its spectrum at point D

Figure 9. Acceleration signal and its spectrum at point A, B, C, and D in the milling of the thin-walled plate.

In addition, in order to more clearly investigate the milling chatter, the morphologies of the machined surface at different points are shown in Figure 10. From Figure 10, we can see that the machining chatter occurs at observation points B and D, which can be observed from the rough surface quality and obvious vibration.

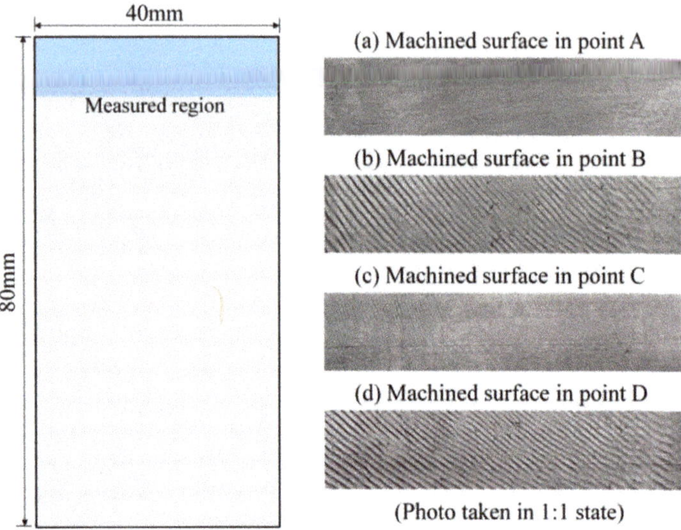

Figure 10. Surface morphology of the thin plate in the milling region.

7. Conclusions

(1) A novel updated FDM is proposed to predict the milling SLD. The cubic-spline interpolation and the Newton interpolation are introduced to approximate the state item and time-delay item, respectively. A discrete map is established between the current state matrix and the previous state matrix, and the SLD is obtained based on the eigenvalue modulus judgement criterion of the transition matrix.

(2) An iterative algorithm is proposed to obviously improve computational efficiency. The calculation of the transition matrix eigenvalues in the chattering region is eliminated. The simulation results of a benchmark example with two different radial immersion ratios show that the algorithm has a faster computational efficiency than other methods, especially when the radial immersion ratio is large.

(3) The proposed method has obvious advantages in terms of computational accuracy and convergence speed than other methods. In terms of calculation accuracy, it already coincides with the reference curve when the discrete number m is small whether the radial immersion ratio is large or small. In addition, it has a faster convergence speed both in the stable or unstable region, and this part will be further studied in the future.

(4) A series of milling experiments under different spindle speeds are designed to verify the accuracy of the proposed method. The experimental results show that the proposed method is in good agreement with the experimental value.

Author Contributions: Conceptualization, J.M.; methodology, J.M. and Y.L.; software, Y.L.; formal analysis, Y.L. and X.P.; writing—original draft preparation, J.M. and Y.L.; investigation, Y.L., X.P. and G.W.; writing—review and editing, D.Z. and B.Z.; resources, J.M. and X.P.; funding acquisition, J.M. All authors have read and agreed to the published version of the manuscript.

Funding: This research was funded by the National Natural Science Foundation of China (no. 52005166), the Postdoc-toral Research Foundation of China (no. 2019M652534), the Henan Postdoc-toral Foundation (no. 19030071), the Foundation of Henan Educational Committee (no. 20A460016), Young Backbone Teachers Foundation Scheme of Henan Polytechnic University (no. 2019XQG-01), the Science and Technology Department of Henan Province (no. 202102210082), and the Doctor Foundation from Henan Polytechnic University (no.B2019-49).

Conflicts of Interest: The authors declare no conflict of interest.

References

1. Insperger, T.; Stepan, G. Updated semi-discretization method for periodic delay-differential equat with discrete delay. *Int. J. Numer. Meth. Eng.* **2004**, *61*, 117–141. [CrossRef]
2. Insperger, T.; Gradišek, J.; Kalveram, M.; Stépán, G.; Winert, K.; Govekar, E. Machine Tool chatter and surface location error in milling processes. *J. Manuf. Sci. Eng.* **2006**, *128*, 913–920. [CrossRef]
3. Insperger, T.; Stepan, G.; Turi, J. State-dependent delay in regenerative turning processes. *Nonlinnear Dynam.* **2007**, *47*, 275–283. [CrossRef]
4. Chandiramani, N.K.; Pothala, T. Dynamics of 2-dof regenerative chatter during turning. *J. Sound Vib.* **2006**, *290*, 448–464. [CrossRef]
5. Eksioglu, C.; Kilic, Z.M.; Altintas, Y. Discrete-Time prediction of chatter stability, cutting forces, and surface location errors in flexible milling systems. *J. Manuf. Sci. Eng.* **2012**, *134*, 061006. [CrossRef]
6. Dun, Y.C.; Zhu, L.D.; Wang, S.H. Multi-modal method for chatter stability prediction and control in milling of thin-walled workpiece. *Appl. Math. Model.* **2020**, *80*, 602–624. [CrossRef]
7. Li, H.; Li, S.; Shao, M.L.; Ling, J.H.; Zhao, C.Y.; Wen, B.C. Analysis of the stability of intermittent cut with energy lost. *Int. J. Mater. Prod. Technol.* **2014**, *48*, 270–285. [CrossRef]
8. Guo, M.X.; Zhu, L.D.; Yan, B.L.; Guan, Z.H. Research on the milling stability of thin-walled parts based on the semi-discretization method of improved Runge-Kutta method. *Int. J. Adv. Manuf. Technol.* **2021**, *115*, 2325–2342. [CrossRef]
9. Ozoegwu, P.; Eberhard, P. Stability analysis of multi-discrete delay milling with helix effects using a general order full-discretization method updated with a generalized integral quadrature. *Mathematics* **2020**, *8*, 1003. [CrossRef]
10. Liu, X.L.; Li, R.Y.; Wu, S.; Yang, L.; Yue, C.X. A prediction method of milling chatter stability for complex surface mold. *Int. J. Adv. Manuf. Technol.* **2016**, *89*, 2637–2648. [CrossRef]
11. Grossi, N.; Scippa, A.; Sallese, L.; Sato, R.; Campatelli, G. Spindle speed ramp-up test: A novel experimental approach for chatter stability detection. *Int. J. Mach. Tools Manuf.* **2015**, *89*, 221–230. [CrossRef]
12. Altintas, Y.; Budak, E. Analytical prediction of stability lobes in milling. *CIRP Ann.-Manuf. Technol.* **1995**, *44*, 357–362. [CrossRef]
13. Merdol, S.D.; Altintas, Y. Multi frequency solution of chatter stability for low immersion milling. *J. Manuf. Sci. Eng.* **2004**, *126*, 459–466. [CrossRef]
14. Faassen, R.P.H.; Wouw, N.; Nijmeijer, H.; Oosterling, J.A.J. An improved tool path model including periodic delay for chatter prediction in milling. *J. Comput. Nonliner Dyn.* **2007**, *2*, 167–179. [CrossRef]
15. Ding, Y.; Zhu, L.M.; Zhang, X.J.; Ding, H. Numerical integration method for prediction of milling stability. *J. Manuf. Sci. Eng.* **2011**, *133*, 031005. [CrossRef]
16. Liang, X.G.; Yao, Z.Q.; Luo, L.; Hu, J. An improved numerical integration method for predicting milling stability with varying time delay. *Int. J. Adv. Manuf. Technol.* **2013**, *68*, 1967–1976. [CrossRef]
17. Li, Z.Q.; Yang, Z.K.; Peng, Y.R.; Zhu, F.; Ming, X.Z. Prediction of chatter stability for milling process using Runge-Kutta-based complete discretization method. *Int. J. Adv. Manuf. Technol.* **2016**, *86*, 943–952. [CrossRef]
18. Dai, Y.N.; Li, H.K.; Xing, X.Y.; Hao, B.T. Prediction of chatter stability for milling process using precise integration method. *Precis. Eng.* **2018**, *52*, 152–157. [CrossRef]
19. Insperger, T.; Stepan, G. Semi-discretization method for delayed systems. *Int. J. Numer. Meth. Eng.* **2002**, *55*, 503–518. [CrossRef]
20. Ding, Y.; Zhu, L.M.; Zhang, X.J.; Ding, H. A full-discretization method for prediction of milling stability. *Int. J. Mach. Tools Manuf.* **2010**, *50*, 502–509. [CrossRef]
21. Ding, Y.; Zhu, L.M.; Zhang, X.J.; Ding, H. Second-order full-discretization method for milling stability prediction. *Int. J. Mach. Tools Manuf.* **2010**, *50*, 926–932. [CrossRef]
22. Quo, Q.; Sun, Y.W.; Jiang, Y. On the accurate calculation of milling stability limits using third-order full-discretization method. *Int. J. Mach. Tools Manuf.* **2012**, *62*, 61–66. [CrossRef]
23. Ozoegwu, C.G.; Omenyi, S.N.; Ofochebe, S.M. Hyper-third order full-discretization methods in milling stability prediction. *Int. J. Mach. Tools Manuf.* **2015**, *92*, 1–9. [CrossRef]
24. Liu, Y.L.; Zhang, D.H.; Wu, B.H. An efficient full-discretization method for prediction of milling stability. *Int. J. Mach. Tools Manuf.* **2012**, *63*, 44–48. [CrossRef]
25. Tang, X.W.; Peng, F.Y.; Yan, R.; Gong, Y.H.; Li, Y.T.; Jiang, L.L. Accurate and efficient prediction of milling stability with updated full-discretization method. *Int. J. Adv. Manuf. Technol.* **2017**, *88*, 2357–2368. [CrossRef]
26. Lakshmikantham, V.; Trigiante, D. *Theory of Difference Equations: Numerical Methods and Applications*; Academic Press: London, UK, 1998; pp. 90–99.

27. Insperger, T. Full-discretization and semi-discretization for milling stability prediction: Some comments. *Int. J. Mach. Tools Manuf.* **2010**, *50*, 658–662. [CrossRef]
28. Altintas, Y. *Manufacturing Automation: Metal Cutting Mechanics, Machine Tool Vibration, and CNC Design*, 2nd ed.; Cambridge University Press: Cambridge, UK, 2000; pp. 44–47.

Article

Influence of Some Microchanges Generated by Different Processing Methods on Selected Tribological Characteristics

Gheorghe Nagîț, Laurențiu Slătineanu, Oana Dodun, Andrei Marius Mihalache, Marius Ionuț Rîpanu and Adelina Hrițuc *

Department of Machine Manufacturing Technology, Gheorghe Asachi Technical University of Iasi, 700050 Iasi, Romania; nagit@tcm.tuiasi.ro (G.N.); slati@tcm.tuiasi.ro (L.S.); oanad@tcm.tuiasi.ro (O.D.); andrei.mihalache@tuiasi.ro (A.M.M.); marius.ripanu@tuiasi.ro (M.I.R.)
* Correspondence: adelina.hrituc@student.tuiasi.ro; Tel.: +40-751640117

Abstract: Different processing methods can change the physical–mechanical properties and the microgeometry of the surfaces made by such processes. In turn, such microchanges may affect the tribological characteristics of the surface layer. The purpose of this research was to study the tribological behavior of a test piece surfaces analyzing the changes on the values of the coefficient of friction and loss of mass that appear in time. The surfaces subjected to experimental research were previously obtained by turning, grinding, ball burnishing, and vibroburnishing. The experimental research was performed using a device adaptable to a universal lathe. Mathematical processing of the experimental results led to the establishment of power-type function empirical models that highlight the intensity of the influence exerted by the pressure and duration of the test on the values of the output parameters. It was found that the best results were obtained in the case of applying ball vibroburnishing as the final process.

Keywords: machining methods; surface layer; microchanges; wear; friction coefficient; erosion; measurement; empirical mathematical model

Citation: Nagît, G.; Slătineanu, L.; Dodun, O.; Mihalache, A.M.; Rîpanu, M.I.; Hrițuc, A. Influence of Some Microchanges Generated by Different Processing Methods on Selected Tribological Characteristics. *Micromachines* **2022**, *13*, 29. https://doi.org/10.3390/mi13010029

Academic Editors: Benny C. F. Cheung and Jiang Guo

Received: 26 November 2021
Accepted: 22 December 2021
Published: 26 December 2021

Publisher's Note: MDPI stays neutral with regard to jurisdictional claims in published maps and institutional affiliations.

1. Introduction

It is known that during the first period of equipment use, the surface roughness decreases gradually to certain values that correspond to the normal use conditions.

Due to the reduction in the surface asperities heights, a change in the friction coefficient values is expected. On the other hand, the surface roughness reduction is accompanied by removal of the material from the two parts involved in relative movement to each other.

All these processes take place at the micro-space level, in a surface layer with a thickness of, at most, several hundred micrometers, as the parameters for evaluating the height of the roughness of the processed surfaces can reach values less than 1 μm.

Many experimental investigations aimed to highlight the influence exerted by the previously applied machining process on the tribological behavior of the surfaces thus obtained. Some of these investigations took into account the processes of ball burnishing and ball vibroburnishing as processes capable of affecting the tribological characteristics of the surfaces of the parts. Other researches were directed at defining the experimental conditions to be used when experimentally assessing the tribological behavior of the burnished or vibroburnished surfaces.

Thus, in the last decades of the previous millennium, in the former Soviet Union, I.G. Schneider and his collaborators developed ample researches concerning the vibroburnishing processes and the functional properties of surfaces obtained using such machining processes. In 1982, Schneider published a monograph in which the functional properties of the so-called surfaces with regular microrelief were analyzed using the methods accessible at that time [1].

Another researcher from the Soviet Union who investigated the vibroburnishing process was L.G. Odintsov. He and his collaborators published papers in which this subject was approached, and a monograph concerning the hardening and finishing surfaces created by plastic deformation processes in surface layers up to several hundred micrometers thick was elaborated [2].

Loh et al. (1990) studied the influence of the ball burnishing process conditions on surface hardening in the case of specimens made of AISI 1045 steel [3]. They noticed the possibility of increasing the hardness by 68% and obtaining the maximum hardness beneath the burnished surface, to depths with values less than 150 μm.

Kim et al. simulated the experimental conditions of a tribological test based on using a reciprocating tribometer and a specimen pneumatically pressed onto the disk counterface sample, corresponding to a block-on-ring type of test [4]. As wear process output parameters, the wear depth and the wear rate were taken into consideration. The research showed good agreement of the experimental results with the results obtained using the simulation, by means of finite element analysis, for the material pair of interest.

El-Tayeb et al. considered using a tribo-test machine. The load was ensured using a first-order lever [5] to investigate the tribological behavior of burnished cylindrical specimens made of aluminum Al-6061. They found a certain combination of values corresponded to the burnishing process input factors, which ensured that the best behavior of the tested material was obtained for the tested friction coefficient values. The experimental research was performed in dry contact and lubricated contact conditions, on specimens made of aluminum 6060 and stainless-steel counterface cups.

Wojciechowski and Nosal addressed the problem of scuffing resistance in the case of test sample surfaces previously obtained by cold burnishing processes [6]. They proposed that the surface layer energy accumulation should be considered. Their experimental tests showed that an increase in surface layer energy accumulation leads to a decrease in scuffing resistance. The surface layer considered in manufacturing engineering is usually less than 1 mm thick, and is frequently even less than 500 μm.

Low took the use of a tribo-test machine, in the presence of a first-order lever, into consideration [7]. He used a 316 stainless steel, 60 mm diameter cup as a counterface, while the specimen was pressed onto the counterface using weights placed at the ends of the lever.

Li and Han investigated how the recrystallization of the Ni_3Al-based single crystal alloy IC6SX is affected when various surface mechanical treatments are applied [8]. Dry sandblasting, wet sandblasting, indentation, shot blasting, and burnishing were used as mechanical treatments. The research showed that the deformation amount was more pronounced when burnishing and dry sandblasting were applied than that generated by applying other superficial mechanical treatments. At the same time, an observation was made, according to which the recrystallization did not affect the mechanical properties of the investigated alloy.

Ovali and Akkurt developed experimental research to compare the results of using burnishing and other hole surface finishing processes in the case of specimens made of brass materials [9]. They noticed that the highest hardness and the best surface quality were obtained when applying the burnishing process.

Al-Saeedi et al. proposed using an adaptive neuro-fuzzy inference system to predict the workpiece hardness and roughness of the surfaces obtained using a roller burnishing process [10]. Their considerations were valid for the parts made of polyoxymethylene.

Lewandowski investigated the wear behavior of cast iron EN-GJSFP-500-7 after applying a burnishing process when using solid lubricant [11]. A tribological test based on the block-on-ring work schema was used. He noticed that better results for the tribological properties and surface roughness parameters were obtained when using machine oil.

Zaleski designed and used a tribological testing stand in which the charging force was applied to the end of a third-order lever using weights of known value [12]. He used such a stand to study the wear behavior of surfaces processed by grinding, and vibratory and shot peening, respectively.

An investigation of the wear of burnished specimens made of low-density polyethylene was conducted by Janczewski et al., using a ball-on-disc test [13]. They noticed improved behavior of the burnished surfaces as compared to the surfaces obtained by milling only. The experimental results proved that there was a decrease in the wear rate by 58%.

Krasnyy et al. investigated the wear and fretting resistance in the case of surfaces previously obtained by applying a vibroburnishing process and presenting certain regular microgroove patterns [14]. They aimed to improve the parts' operation performances acting on the values of the parameters that characterize the groove patterns. In another paper, Krasnyy and Maksarov presented the results of research that aimed to highlight the tribotechnical characteristics of surfaces obtained by vibroburnishing [15]. The investigated surfaces presented regular microgeometric relief, and, in certain situations, they proved to have high wear resistance compared to the wear resistance of another finishing process.

Takada and Sasahara aimed to investigate the influence exerted by the shape of the frictional stir burnishing tool on the hardness of the surface layer, on the residual stress, and on the values of the surface roughness parameter Ra [16]. It was confirmed that a lower roughness value could be obtained when the tool diameter was larger. However, an increase in the thickness of the microhardened layer was observed for lower values of the tool tip radius.

Dzierwa and Markopoulos addressed the issue of the influence of the ball burnishing process on the topography and tribological properties of hardened steel [17]. Experimental research has confirmed the possibility of reducing the root mean square height of the surface Sq from 0.522 μm to 0.051 μm and increasing the wear resistance by using a ball burnishing process.

The integrity of the surface achieved when using micro-textured ball-end milling was studied by Yang et al. [18]. Empirical mathematical models were determined to highlight the influence exerted by different input factors of the ball-end milling process on the hardness of HV and the values of the roughness parameter Ra of the processed surfaces.

Hou and Li investigated the influence of micro-hardness changes on a titanium alloy's friction and wear characteristics [19]. For the experimental research, they used the characteristics of the ball-on-disc contact test. According to one of the conclusions of their research, the effects of previous processing can be observed on a surface layer thickness of less than 100 μm, and these microchanges can significantly affect the wear and friction behavior of the specimen material.

The above information, and the information included in other works [20–25], refer to changes in some operating properties when a surface layer with a thickness of, at most, several hundred micrometers is affected by the processing methods by which the analyzed surfaces were made.

The results obtained by the researchers for the previously mentioned applied processing methods regarding the influence exerted on the obtained surface layer of the tribological behavior of the cylindrical surfaces, showed less information concerning the influence exerted by the burnishing and vibroburnishing processes, and these results were obtained when many actual research methods were not known. In this way, it can be considered that experimental research could highlight the evolution, over time, of the friction coefficient μ and lost mass Δm when using certain tribological tests. Improved solutions for mathematical processing of the theoretical and experimental results could be applied.

The aim of the research, the results of which were included in the article, was to identify empirical mathematical models that highlight how distinct processing methods influence the variation, over time, in the friction coefficient and the loss of mass under the tribological conditions specified for the experimental test.

2. Materials and Methods

2.1. Theoretical Premises

It is expected that the distinct processing method affects the structure, chemical composition, and mechanical properties of the surface layer (which has a thickness of several hundred micrometers) in distinct ways.

Thus, in the case of cylindrical turning using a lathe tool, the grains from the surface layer change their shape due to the pressure exerted by the tool during the process of material removal from the workpiece. The tool used for the experiments was a SN 400 × 1000 lathe, manufactured at the Lathe Enterprise of Arad (Romania). An increase in the microhardness of the surface layer with a certain thickness can be observed, compared to the microhardness of the material that was not affected by the cutting process (Figure 1a).

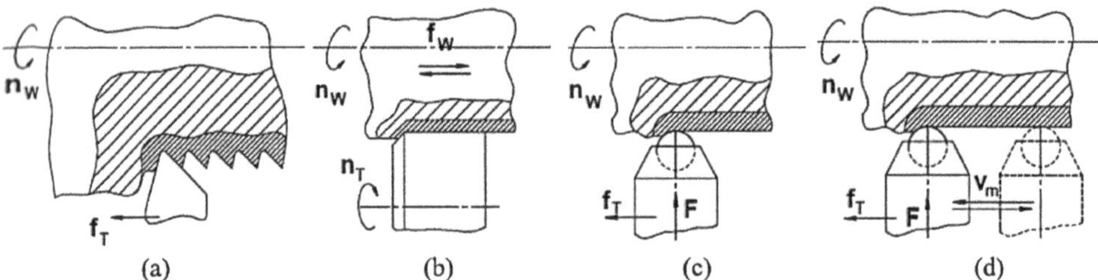

Figure 1. Microchanges in the surface layers after applying various processing methods: (**a**) after turning; (**b**) after grinding; (**c**) after burnishing; (**d**) after vibroburnishing.

The thickness of the hardened surface layer could be lower when applying a cylindrical grinding process (Figure 1b) since the forces generated by such a machining process are lower than the forces occurring during the turning process. If high cutting speeds are used in the case of turning or grinding, a possible heat-affected zone can also be generated in the surface layer of the machined part.

Considering the chemical composition of the surface layer of the test piece, the temperature reached in the processing zone, and the conditions of heat dissipation, some thermal changes could affect the microstructure of the surface layer. In this surface layer, phenomena of overheating, normalization, recrystallization, and hardening could develop, changing its mechanical properties and, as a consequence, the wear behavior of the surface layer.

In the case of the burnishing process, the burnishing tool (ball or roll) is forced to roll under a certain pressure over the entire surface that is intended to be processed, due to the relative movements performed by the workpiece and the burnishing tool, one to the other. The machining scheme presented in Figure 1c corresponds to the burnishing of a cylindrical surface when the workpiece performs a rotation movement, and the burnishing ball materializes a feed movement along the workpiece rotation axis. Generally, when applying a burnishing process, the objectives are to improve the surface roughness, increase the surface layer microhardness, and sometimes even obtain certain microrelief of the processed surface. Due to the high value of the radial force F exerted by the burnishing tool on the workpiece surface layer, a higher value of the hardened surface layer is expected (Figure 1c).

Compared to the common burnishing process, in the case of the vibroburnishing process, an additional vibratory movement is performed by the vibroburnishing tool, with a small amplitude and a relatively low frequency [19]. Due to the more complex movements performed by the vibroburnishing tool on the workpiece surface (Figure 1d), it is expected that certain differences concerning the microstructure and the thickness of the layer affected by the vibroburnishing process could occur, compared to the same aspects in the case of the burnishing process.

Certain microchanges develop in the surface layer due to the processing methods that involve removal of the material (e.g., the turning and grinding methods) or to the processing methods that aim to increase the microhardness and modify the surface asperities geometry. It is expected that distinct behaviors of the processed surface layers will be obtained under the conditions of tribological tests.

2.2. Experimental Conditions

To investigate the evolution of the friction coefficient value and the mass lost by the test piece during a tribological test, equipment presented in Figure 2 was used. The equipment was adapted on a universal lathe of medium size.

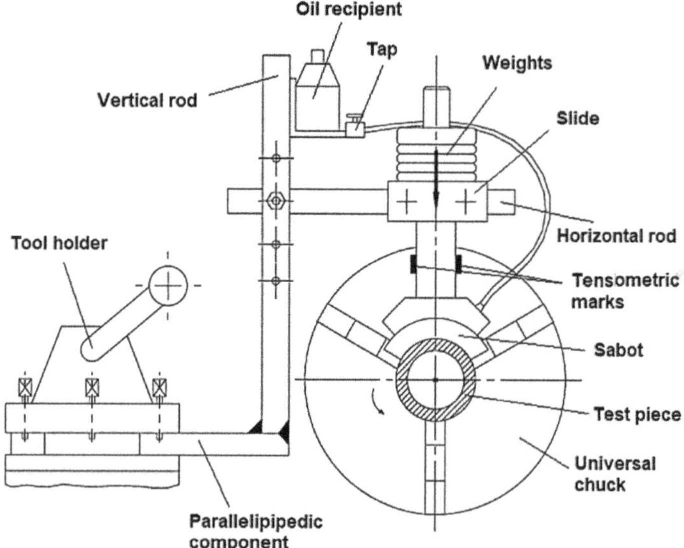

Figure 2. Equipment for testing the wear behavior of surfaces obtained by distinct processing methods.

It can be observed that the conditions of the tribological test present certain similitudes with the conditions specific to the processes that develop in the first period of operation, when there are relative movements between the surfaces of two parts found in contact. It is known that regardless of the initial roughness, after a certain period, the values of surface roughness correspond to the local tribological conditions (local operation conditions) and these values are maintained along with the normal service period of the part.

In the case of equipment for investigating the tribological behavior of the surface layer previously obtained by different processing methods, in the lathe tool holder, a rigid frame obtained by welding was clamped. The rigid frame has a parallelepipedal component for clamping in the lathe tool holder instead of the common lathe tool. A horizontal rod could be attached, using a cylindrical joint, to the vertical component of the rigid frame. In the vertical component of the rigid frame, there are many holes where the bolt of the cylindrical joint could be placed to adapt equipment in the cases of distinct diameters of the test pieces. Along the horizontal rod, a slide could be moved and clamped in an adequate position to the vertical plane of the lathe main shaft axis. A vertical rod could move vertically in a hole that exists in the slide. At the lower end of the vertical rod, there is a part to which sabot support is assembled using screws. A proper sabot made of a material characterized by high hardness, high wear resistance, and certain roughness of the active surface is attached to the sabot support. The sabot is pressed onto the cylindrical surface of the test piece using weights of known values and placed on the vertical rod. The slide is supported by a nut screwed into the vertical rod.

A recipient containing mineral oil was attached to the vertical rod to ensure lubrication conditions for the zone between the sabot and the test piece. A tap ensures oil flow adjustment to the zone found between the sabot and test piece.

At its end, the test piece that has a bushing shape is positioned and clamped onto a mandrel clamped in the lathe universal chuck and the rotating live center.

On the vertical rod, tensometric marks are attached. These tensometric marks are connected in the electric circuit of a tensometric bridge, and they ensure the evaluation of the friction moment M_f.

Calibration of the equipment, including the tensometric bridge for obtaining distinct values of the friction moment M_f, was performed using the working scheme from Figure 3. It can be observed that a cylindrical stud was used to solidarize the sabot with a special test piece and an initially horizontal threaded rod was attached to the special test piece; the calibration operation was then performed. Different weights of known mass values were attached to the free end of the threaded rod. Thus, they generated different torsion moments, M_t. The value of the torsion moment M_t applied to the test piece was measured using an adequate apparatus based on the use of a tensometric bridge.

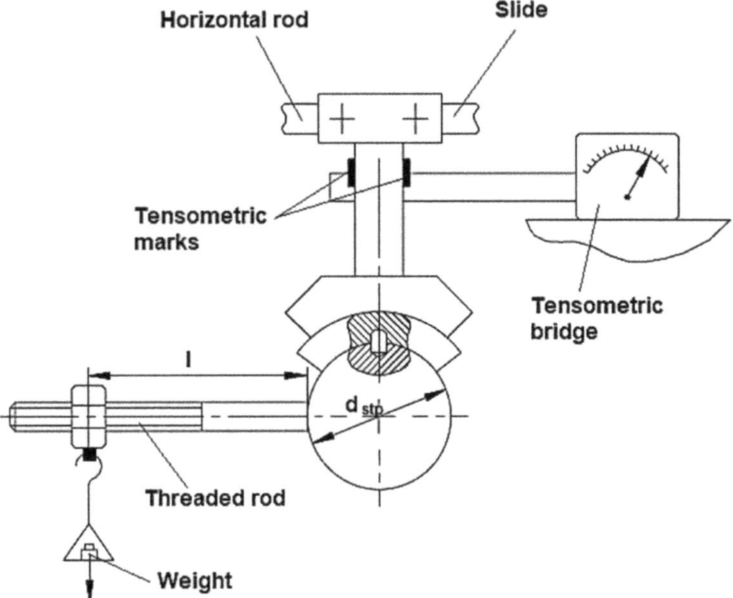

Figure 3. Calibration of equipment for studying the wear behavior of surfaces obtained by different processing methods.

The resulting calibration diagram is presented in Figure 4. In the diagram, the average values of the obtained results upon charging and discharging the equipment were taken into consideration. The diagram was made using Excel software, which made it possible to inscribe the linear function and the value of the coefficient of determination R^2 corresponding to that function on the diagram.

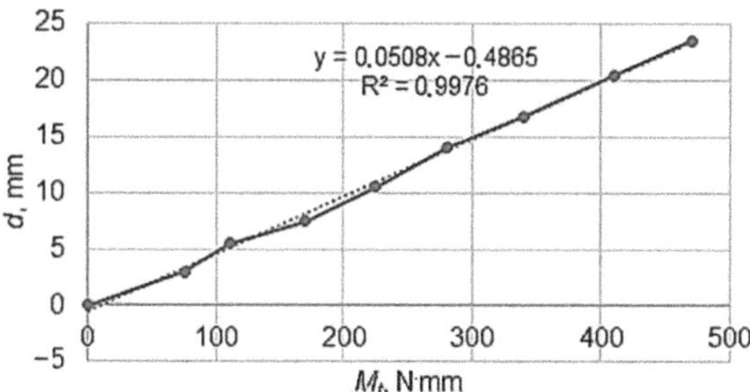

Figure 4. Calibration diagram for the indication *d* of the pressure measuring apparatus depending on the torsion moment M_t.

The torsion moment generated by the mass *m* of the weight attached to the free end of the threaded rod is given by the following equation:

$$M_t = mg\left(l + \frac{d}{2}\right) \qquad (1)$$

where *g* is the gravitational acceleration (mm/s^2), *m* is the mass of the weight used in each test, *l* is the length of the threaded rod (mm), and *d* is the diameter of the special test piece, expressed in mm.

Knowing the values of the torsion moment M_t, the values of the friction coefficient μ could be calculated as follows:

$$\mu = \frac{2M_t}{mgd} \qquad (2)$$

During the experimental tests, weights of known mass values, *m*, were placed on the vertical rod to generate different pressures exerted by the sabot on the test piece.

The values of pressure were evaluated using the following equation:

$$p = \frac{F}{A} \qquad (3)$$

where *A* is the projection of the work surface area of the sabot in a horizontal plane.

Values of the pressure, 14,400–143,300 Pa, were obtained using weights with masses of 1–10 kg.

The test piece had an 800 rev/min rotation speed, corresponding to a peripheral speed of $v = 75$ m/min. The test pieces were made of steel 1.0060 (characterized by a tensile strength of 650–880 MPa). The test piece dimensions were as follows: the external diameter $d_{ext} = 30$ mm, the internal diameter $d_{int} = 22$ mm, and length $l_{tp} = 22$ mm. The sabot material was X 210 Cr 12 (1.2080). The test pieces were weighted after each experimental test, which had a duration of 30 min. The total duration of a test that used a certain combination of process input factors was 210 min.

The following four processing methods were used to generate the external cylindrical surfaces, aiming to obtain information concerning the tribological behavior of surfaces generated by different machining methods:

- Turning with a rotation speed of 500 rev/min ($v = 47.1$ m/min), a longitudinal feed $f_l = 0.1$ mm/rev and a depth of cut $a_p = 0.2$ mm. As a result, roughness characterized by values $Ra = 1.23–1.35$ μm was generated on the turned surface;
- Grinding (using parts prepared previously by turning), which led to a surface roughness $Ra = 0.74–0.7$ μm;
- Ball burnishing applied after turning with a burnishing force $F_b = 400$ N, rotation speed $n = 500$ rev/min ($v = 47.1$ m/min), burnishing feed $f_l = 0.096$ mm/rev, and ball diameter

d_b = 15.85 mm. As a result, surface roughness characterized by Ra = 0.45–0.51 µm was obtained;

A so-called microrelief of 3rd category was obtained by vibroburnishing.

- Ball vibroburnishing occurs when an additional vibration characterized by the amplitude A = 2.6 mm and frequency f = 470 strokes/min (7.83 Hz) is applied to the burnishing process, as mentioned above. The other processing parameters valid in the case of the ball vibroburnishing process had the same values as those used in the case of the ball burnishing process. Microrelief of the 3rd category is characterized by a partial overlap of the grooves generated by the ball on the vibroburnished test piece surface.

For ball burnishing and ball vibroburnishing, equipment schematically presented in Figure 5 was used. The processing scheme involved the rotation of the test piece, while the ball used as a vibroburnishing tool performed a vibration movement in a parallel direction to the test piece axis. This vibration movement was obtained by using an electric motor whose rotation movement was transformed into a rectilinear alternative movement of the ball support using a mechanism-type rod-connecting rod. The pressure necessary in the vibroburnishing process was ensured by using a spring whose compression could be set by means of a nut that moves along a threaded screw.

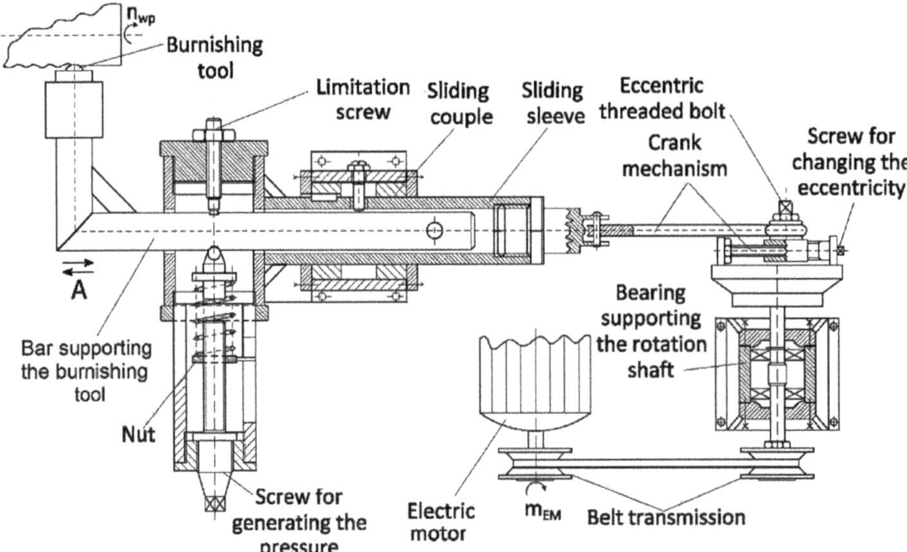

Figure 5. Device adapted on universal lathes to materialize vibroburnishing and oscillatory burnishing of the workpiece surface layer [26].

Belt wheels with different external diameters were used to change the vibratory movement frequency. The amplitude of the vibratory movement was set, acting on the length of the rod included in the rod-connecting rod mechanism.

Seven values of the pressure exerted on the sabot, found to be between 14,400 Pa and 143,300 Pa, were considered to obtain a larger image concerning the influence exerted by the process input factors on the values of the friction coefficient μ and lost mass Δm in the cases of surfaces generated by previously applying four different machining processes. The range of surface pressures for experimental research was established based on some information identified in the consulted literature. The results of preliminary experimental research and the possibilities of the equipment on which the experimental research was carried out were also considered, with the aim of obtaining sufficiently differentiated values of the output parameters, to increase the confidence level of the proposed mathematical models.

Appreciating that the test duration t could significantly affect the values of the process output parameters, the values of the friction coefficient μ and lost mass Δm were measured every 30 min for the total test duration, which was 210 min. When investigating the influence of the test duration t on the variation in the friction coefficient μ in the preliminary experimental tests, it was found that the values of the friction coefficient μ stabilized after about 90 min. However, to increase the confidence level of the experimental results, longer test durations were used, i.e., 210 min.

Two experimental tests were established for each experiment characterized by the imposed values of the sets of process input factors. Only the average value obtained for the process output parameters in the two experiments are included in Table 1.

Table 1. Experimental conditions and results.

| Line no.1 | Applied Processing Method | Testing Pressure, | Friction Coefficient, μ, for a Duration Test, t (min): | | | | | | | | Mass Lost by the Test Piece, Δm (mg), for a Test Duration t (min) | | | | | | | |
| 2 | | p, Pa | 0 | 30 | 60 | 90 | 120 | 150 | 180 | 210 | 0 | 30 | 60 | 90 | 120 | 150 | 180 | 210 |
Column no.1	2	3	4	5	6	7	8	9	10	11	12	13	14	15	16	17	18	19
4		14,400	0.212	0.217	0.220	0.223	0.226	0.229	0.233	0.234	0	27.5	41.2	49.0	57.8	62.8	66.2	72.0
5		33,120	0.219	0.222	0.225	0.227	0.230	0.233	0.237	0.238	0	28.3	43.3	51.2	58.5	63.4	67.3	72.5
6		59,300	0.225	0.229	0.230	0.233	0.236	0.238	0.243	0.243	0	29.1	44.2	52.0	59.5	64.0	68.4	73.0
7	Turning	76,600	0.233	0.238	0.241	0.242	0.243	0.249	0.255	0.255	0	30.0	45.0	53.0	60	64.5	70.1	73.5
8		103,800	0.237	0.243	0.244	0.245	0.247	0.253	0.260	0.261	0	31.2	45.8	53.8	61.0	64.9	70.9	73.9
9		123,900	0.241	0.247	0.248	0.250	0.251	0.255	0.263	0.264	0	33.0	46.7	55.0	61.8	65.4	71.5	74.3
10		143,300	0.244	0.252	0.253	0.254	0.255	0.258	0.265	0.266	0	33.8	47.9	55.9	62.4	66.0	71.9	74.7
11		14,400	0.198	0.203	0.207	0.210	0.214	0.217	0.220	0.221	0	24.1	38.2	47.0	55.1	59.5	63.7	66.1
12		33,120	0.205	0.207	0.211	0.216	0.218	0.222	0.226	0.227	0	24.7	39.3	47.8	55.8	59.9	64.6	66.5
13		59,300	0.211	0.213	0.215	0.223	0.226	0.229	0.233	0.234	0	25.4	39.7	49.1	56.3	60.5	65.0	66.9
14	Grinding	76,600	0.216	0.218	0.219	0.227	0.230	0.232	0.239	0.241	0	26.0	40.0	50.0	57.0	61.0	65.6	67.2
15		103,800	0.220	0.225	0.224	0.230	0.234	0.238	0.244	0.245	0	26.7	40.6	50.7	57.8	61.4	66.2	67.5
16		123,900	0.226	0.229	0.231	0.233	0.237	0.241	0.245	0.248	0	27.3	41.5	51.2	58.2	61.7	66.9	67.8
17		143,300	0.229	0.233	0.236	0.239	0.242	0.245	0.248	0.250	0	28.2	42.4	51.9	59.2	62.0	67.5	68
18		14,400	0.164	0.167	0.170	0.172	0.173	0.174	0.175	0.175	0	19.0	32.1	41.0	45.7	48.1	50.1	51.0
19		33,120	0.168	0.171	0.174	0.176	0.177	0.178	0.178	0.179	0	19.5	32.9	41.7	46.4	48.3	50.5	51.3
20		59,300	0.172	0.175	0.178	0.180	0.181	0.181	0.182	0.183	0	20.2	33.9	42.5	47.0	48.6	50.9	51.6
21	Burnishing	76,600	0.175	0.178	0.181	0.183	0.185	0.185	0.186	0.186	0	21.0	35.0	43.0	47.5	49.0	51.3	51.9
22		103,800	0.178	0.181	0.184	0.186	0.187	0.188	0.190	0.191	0	21.3	35.6	43.4	47.8	49.3	51.6	52.1
23		123,900	0.181	0.184	0.187	0.189	0.191	0.192	0.192	0.193	0	21.9	35.9	43.8	48.0	49.5	51.9	52.3
24		143,300	0.184	0.187	0.190	0.192	0.193	0.194	0.194	0.195	0	23.1	37.8	44.0	48.2	49.7	52.2	52.4
25		14,400	0.124	0.127	0.129	0.130	0.130	0.131	0.131	0.131	0	10.0	17.6	27.0	35.9	36.6	37.0	37.5
26		33,120	0.127	0.130	0.132	0.133	0.133	0.133	0.133	0.133	0	10.3	18.2	27.3	36.3	36.8	37.3	37.7
27	Vibroburni-	59,300	0.131	0.134	0.136	0.137	0.137	0.137	0.137	0.137	0	10.6	18.8	27.5	36.6	37.0	37.6	37.9
28	shing	76,600	0.135	0.136	0.138	0.139	0.139	0.139	0.139	0.139	0	11.0	19.0	28.0	36.9	37.2	37.8	38.0
29		103,800	0.138	0.141	0.143	0.144	0.144	0.144	0.144	0.144	0	11.3	19.3	28.3	37.2	37.3	37.9	38.1
30		123,900	0.140	0.143	0.147	0.148	0.148	0.149	0.149	0.149	0	11.4	19.4	28.6	37.3	37.5	38.1	38.2
31		143,300	0.141	0.144	0.148	0.149	0.149	0.150	0.150	0.150	0	11.9	19.6	28.9	37.4	37.6	38.2	38.3

2.3. Modeling the Tribological Test Using the Finite Element Method

Starting from the schematic representation of the equipment used to test the wear behavior in Figure 2, a simulation based on ANSYS software for finite element analysis was considered. The aim was to highlight some aspects that characterize the wear test process. The materials of the two components directly participating in the wear process were those used in the experimental research (steel 1.0060 as the test piece and steel X210Cr12 for the sabot). The two surfaces in contact were considered under ideal conditions, so the heights of the roughness in the process were not considered.

We started by evaluating the amount of material removed from the test piece under the pressure p = 14,400 Pa and coefficient of friction μ = 0.127. The sabot was considered to be pushed down by a force of 162.86 N, which corresponds to a uniform pressure of 14,400 Pa. The results obtained correspond to a transient analysis for a full revolution time stepped in 18 equal intervals.

The amount of material removed from the test piece was determined by taking into account the volume of the test piece before the start of the process and the volume of the test piece after one rotation. In the simulation process, APDL-type POST commands were used for this purpose.

Limitations in computing power forced us to set just a 1 mm element size for meshing purposes. Considering a tensile yield strength of 310 MPa in the case of the test piece material, equivalent (von Mises) stresses were set to all parts.

As a result of applying the finite element analysis method, the graphical representations from Figures 6–8 were developed. Thus, in Figure 6, the variation in the pressure on the contact surface between the sabot and the test piece with a bushing shape was highlighted. Higher pressure values can be observed near the axis of the rod through which the sabot is pressed on the outer cylindrical surface of the test piece. The contact pressure peaked at just 10% of the yielding value.

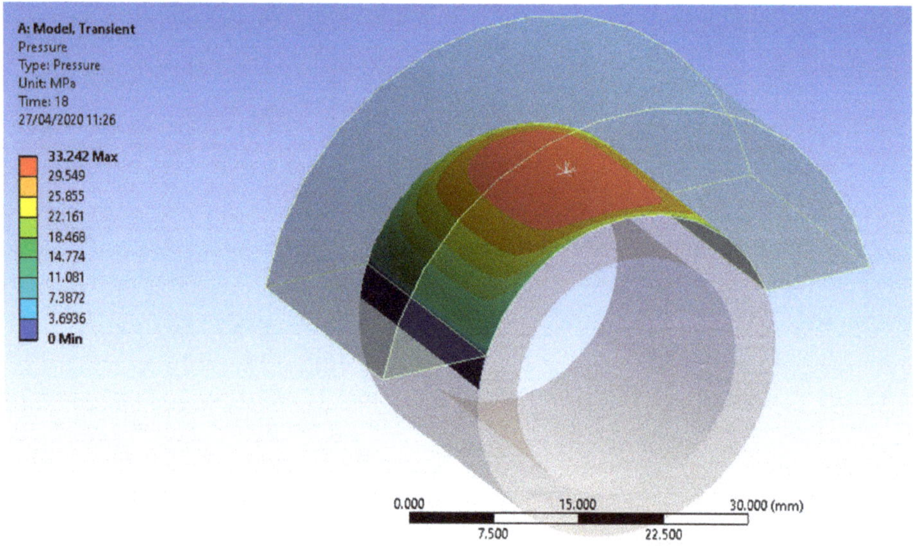

Figure 6. The pressure exerted by the sabot on the test piece surface.

Figure 7 took into account a survey of the penetration of the sabot into the test piece material due to the evolution of the wear process. The APDL POST commands have given a new microvolume value for the test piece, which is about 0.003 mm^3 smaller than the initial microvolume.

In Figure 8, the variation in the equivalent stresses (von Mises) can be observed, appreciating that they could affect, possibly even to a small extent, the removal of material from the test piece due to the wear process. The yielding value was not reached for the steel 1.0060 ST 60-2. This essentially means that the elastic state of the material was not exceeded, and there was only a small amount of material flow evaluated as being below 0.1%.

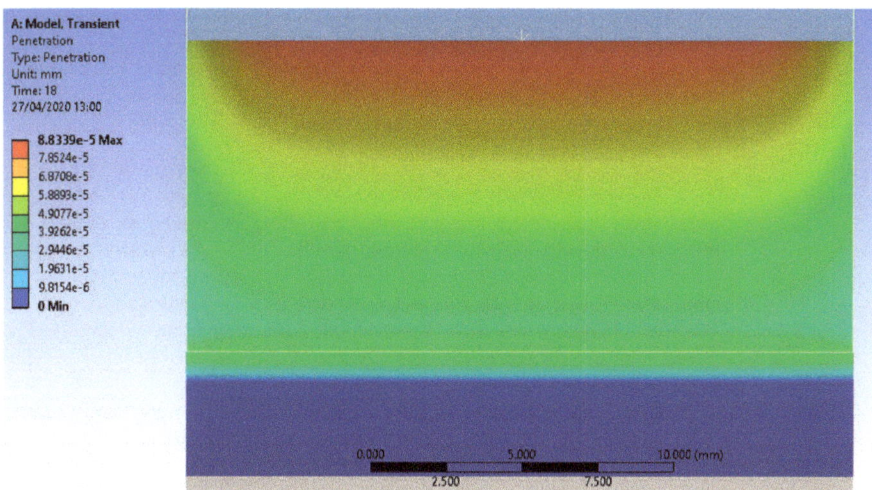

Figure 7. Lowering the sabot (penetration) as a consequence of the wear process that affects the test piece.

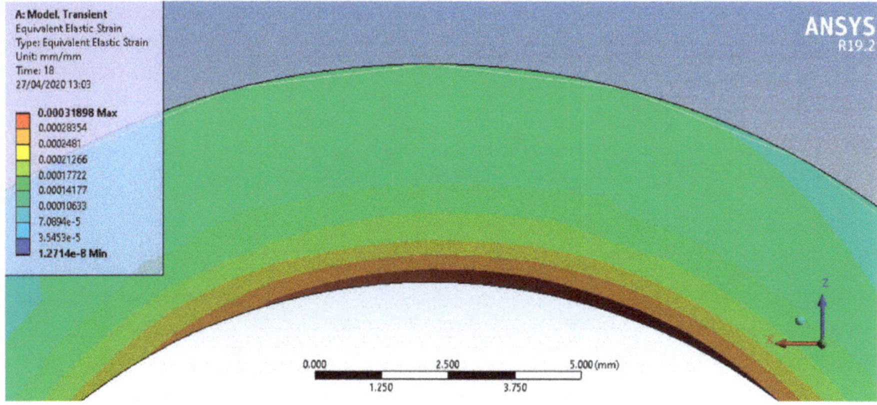

Figure 8. Equivalent stresses (von Mises) developed in the test piece.

3. Results

The experimental results included in Table 1 were mathematically processed using specialized software, based on the method of least squares [27]. The software facilitates the selection of the most convenient empirical mathematical model using Gaussian criterion. This criterion considers the value of the so-called Gauss sum, which is determined as the sum of the squares of the differences between the ordinates of the points that correspond to the selected empirical mathematical model, and those determined experimentally for the same values of the abscissas. The lower the value of Gauss's criterion is, the more convenient the considered empirical mathematical model is to the experimental results.

As mentioned above, the software ensures the selection of an adequate empirical mathematical model, taking into consideration five such models (polynomial of the first and second degree, power-type function, exponential-type function, and hyperbolic function). For all eight groups of experimental results (that correspond to the sizes of the friction coefficient μ and the lost mass Δm in the cases of the four previously applied processing methods), the software showed that the most appropriate empirical mathematical model is a polynomial of the second degree (for which the minimum value of the Gauss's criterion was determined).

For example, in the case of turning, such an empirical mathematical model is as follows:

$$\mu = 0.209 + 2.766 \times 10^{-7} p - 1.304 \times 10^{-13} p^2 + 8.117 \times 10^{-5} t + 7.970 \times 10^{-9} t^2 \tag{4}$$

for which the Gauss's coefficient has the value $S_G = 4.336001 \times 10^{-6}$.

On the other hand, it can be noticed that the empirical mathematical model of a polynomial of the second degree does not allow an immediate image, concerning the influence exerted by each of the process input factors on the values of the friction coefficient μ and lost mass Δm, to be formed. For this reason, the power-type functions were preferred, since, in the case of such functions, the values of the exponents attached to the sizes corresponding to the process input factors offer immediate information regarding the intensity of the influence exerted by each of the considered process input factors on the values of the output parameters.

In this way, the following empirical mathematical model was determined for the friction coefficient μ corresponding to the tested turned surface:

$$\mu = 0.123 p^{0.0593} t^{0.00335} \tag{5}$$

for which the value of the Gauss's criterion is $S_G = 3.825921 \times 10^{-5}$.

It can be observed that the value of Gauss's criterion determined when using the power-type function is higher than the value of the same criterion when using a polynomial-type function of the second degree. Still, we considered that the advantages of the immediate obtainment of an image concerning the influence exerted by the process input factors on the parameter of technical interest (in this case, the friction coefficient μ) are more important. Subsequently, only the power-type functions were used.

By mathematical processing of the experimental results included in Table 1, the following empirical mathematical models have been determined thus far:

- In the case of surfaces obtained by turning, the following model was determined:

$$\Delta m = 4.285 p^{0.0444} t^{0.442} \tag{6}$$

($S_G = 5.982294$);

- In the case of the surfaces obtained by grinding, the following models were determined:

$$\mu = 0.121 p^{0.0554} t^{0.00387} \tag{7}$$

($S_G = 4.357229 \times 10^{-5}$);

$$\Delta m = 3.632 p^{0.0335} t^{0.486} \tag{8}$$

($S_G = 8.259409$);

- In the case of surfaces obtained by burnishing, the following models were determined:

$$\mu = 0.107 p^{0.0471} t^{0.00285} \tag{9}$$

($S_G = 7.776968 \times 10^{-6}$);

$$\Delta m = 3.308 p^{0.0348} t^{0.459} \tag{10}$$

($S_G = 11.31938$);

- In the case of vibroburnished surfaces, the following models were determined:

$$\mu = 0.0634 p^{0.0708} t^{0.00295} \tag{11}$$

($S_G = 1.884871 \times 10^{-4}$);

$$\Delta m = 0.876 p^{0.0292} t^{0.672} \tag{12}$$

(S_G = 15.76152).

The empirical mathematical model defined by Equation (6) was determined by taking into account the experimental values of the friction coefficient μ recorded in columns 4, 5, 6, 8, 9, and 10 in Table 1 (except, therefore, when determining the model mathematically; the results are in column 7 in this case). The graphical representation from Figure 9 was developed to verify the validity of the empirical mathematical model defined by Equation (6). In this graphical representation, the continuous curved line corresponds to the mathematical model defined by Equation (6). The points in the diagram in Figure 6 correspond to the values of the experimental results recorded in column 10 in Table 1. It can be observed that the values of the coefficient of friction μ are near to the curved line, both above and below it. This confirms the validity of the empirical model determined by mathematical processing of the experimental results, since the points in the diagram represent results that were not used in determining the empirical mathematical model.

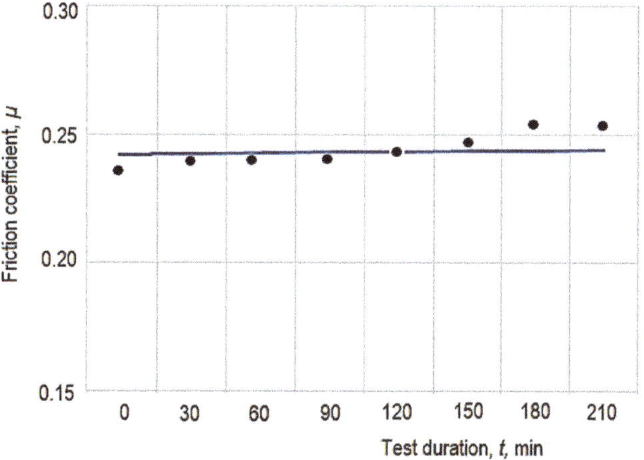

Figure 9. Testing the validity of the empirical mathematical model defined by Equation (6) (p = 76,600 Pa).

The graphical representations in Figures 10–16 were achieved by considering the empirical mathematical models defined by Equations (5)–(12).

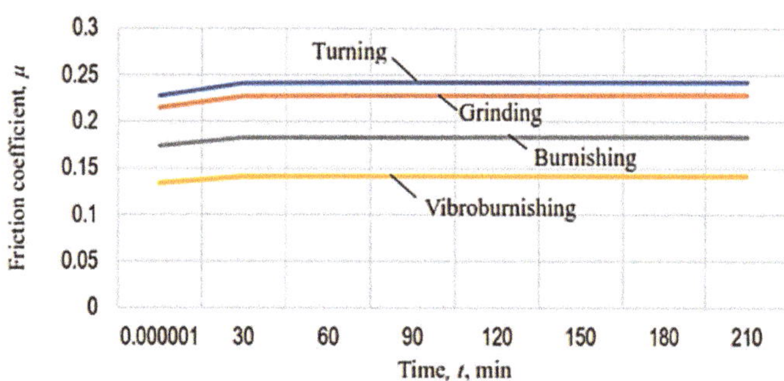

Figure 10. The influence exerted by the process duration on the value of the friction coefficient μ for different machining processes applied to the tested surface (p = 70,000 Pa).

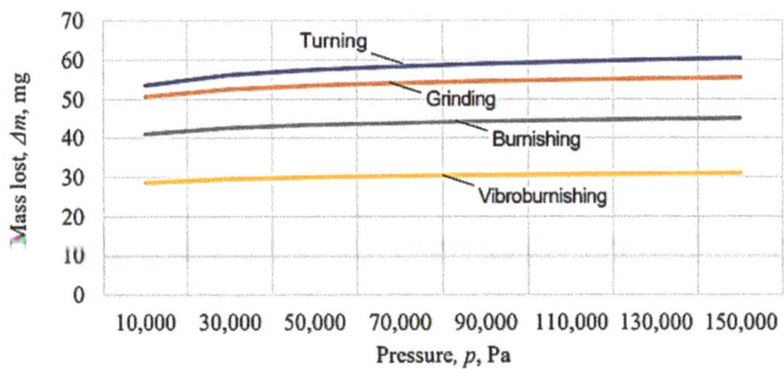

Figure 11. Influence exerted by the pressure on the mass lost Δ*m* for different machining processes applied to the tested surface (*t* = 150 min).

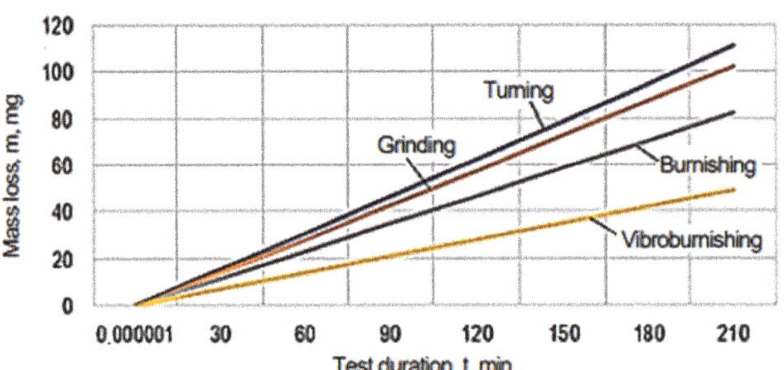

Figure 12. The influence exerted by the process duration *t* on the mass lost Δ*m* for different machining processes applied to the tested surfaces (*p* = 70,000 Pa).

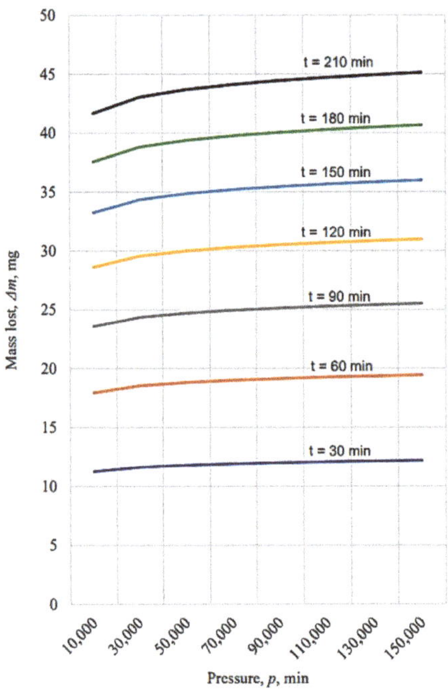

Figure 13. The influence exerted by the pressure on the mass lost Δ*m* for different durations of the process applied to the tested surface in the case of the vibroburnishing process.

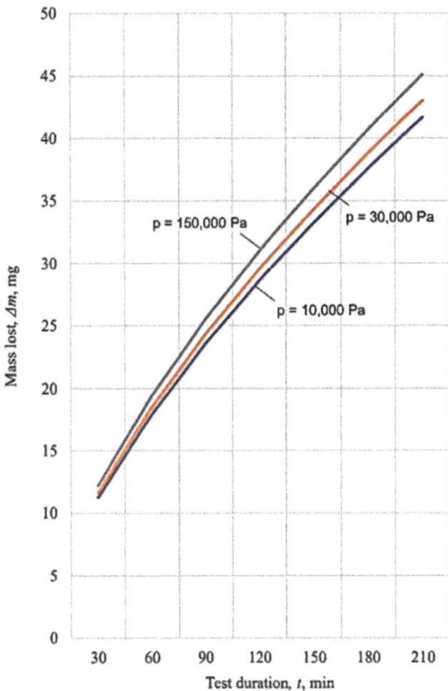

Figure 14. Influence exerted by the process duration *t* on the mass lost Δ*m* for different pressures *p* applied to the tested surface in the case of the vibroburnishing process.

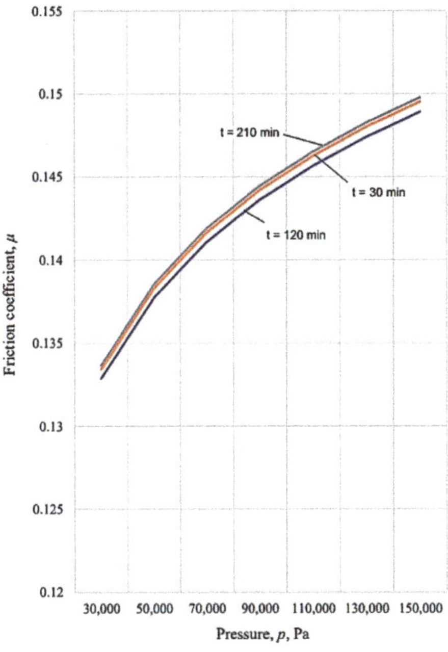

Figure 15. Influence exerted by the pressure *p* on the friction coefficient μ for different process durations *t* in the case of vibroburnishing applied to the tested surface.

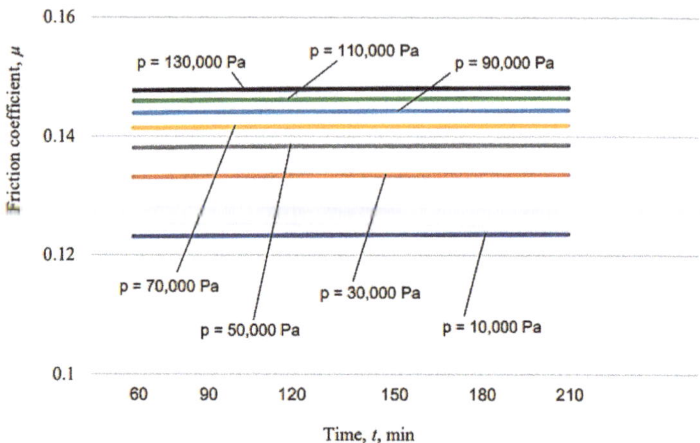

Figure 16. Influence exerted by the process duration t on the friction coefficient μ for different pressures p in the case of vibroburnishing applied to the tested surface.

4. Discussion

Some remarks could be formulated by examining the empirical mathematical models from Equations (5–12) and the graphical representations in Figures 10–16.

Thus, the similitude of the empirical mathematical models to the empirical mathematical models that represent the variation in the friction coefficient μ and lost mass Δm could be observed for all the test pieces obtained by different processing methods. In all eight empirical mathematical models, the low values of the exponents attached to the pressure p show that this process input factor (pressure, p) exerts a certain influence on the friction coefficient μ and lost mass Δm, but this influence is low enough. As expected, the increase in pressure, p, is accompanied by an increase in the friction coefficient μ, due to the increase in the normal force F exerted on the contact surface between the test piece and sabot.

On the other hand, the influence exerted by the test duration t on the friction coefficient μ is also low, and is even less than the influence exerted by the pressure p, since the values of the exponents attached to the duration test t in the empirical mathematical models defined by Equations (5)–(12) are lower than the values of the exponents attached to the pressure p in the same mathematical equations.

In contrast to the influence exerted on the friction coefficient μ, in the case of the lost mass Δm, it can be noticed that the duration test t exerts a stronger influence, since the values of the exponents attached to the test duration t are higher than the values of the exponents attached to the pressure p in the same mathematical equations. As expected, the increase in the test duration t causes an increase in the lost mass Δm.

The diagrams presented in Figures 10–16 show that the decrease in the initial values of the surface roughness parameter Ra, which corresponds to the four surfaces obtained using different processing methods, leads to a decrease in the values of the friction coefficient μ and lost mass Δm. In the case of the friction coefficient μ, this aspect is justified as it is known that, generally, the value of the coefficient μ diminishes when the heights of the surface micro asperities are lower.

Similar arguments could also be valid in the case of the change in the value of the lost mass Δm when the height of the surface asperities is lower. Indeed, when the heights of the surface micro asperities are higher, the local pressure on the micro asperities found in contact is higher, and a higher quantity of the test piece materials could be removed.

5. The Correlation between the Values of the Friction Coefficient and Lost Mass When Applying Tribological Tests

The inverse evolution of the two process output parameters analyzed in this paper showed a correlation between them. To find the answer to this question, the correlation coefficient value was determined, considering that the values of both output parameters (friction coefficient μ and lost mass Δm) were established within the same sets of experiments.

It is known that the value of the coefficient of the correlation (the Pearson's correlation coefficient r_{xy} for the uncorrected, non-standard form) is given by the following [28]:

$$r_{xy} = \frac{n\sum x_i y_i - \sum x_i \sum y_i}{\sqrt{n\sum x_i^2 - (\sum x_i)^2}\sqrt{n\sum y_i^2 - (\sum y_i)^2}} \tag{13}$$

where n corresponds to the number of measurements found in each of the two comparison sets of the measured values x_i and y_i when $i = 1, 2, \ldots, n$. It is usual to consider a strong correlation between the two series of considered values if the Pearson's coefficient is close to 1.00 or -1.00, and a low correlation when the Pearson's coefficient has a value close to zero.

The function CORRELATION from the Excel software was used to evaluate the correlation between the values of the friction coefficient μ and the lost mass Δm in the cases of the surfaces previously obtained by applying different processing methods. The determined Pearson's coefficient values when applying the four different processing methods applied previously to the tested surface were the following:

$r_{xy} = 0.534322$ for turning;
$r_{xy} = 0.617232$ for grinding;
$r_{xy} = 0.531417$ for ball burnishing;
$r_{xy} = 0.34025$ for ball vibroburnishing.

Following the existing conventions concerning the evaluation of the correlation using the Pearson's coefficient, this means that, in the cases of all the different processes applied to the test piece surfaces before testing their tribological behavior, there is an average correlation between the compared sizes in the case of the ball vibroburnishing process, and a strong correlation in the cases of turning, grinding, and ball burnishing. The maximum value of Pearson's coefficient corresponds to the grinded surface, while the minimum value of Pearson's coefficient was found for the ball vibroburnished surface.

6. Conclusions

The tribological characteristics of the surfaces can be influenced, to a significant extent, by the processing technique by which the respective surfaces were obtained. Testing equipment, adapted on a universal lathe, was designed and materialized to mathematically model the variation in the friction coefficient and the lost mass. The equipment involves pressing a sabot made of hard material onto the external cylindrical surface of a rotating test piece made of the material to be tested. The experimental research was carried out on steel 1.0060 test pieces; the mechanical treatment by burnishing and vibroburnishing could improve the tribological characteristics more compared with the surfaces obtained by turning and grinding. Mathematical models using the power-type function have been determined to highlight the influence of pressure and test duration on the size of the friction coefficient and the microamount of material removed from the test piece due to the tribological test. It was considered that the change in the tribological behavior of the test piece material is a result of the change in the microstructure of the material in the surface layer; the shape and dimensions of the micro asperities result from the previous application of certain processing methods. Information was obtained regarding the intensity of the variation in the friction coefficient and mass loss depending on the applied pressure and the experiment test duration. It can be noticed that, essentially, the influence exerted by the pressure and test duration is relatively low in the case of the friction coefficient. If the mass loss is analyzed, it can be noticed that the test duration has a stronger influence than the influence exerted by the pressure for the considered experimental conditions. Among the investigated processing techniques, burnishing and vibroburnishing had the most convenient results for the friction coefficient and mass microamount removed from the test piece.

Author Contributions: Conceptualization, G.N. and L.S.; methodology, O.D.; software, A.M.M. and M.I.R.; validation, O.D., A.H. and L.S.; formal analysis, A.H.; investigation, G.N.; resources, O.D.; writing—original draft preparation, L.S.; writing—review and editing, O.D.; visualization, G.N.; supervision, L.S.; All authors have read and agreed to the published version of the manuscript.

Funding: This research received no external funding.

Institutional Review Board Statement: Not applicable.

Informed Consent Statement: Not applicable.

Data Availability Statement: Not applicable.

Conflicts of Interest: The authors declare no conflict of interest.

References

1. Shneider, I.G. *Performance Properties of Parts with Regular Microrelief*; Mashinostroenie: Leningrad, Russia, 1982. (In Russian)
2. Odintsov, L.G. *Hardening and Finishing the Parts Surfaces Using the Surface Plastic Deformation*; Handbook; Maschinostroenie: Moscow, Russia, 1987. (In Russian)
3. Loh, N.H.; Tam, S.C.; Miyazawa, S. Surface hardening by ball burnishing. *Tribol. Int.* **1990**, *23*, 413–417. [CrossRef]
4. Kim, N.H.; Won, D.; Burris, D.; Holtkamp, B.; Gessel, G.R.; Swanson, P.; Sawyer, W.G. Finite element analysis and experiments of metal/metal wear in oscillatory contacts. *Wear* **2005**, *258*, 1787–1793. [CrossRef]
5. El-Tayeb, N.S.M.; Low, K.O.; Brevern, P.V. On the surface and tribological characteristics of burnished cylindrical Al-6061. *Tribol. Int.* **2009**, *42*, 320–326. [CrossRef]
6. Wojciechowski, L.; Nosal, S. The preparation of surface layer of non-alloyed steel to cooperation with lubricating medium. *J. KONES* **2009**, *16*, 533–540. Available online: http://yadda.icm.edu.pl/yadda/element/bwmeta1.element.baztech-article-BUJ7-0018-0052 (accessed on 20 October 2021).
7. Low, K.O.; Wong, K.J. Influence of ball burnishing on surface quality and tribological characteristics of polymers under dry sliding conditions. *Tribol. Int.* **2011**, *44*, 144–153. [CrossRef]
8. Li, Y.; Han, Y. Recrystallization of Ni3Al Base Single Crystal Alloy IC6SX with different surface mechanical processes. *J. Mater. Sci. Technol.* **2010**, *26*, 883–888. [CrossRef]
9. Ovali, I.; Akkurt, A. Comparison of burnishing process with other methods of hole surface finishing processes applied on brass materials. *Mater. Manuf. Process.* **2011**, *26*, 1064–1072. [CrossRef]
10. Al-Saeedi, S.; Sarhan, A.A.D.; Bushroa, A.R. Investigating the tribological characteristics of burnished polyoxymethylene—ANFIS and FE modeling. *Tribol. Trans.* **2011**, *61*, 880–888. [CrossRef]
11. Lewandowski, A. Tribological properties of cast iron EN-GJSFP-500-7 after a burnishing process modified by exploration preparations based on solid lubricant and chemical action. *Tribologia* **2013**, *4*, 67–84. (In Polish)
12. Zaleski, K. The effect of vibratory and rotational shot peening and wear on fatigue life of steel. *Eksploat. I Niezawodn.–Maint. Reliab.* **2017**, *19*, 102–107. [CrossRef]
13. Janczewski, Ł.; Toboła, D.; Brostow, W.; Czechowski, K.; Hagg Lobland, H.E.; Kot, M.; Zagórski, K. Effects of ball burnishing on surface properties of low density polyethylene. *Tribol. Int.* **2016**, *93*, 36–42. [CrossRef]
14. Krasnyy, V.A.; Maksarov, V.V. Tribological characteristics of mining machines parts with regular surface microgeometry (in Russian). *Metallobrabotka* **2016**, *91*, 2135.
15. Krasnyy, V.; Maksarov, V.; Olt, J. Increase of wear and fretting resistance of mining machinery parts with regular roughness patterns. In Proceedings of the 27th DAAAM International Symposium, Vienna, Austria, 26–29 October 2016; Katalinic, B., Ed.; DAAAM International: Vienna, Austria, 2016; Volume 27, pp. 151–156. [CrossRef]
16. Takada, Y.; Sasahara, H. Effect of Tip Shape of Frictional Stir Burnishing Tool on Processed Layer's Hardness, Residual Stress and Surface Roughness. *Coatings* **2018**, *8*, 32. [CrossRef]
17. Dzierwa, A.; Markopoulos, A.P. Influence of Ball-Burnishing Process on Surface Topography Parameters and Tribological Properties of Hardened Steel. *Machines* **2019**, *7*, 11. [CrossRef]
18. Yang, S.; Yu, S.; He, C. The surface integrity of titanium alloy when using micro-textured ball-end milling cutters. *Micromachines* **2019**, *10*, 21. [CrossRef] [PubMed]
19. Hou, G.; Li, A. Effect of surface micro-hardness change in multistep machining on friction and wear characteristics of titanium alloy. *Appl. Sci.* **2021**, *11*, 7471. [CrossRef]
20. Nagîț, G.; Slătineanu, L.; Dodun, O.; Rîpanu, M.I.; Mihalache, A.M. Surface layer microhardness and roughness after applying a vibroburnishing process. *J. Mater. Res. Technol.* **2019**, *8*, 4333–4346. [CrossRef]

21. Attabi, S.; Himour, A.; Laouar, L.; Motallebzadeh, A. Mechanical and wear behaviors of 316L stainless steel after ball burnishing treatment. *J. Mater. Res. Technol.* **2021**, *15*, 3255–3267. [CrossRef]
22. Capilla-González, G.; Martínez-Ramírez, I.; Díaz-Infante, D.; Hernández-Rodríguez, E.; Alcántar-Camarena, V.; Saldaña-Robles, A. Effect of the ball burnishing on the surface quality and mechanical properties of a TRIP steel sheet. *Int. J. Adv. Manuf. Technol.* **2021**, *116*, 3953–3964. [CrossRef]
23. Slavov, S.; Dimitrov, D.; Konsulova-Bakalova, M.; Vasileva, D. Impact of ball burnished regular reliefs on fatigue life of AISI 304 and 316L austenitic stainless steels. *Materials* **2021**, *14*, 2529. [CrossRef]
24. Travieso-Rodríguez, J.A.; Jerez-Mesa, R.; Gómez-Gras, G.; Llumà-Fuentes, J.; Casadesús-Farràs, O.; Madueño-Guerrero, M. Hardening effect and fatigue behavior enhancement through ball burnishing on AISI 1038. *Mater. Res. Technol.* **2019**, *8*, 5639–5646. [CrossRef]
25. Kluz, R.; Antosz, K.; Trzepieciński, T.; Bucior, M. Modelling the Influence of slide burnishing parameters on the surface roughness of shafts made of 42CrMo4 heat-treatable steel. *Materials* **2021**, *14*, 1175. [CrossRef] [PubMed]
26. Nagîţ, G.; Braha, V.; Slătineanu, L. Vibration Rolling Device. Romanian Patent 116954, 10 January 1996. Available online: https://worldwide.espacenet.com/patent/search/family/064360902/publication/RO116954B1?q=Nag%C3%AE%C5%A3 (accessed on 20 October 2021).
27. Creţu, G. *The Basis of Experimental Research*; Laboratory Guide; Gheorghe Asachi Technical University: Iasi, Romania, 1992. (In Romanian)
28. Statistics How To. 2018. Available online: http://www.statisticshowto.com/probability-and-statistics/correlation-coefficient-formula (accessed on 12 November 2021).

 *micromachines*

Article

Study on the Mechanism of Solid-Phase Oxidant Action in Tribochemical Mechanical Polishing of SiC Single Crystal Substrate

Wanting Qi, Xiaojun Cao, Wen Xiao, Zhankui Wang and Jianyun Su *

School of Mechanical and Electrical Engineering, Henan Institute of Science and Technology, Xinxiang 453003, China; qwt245147@163.com (W.Q.); cxj18303659239@163.com (X.C.); xw68886@163.com (W.X.); luckywzk@126.com (Z.W.)
* Correspondence: dlutsu2004@126.com; Tel.: +86-373-3693180

Abstract: Na_2CO_3—1.5 H_2O_2, $KClO_3$, $KMnO_4$, KIO_3, and NaOH were selected for dry polishing tests with a 6H-SiC single crystal substrate on a polyurethane polishing pad. The research results showed that all the solid-phase oxidants, except NaOH, could decompose to produce oxygen under the frictional action. After polishing with the five solid-phase oxidants, oxygen was found on the surface of SiC, indicating that all five solid-phase oxidants can have complex tribochemical reactions with SiC. Their reaction products are mainly SiO_2 and $(SiO_2)x$. Under the action of friction, due to the high flash point temperature of the polishing interface, the oxygen generated by the decomposition of the solid-phase oxidant could oxidize the surface of SiC and generate a SiO_2 oxide layer on the surface of SiC. On the other hand, SiC reacted with H_2O and generated a SiO_2 oxide layer on the surface of SiC. After polishing with NaOH, the SiO_2 oxide layer and soluble Na_2SiO_3 could be generated on the SiC surface; therefore, the surface material removal rate (MRR) was the highest, and the surface roughness was the largest, after polishing. The lowest MRR was achieved after the dry polishing of SiC with $KClO_3$.

Keywords: 6H-SiC; fixed abrasive; tribochemical mechanical polishing; solid-phase oxidant; dry polishing

Citation: Qi, W.; Cao, X.; Xiao, W.; Wang, Z.; Su, J. Study on the Mechanism of Solid-Phase Oxidant Action in Tribochemical Mechanical Polishing of SiC Single Crystal Substrate. *Micromachines* **2021**, *12*, 1547. https://doi.org/10.3390/mi12121547

Academic Editors: Jiang Guo and Benny C. F. Cheung

Received: 14 November 2021
Accepted: 8 December 2021
Published: 12 December 2021

Publisher's Note: MDPI stays neutral with regard to jurisdictional claims in published maps and institutional affiliations.

1. Introduction

As a third-generation semiconductor material, SiC has excellent chemical and physical properties [1,2] and widely used in satellite communications, integrated circuits and consumer electronics [3–5]. However, SiC is characterized by high hardness, high brittleness, and good physical and chemical stability, therefore, it is typically a difficult material to machine [6].

At present, the common method for ultra-smooth processing of SiC is free abrasive chemical mechanical polishing (CMP), which achieves the ultra-smooth, damage-free, and ultra-flat surface processing of the workpiece through acombination of chemical etching of polishing solution and mechanical action of abrasive, and is one of the more effective global flattening processing methods for semiconductor materials [7,8]. However, CMP has the following disadvantages, low processing efficiency, poor environmental friendliness, poor surface consistency, and poor process engineering controllability [9]. Fixed abrasive chemical machining can effectively avoid the above disadvantages of free abrasive machining and has become one of the emerging technologies in the field of ultra-precision machining [10–12]. Fixed abrasive tribochemical mechanical polishing is a fixed abrasive chemical-mechanical processing technology, which can use the abrasive and chemical additives in the polishing pad and the surface of the workpiece in a tribochemical reaction to change the surface of the workpiece material and chemical organization. This mechanism achieves the efficient removal of its material; therefore, is the process increasingly gaining the attention of researchers [13].

The core of fixed abrasive tribochemical mechanical polishing is the tribochemical reaction produced on the surface of the workpiece in the polishing process. Therefore, understanding its chemical reaction mechanism is the key to studying its processing mechanism. Usually, there are two kinds of tribochemical reaction, one is the tribochemical reaction in the dry friction state, and the other is the tribochemical reaction in the lubricated state [14]. Tribochemical mechanical polishing does not use apolishing solution containing free abrasive, but instead uses tribochemical mechanical polishing tools. The polishing slurry is trace deionized water, or chemical solution, without adding any free abrasive, its tribochemical reaction is mostly wet tribochemical action, therefore, the workpiece material removal method is mostly tribochemical wear removal [15]. Since the workpiece obtained by this method has the advantages of low residual stress, flatness, and ultra-smoothness without damage, it has gained the attention of a large number of researchers. Z. Zhu et al. [16] conducted tribochemical mechanical polishing tests on SiC using abrasive-free oxidants H_2O_2, CrO_3, and $KMnO_4$, respectively, and excellent surfaces with surface roughness less than 50 nm and residual stress less than 50 MPa were obtained. S. Kitaoka et al. [17] proposed the theory of tribochemical wear based on a hydrothermal reaction for SiC and Si_3N_4 anaerobic ceramics. It is believed that SiC undergoes a tribochemical reaction with water at 120 °C and a SiO_2 reaction layer is generated on its surface, which achieves ultra-precision machining after the removal of abrasive particles. Yusuke Ootani et al. [18] studied the kinetics of the tribochemical reactions of Si_3N_4 and SiC under an aqueous environment and analyzed different tribochemical reaction mechanisms during the lapping of Si_3N_4 and SiC. Zhou F et al. [19] proposed the wear mechanism of SiC/SiCin water related to the microfracture of the ceramic and the instability of the tribochemical reaction layer.

In summary, the current tribochemical mechanical polishing for SiC is mainly used in water-based polishing solutions for tribochemical mechanical polishing, which contains chemical additives in the polishing slurry, and the discharge of the polishing slurry will bring about environmental pollution and an increase in processing costs. Therefore, the use of fixed abrasive dry tribochemical mechanical polishing can reduce its chemical pollution and production cost. However, there are few studies on the dry tribochemical mechanical polishing of fixed abrasives regarding SiC, and the mechanism of its action arestill not clear enough, particularly the oxidation mechanism of the dry friction of SiC workpieces is still unclear and needs further research and exploration.

In this paper, five solid-phase oxidants were selected for dry polishing with a 6H-SiC single crystal substrate.The changes to their 3D morphology, compounds, and elements on the workpiece surface before and after polishing were examined to analyze their surface tribochemical reactants and to study the oxygen production mechanism of their solid-phase oxidants during the tribochemical mechanical polishing of the fixed abrasive. The study can provide help to understand the mechanism of oxygen production and the tribochemical reaction mechanisms of SiC during fixed abrasive tribochemical mechanical polishing, providing aid regarding the selection of a solid-phase oxidant for the fixed abrasive tribochemical mechanical polishing pad and the formulation of a solid-phase oxidant for a fixed abrasive.

2. Experiment and Characterization

2.1. Polishing Test

The test sample was the n-type of a 6H-SiC single crystal substrate (Tianke Heda, Beijing, China) with a thickness of 0.4 mm, a diameter of 50.8 mm, and an initial surface roughness of 6–7 nm. The SiC was pasted to the center of the carrier table with paraffin wax before the test, and was dry polished on the C side (0001) at room temperature using the ZYP230 rotary oscillating gravity lapping and polishing machine (Kemai, Shenyang, China). The processing principle is shown in Figure 1, and the polishing process parameters are shown in Table 1. The polishing pad used for the test was a polyurethane polishing pad, and the polishing medium was five typesof solid-phase oxidant. The specific compositions are shown in Table 2.

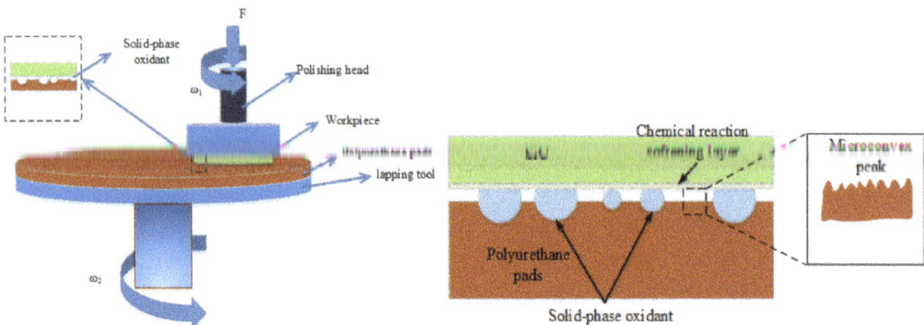

Figure 1. The schematic of tribochemica lmechanical polishing.

Table 1. Polishing process parameters.

Factors	Speed of Polishing Tool n_1 (r/min)	Speed of Polishing Head n_2 (r/min)	Polishing Pressure P (psi)	Time t (h)
Parameters	60	45	2	1.5

Table 2. Solid-phase oxidant and its composition used in the test.

	1	2	3	4	5
6H-SiC	Na_2CO_3—1.5 H_2O_2	NaOH	KIO_3	$KClO_3$	$KMnO_4$

The solid-phase oxidant was spread evenly on the polishing pad, as shown in Figure 2a, and the dry polishing process had a dosing rate of 20g/h. Figure 2b,c show the beginning of the dry polishing process, during the dry polishing process, and after the completion of dry polishing, respectively. The single-factor method was used for the experiments and analysis to explore the oxygen production mechanisms of different solid-phase oxidant polishing.

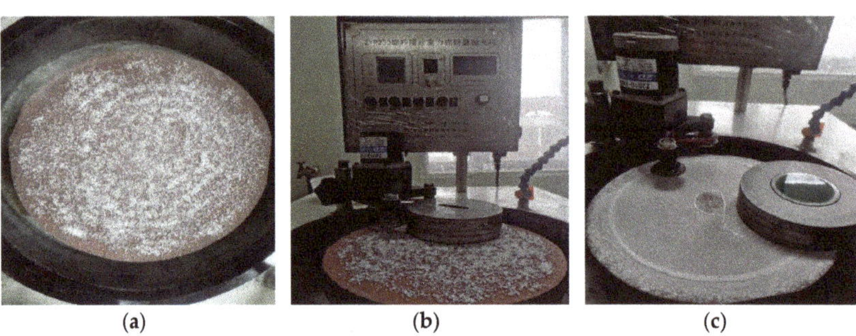

(a)   (b)   (c)

Figure 2. Test process. (**a**) The oxidant is evenly distributed on the polishing pad. (**b**) Start polishing. (**c**) Finish polishing.

The mass of each sample was measured using a precision electronic balance.Before and after its processing, the difference was calculated, and the material removal rate (MRR, nm/min) for polishing was calculated using Equation (1). The surface roughness and 3D morphology of the sample before and after polishing were measured on a ContourGTk-1 3D profile inspection system (Bruker, Billerica, MA, USA).

$$MRR = \frac{\Delta m}{\rho r^2 \pi t} \times 10^7 \tag{1}$$

where, Δm is the mass difference before and after polishing, g, t is the processing time, min, ρ is the density of SiC, g/cm^3, which is taken as 3.2 g/cm^3, r is the radius of the test sample, mm.

2.2. Workpiece Surface Composition Testing

In order to explore the solid-phase chemical reaction between different solid-phase oxidants and SiC and their oxygen production mechanism, the chemical elemental composition of the sample surface before and after polishing was examined by Quanta 200 SEM and the accompanying OXFORDINCA250 energy spectrometer system (FEI, Hillsboro, OR, USA). In addition, the chemical structure composition of the sample surface before and after polishing was examined by Bruker D8 Advance A25 XRD (Bruker, Billerica, MA, USA).

3. Analysis of Results

3.1. Elements and Content of SiC Surface after Polishing

The percentage of surface oxygen element content on the surface of SiC before and after dry polishing with five solid-phase oxidants was detected by SEM, and the results showed that the initial surface of SiC didnot contain oxygen before polishing. The atomic percentages of surface C and Si are shown in Table 3. After testing, oxygen appeared on the surface of SiC after dry polishing with five solid-phase oxidants.The percentage of oxygen atoms is shown in Figure 3.

Table 3. Initial surface elements and content of SiC.

Element	C (%)	Si (%)
Atomic percentage of SiC surface elements	45.67	54.33

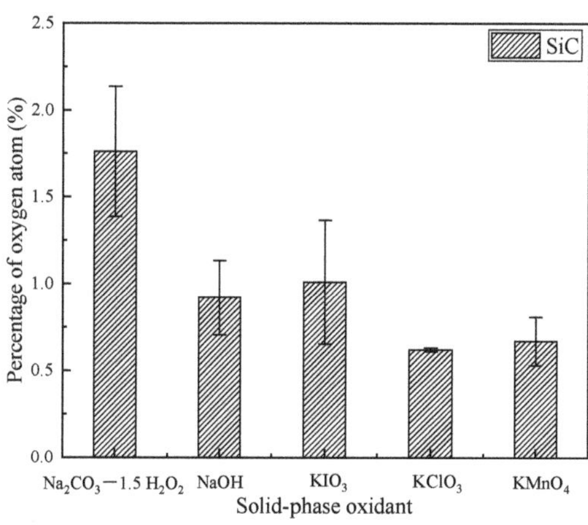

Figure 3. Percentage of oxygen atoms on SiC surface after dry polishing.

The occurrence of the tribochemical reaction of SiC can be reflected by the change of the atomic percentage on its surface, as shown in Figure 3. After polishing with five solid-phase oxidants, although oxygen was produced on the surface, the oxygen atomic percentage content varied, indicating that different degrees and mechanisms of tribochemical reactions occur between the five solid-phase oxidants and SiC. The highest percentage of oxygen atomson the surface reaction layer was observed after the dry polishing of SiC with the solid-phase oxidant Na$_2$CO$_3$—1.5 H$_2$O$_2$, and the lowest percentage of oxygen atoms in the surface of SiC was observed after dry polishing with the solid-phase oxidant KClO$_3$.

According to the SEM analysis, the appearance of oxygen on the surface of SiC indicated that the solid-phase oxidant could generate oxygen to oxidize the SiC surface under the action of frictional heat to produce an oxidation reaction film on the SiC surface.

Therefore, it has been shown that SiC can generate a more shearable reaction film by tribochemical reaction at room temperature [20,21].

3.2. Physical Phase Analysis of SiC Surface after Polishing

The XRD results of SiC after dry polishing were compared with those of SiC before dry polishing, as shown in Figure 4. After importing the XRD data before and after dry polishing into Jade software, it wasfound that the same peaks appeared between 30° and 40° and between 70° and 90°. The peak at 32° may be SiO_4 after software comparison analysis, and the intensity of the detected peak on the surface of the initial SiC was small. No oxygen appears, indicating that the content on the surface of SiC is small and not easy to detect. The intensity of the peak increased after dry polishing, indicating that silicon oxides were generated on the surface.

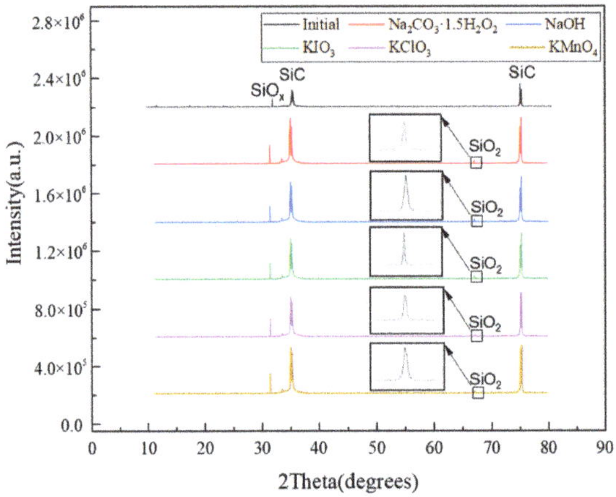

Figure 4. XRD of SiC after dry polishing with solid-phase oxidant.

The appearance and change of some micropeaks in the detection results may be $(SiO_2)_x$, a class of microporous silicate inclusion compounds with a (4,2)-3D structure [22]. In the dry polishing test, a surface tribochemical reaction generated a layered structure.This layered structure may be due to oxidation during the solid-phase oxidant and SiC test period under the thermal effect of tribochemical reaction transformation of oxidation substances.This layered structure includes a multi-functional layer useful for redox, friction reduction, and anti-wear functions [23,24].

3.3. Material Removal Rate after Polishing

Figure 5 show the material removal rates of SiC after dry polishing with five different solid-phase oxidants. The results showed that the five different solid-phase oxidants used for the tribochemical mechanical polishing tests all produce material removal from the SiC, indicating that the five solid-phase oxidants used in the tests may have experienced tribochemical reactions with the workpiece material. Among them, the highest material removal rate was achieved with the solid-phase oxidant NaOH and the lowest with the solid-phase oxidant $KClO_3$.

The highest material removal rate after dry polishing of SiC with NaOH is due to the tribochemical reaction between SiC and NaOH during the dry polishing process to generate CO and CO_2 released in the air. On the other hand, because silicon oxides and water-soluble silicates are easily removed by mechanical action, they are also generated on the surface of SiC.

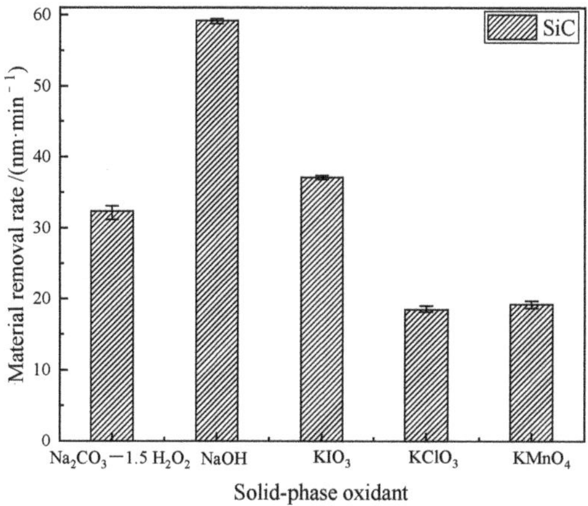

Figure 5. Material removal rate after dry polishing.

3.4. Surface Roughness and Surface Morphology after Polishing

Figure 6 show the changes in surface roughness before and after the dry polishing of SiC using five different solid-phase oxidants. The results showed that the tribochemical polishing tests using five different solid-phase oxidants all affect the surface roughness of the SiC, indicating that the tribochemical interaction between the five solid-phase oxidants used in the tests and the workpiece material all cause the removal of some material from the workpiece surface, thus changing its surface roughness. Among them, the surface roughness of SiC after the action of NaOH increased significantly, while the surface roughness of SiC after the action of other solid-phase oxidants increased slightly but not significantly.

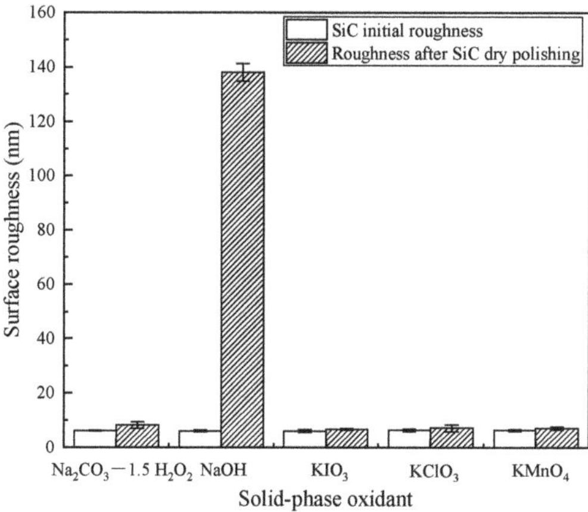

Figure 6. Surface roughness.

The comparison of SEM before and after the dry polishing of SiC with solid-phase oxidant is shown in Figure 7. The surface of SiC after dry polishing with NaOH showed a significant change in pits and scratches compared to the initial morphology.

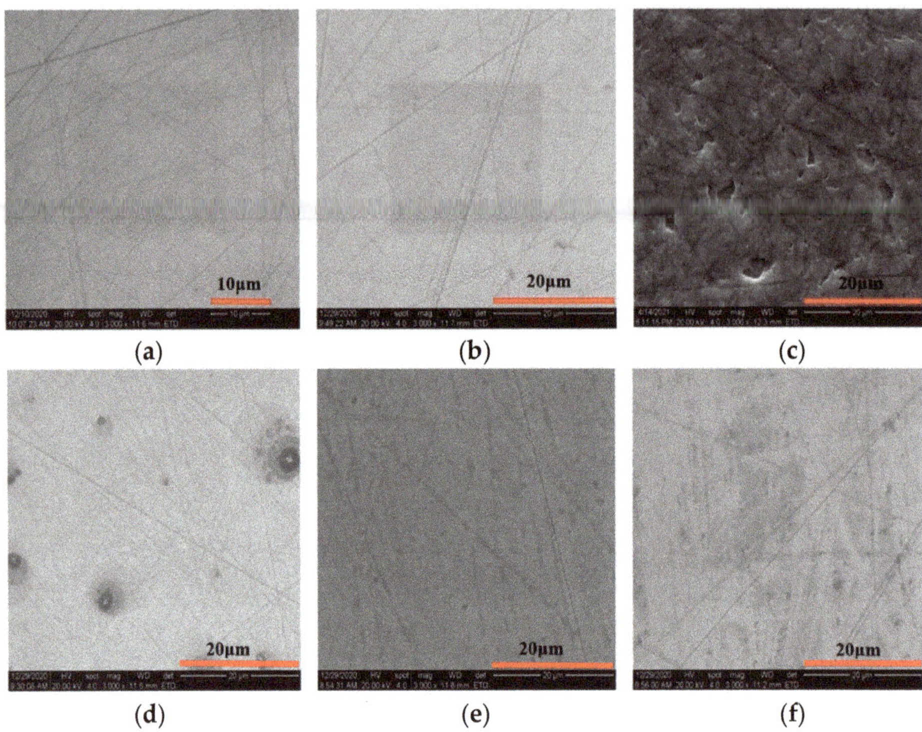

Figure 7. SEM of SiC surface after dry polishing with solid-phase oxidant. (**a**) Initial. (**b**) Na$_2$CO$_3$—1.5 H$_2$O$_2$. (**c**) NaOH. (**d**) KIO$_3$. (**e**) KClO$_3$. (**f**) KMnO$_4$.

4. Discussion

4.1. Solid-Phase Oxidant Tribochemical Reaction Oxygen Generation Mechanism

From Figure 2, it can be seen that the solid-phase oxidant fills between the SiC and the polishing pad during the polishing process, but under the polishing pressure, the SiC specimen and the polishing pad or solid-phase oxidant can be in contact at the micro-convex body [25]. Friction, local compression, or micro-collisions may occur on the micro-convex body at the polishing interface, which will generate concentrated local stresses at the point of contact (several gigapascals [26])and high flash point temperatures(up to 1000 degrees Celsius [27,28]). Then, under the action of friction and high flash point temperatures, etc., the solid-phase oxidant decomposes oxygen and oxidizes SiC or reacts with SiC by friction chemistry with other media [25,29–31].

Sodium percarbonate (Na$_2$CO$_3$—1.5 H$_2$O$_2$) is an inorganic substance and white granular solid commonly known as solid H$_2$O$_2$; it is a strong oxidant. It is easy to separate out oxygen when exposed to moisture to obtain Na$_2$CO$_3$, H$_2$O, and O$_2$.In addition, sodium percarbonate is a heat-sensitive substance, dry Na$_2$CO$_3$—1.5 H$_2$O$_2$ at 120 °C decomposition. However, in the presence of water, heat, or if mixed with heavy metal and organic material, it is very easy to decompose into Na$_2$CO$_3$, H$_2$O, and O$_2$, and its stability decreases with the rise of temperature [32,33]. See Equation (2).

$$\mathrm{Na_2CO_3-1.5\ H_2O_2(2Na_2CO_3-3\ H_2O_2)} \rightarrow 4\mathrm{Na_2CO_3} + 6\mathrm{H_2O} + 3\mathrm{O_2}\uparrow\ (120\ ^\circ\mathrm{C}) \qquad (2)$$

Studies have shown that the decomposition of sodium percarbonate is an autocatalytic mechanism. In the decomposition of sodium percarbonate, H$_2$O is the main catalyst [34]. The product of sodium percarbonate decomposition diffuses to the reaction interface to form intermediates with the reactants, which reduces the activation energy of the reaction and accelerates the reaction. It can be considered that the autocatalytic decomposition of sodium percarbonate proceeds in the following steps.

$$Na_2CO_3{-}1.5\,H_2O_2 \rightarrow [Na_2CO_3 \cdots 1.5\,H_2O_2{-}H_2O] \tag{3}$$

$$[Na_2CO_3 \cdots 1.5\,H_2O_2{-}H_2O] \rightarrow Na_2CO_3 + [1.5\,H_2O_2{-}H_2O] \tag{4}$$

$$[1.5\,H_2O_2{-}H_2O] \rightarrow 2.5H_2O + 0.75O_2\uparrow \tag{5}$$

The $Na_2CO_3{-}1.5\,H_2O_2$ molecule first combines with H_2O to form the activation complex $[Na_2CO_3{-}1.5\,H_2O_2{-}H_2O]$, which is unstable and quickly decomposes into Na_2CO_3. The reactive intermediate $[1.5\,H_2O_2{-}H_2O]$, $[1.5\,H_2O_2{-}H_2O]$ is also unstable and decomposes into H_2O and O_2. Where Equation (3) proceeds slowly, Equations (4) and (5) proceed more rapidly and reach an equilibrium quickly [33].

Sodium hydroxide (melting point is 318.4 °C, the boiling point is 1390 °C) powder will turn into molten sodium hydroxide under the action of frictional heat. In addition, the oxidant sodium hydroxide is easily deliquesced in air and reacts with CO_2 to form Na_2CO_3 and H_2O [35]. See Equation (6).

$$2NaOH + CO_2 \rightarrow Na_2CO_3 + H_2O \tag{6}$$

Potassium chlorate ($KClO_3$) is an inorganic compound, a colorless or white crystalline powder, that is a strong oxidantandis stable at room temperature. When heated to approximately 360 °C (the melting point of potassium chlorate), oxygen is released, and the reaction mechanism can be expressed in Equation (7). At continuous heating to 610 °C, the rate of oxygen release becomes slower, and the viscosity of the system thickens. At this point, the reaction is as in Equation (8); that is, potassium chlorate is oxidized to potassium perchlorate ($KClO_4$) by self-disproportionation. Equations (7) and (8) occur objectively at the same time, and when further heating to 800 °C is conducted, oxygen is released again until the system is completely changed to potassium chloride, such as Equation (9) [36].

$$2KClO_3 \rightarrow 2KCl + 3O_2\uparrow \; (365 \pm 5\,°C) \tag{7}$$

$$KClO_3 \rightarrow KClO_4 + KCl \tag{8}$$

$$KClO_4 \rightarrow KCl + O_2\uparrow \tag{9}$$

Potassium permanganate ($KMnO_4$) is a strong oxidant with a melting point of 240 °C. Its thermal decomposition process is complex, and within 190–700 °C, the following decomposition reactions are produced [37–39].

$$6KMnO_4 \rightarrow 2K_2MnO_4 + K_2Mn_4O_8 + 4O_2\uparrow \tag{10}$$

$$2KMnO_4 \rightarrow KMnO_2 + O_2\uparrow \tag{11}$$

$$3K_2MnO_4 \rightarrow 2K_3MnO_4 + MnO_2 + O_2\uparrow \tag{12}$$

In addition, light has a catalytic effect on the decomposition of potassium permanganate, $KMnO_4$ is not very stable in sunlight, and $KMnO_4$ can spontaneously undergo redox reactions with H_2O [40].

$$4KMnO_4 + 2H_2O \rightarrow 4KOH + 4MnO_2 + 3O_2\uparrow \tag{13}$$

Potassium iodate (KIO_3) is an inorganic substance. It is a colorless crystal, and its melting point is 560 °C (decomposition). It can be decomposed into KI by heat; KI reacts with O_2 and H_2O in moist air to form KOH [41,42].

$$2KIO_3 \rightarrow 2KI + 3O_2\uparrow \; (525\,°C) \tag{14}$$

$$4KI + O_2 + 2H_2O \rightarrow 2I_2 + 4KOH \tag{15}$$

4.2. Mechanism of Tribochemical Oxidation Reaction on the Surface of SiC

(1)   Solid-phase oxidant NaOH

See Equation (6), the solid-phase oxidant NaOH readily deliquesces in air and reacts with CO_2 to form Na_2CO_3 and H_2O [35]. In addition, under the action of friction, the suspended silicon bond in SiC will also undergo the following tribochemical reaction. The main chemical equation is as follows [42,43]:

$$Si + 2NaOH + H_2O \rightarrow Na_2SiO_3 + 2H_2\uparrow \tag{16}$$

The Na_2SiO_3 produced by the reaction is soluble and can be easily removed from the SiC surface by mechanical action.

(2)   Other solid-phase oxidants

From Section 3.1, Na_2CO_3—1.5 H_2O_2, KIO_3, $KClO_3$, and $KMnO_4$ can produce O_2 by decomposition under the action of friction heat, then the surface of the SiC undergoes a tribochemical oxidation reaction under the action of frictional heat and other media [20,21].

$$SiC + 2O_2 \rightarrow SiO_2 + CO_2\uparrow \tag{17}$$

(3)   Tribochemical hydration reaction on SiC surface

During the polishing process, due to the high flash point temperature at the polishing interface, SiC reacts with H_2O to produce SiO_2 on the SiC surface. The main chemical reactions are as follows [44–47]:

$$SiC + 2H_2O \rightarrow SiO_2 + C + 2H_2\uparrow \tag{18}$$

$$SiC + 4H_2O \rightarrow SiO_2 + CO_2 + 4H_2\uparrow \tag{19}$$

$$SiC + O_2 + H_2O \rightarrow SiO_2 + CO\uparrow + H_2\uparrow \tag{20}$$

The flash point temperature during polishing excites the oxidation reaction in Equations (18)–(20). Therefore, the temperature of the test environment is so low that it does not affect the occurrence of thetribochemical reaction and has little effect on friction behavior [48,49].

Thus, as described above, the surface of SiC transforms into SiO_2, Na_2SiO_3, or a surface film composed of SiO_2 and Na_2SiO_3 [26,42,43]. The resulting product, regardless of the state in which the generated product exists, is less hard than SiC. This oxide layer is easily removed using abrasive.

4.3. Material Removal Mechanism of Solid-Phase Oxidant

From Figures 3 and 4, after polishing SiC with five solid-phase oxidants, it was found that the surface of SiC contained oxygen, and the surface products of SiC are SiO_2 and silicon oxides. This illustrates the complex tribochemical reactions generated at the polishing interface during the polishing process, see Equations (2)–(20).Moreover, SiO_2 on the surface of SiC is obtained by the tribochemical reaction shown in Equations (17)–(20).

4.3.1. Solid-Phase Oxidant Sodium Hydroxide (NaOH)

(1)   The surface phases of SiC after polishing all contain oxygen, and the surface compounds are SiO_2 and silicon oxides. On the one hand, the solid-phase oxidant NaOH reacts with CO_2 in the air under the action of friction to form Na_2CO_3 and H_2O, see Equation (6).On the other hand, the SiC reacts with H_2O in the air to form SiO_2 on its surface, see Equations (18)–(20).The CO and CO_2 generated by the reaction escape into the air, and the generated C is removed by friction; the SiO_2 generated is attached

to the SiC surface. Figures 8 and 9 show the surface morphology of SiC after dry polishing with NaOH.

(2) Under the action of friction, the suspended silicon bonds of SiC also undergo a tribochemical reaction with NaOH, see Equation (16). The reaction produces soluble Na_2SiO_3, which is removed from the SiC surface. As can be seen from the SEM inspection of the enlarged Figures 7c and 9b, after polishing, more small pits appear on the SiC surface with a smoother edge, not just brittle fracture removal. It can be shown that the Na_2SiO_3 produced by the reaction is removed by dissolution.

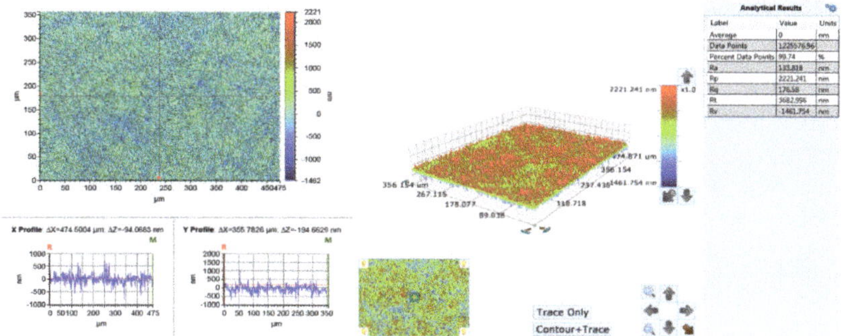

Figure 8. Surface morphology after dry polishing with NaOH.

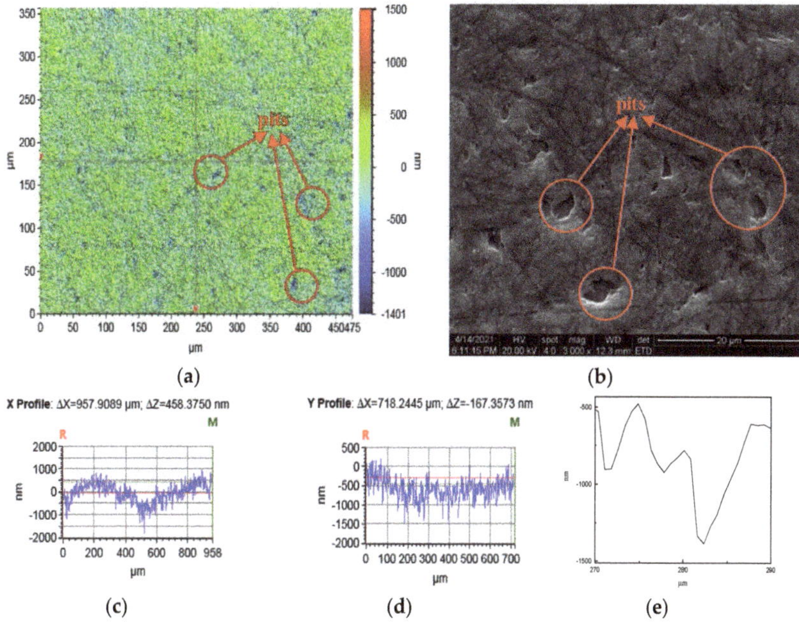

Figure 9. Surface pits after dry polishing with NaOH. (**a**) White light interferometer detection of surface pits. (**b**) SEM inspection of surface pits. (**c**) Partial X coordinate shape. (**d**) Local Y coordinate shape. (**e**) Local enlargement of the pit.

4.3.2. Solid-Phase Oxidant Sodium Percarbonate (Na_2CO_3—1.5 H_2O_2)

The solid-phase oxidant Na_2CO_3—1.5 H_2O_2 decomposes into Na_2CO_3, H_2O, and O_2 under the action of friction, see Equation (2), and the products, in turn, produce a tribochemical reaction with SiC, as shown in Equations (17)–(20).The generated SiO_2 adheres to the surface of SiC, the generated CO and CO_2 escape into the air, and the generated C is removed by friction.The material removal rate consists of the products C, CO, and CO_2; however, the Si atoms in the SiC are not lost, but partly oxidized to SiO_2.Therefore, the removal rate is lower than that of the solid-phase oxidant NaOH, see Figure 5.

4.3.3. Solid-Phase Oxidant Sodium Iodate (KIO_3)

Under frictional heat, KIO_3 produces O_2 and generates KI, see Equation (14), which in turn reacts with the SiC surface in an oxidation reaction, see Equation (17). Furthermore, KOH is generated from the reaction of KI with O_2 and H_2O in the air, which can also provide an alkaline environment to induce the SiC to react with O_2, see Equation (15). Since there is no H_2O in the KIO_3 decomposition reaction equation, SiC may also react with H_2O in the air by frictional chemistry under the action of frictional heat, see Equations (18)–(20). However, the Si atoms in SiC are also not lost but partially oxidized to SiO_2.

The SiO_2 generated by the tribochemical reaction adheres to the SiC surface, and the generated gas escapes into the air, thus creating a material removal rate.

4.3.4. Solid-Phase Oxidant Sodium Chlorate ($KClO_3$)

Under frictional heat, $KClO_3$ decomposes to produce O_2 and generates KCl, see Equations (7)–(9).The resulting O_2 reacts with the SiC surface in an oxidation reaction, see Equation (17). The SiO_2 generated by the tribochemical reaction adheres to the SiC surface, and the generated CO_2 escapes into the air, thus creating a material removal rate.

Since there is no H_2O in the $KClO_3$ decomposition reaction equation, SiC may also react with H_2O in the air by frictional chemistry under the effect of frictional heat, see Equations (18)–(20). However, the Si atoms in the SiC are not lost, but partially oxidized to SiO_2.

4.3.5. Solid-Phase Oxidant Sodium Permanganate ($KMnO_4$)

Under the action of frictional heat, $KMnO_4$ decomposes to produce O_2 and generates KCl, see Equations (10)–(12). In turn, this reacts with the SiC surface in an oxidation reaction, see Equation (17). At the same time, $KMnO_4$ can spontaneously react with H_2O in the air to produce O_2 via a redox reaction.

The SiO_2 generated by the tribochemical reaction adheres to the SiC surface, and the generated CO_2 escapes into the air, thus creating a material removal rate.

Since there is no H_2O in the $KMnO_4$ decomposition reaction equation, SiC may also produce a tribochemical reaction with H_2O in the air under the effect of friction heat, see Equations (18)–(20). However, Si atoms in SiC are not lost, but partly oxidized to SiO_2.

To sum up, under the action of friction, a more complex tribochemical reaction was produced between the solid-phase oxidant NaOH and SiC and air medium, not only the removal of C atoms, but also Si atoms. Comparatively, in the other reactions, only the C atoms were removed. Therefore, the largest removal rate wasproduced when polishing with the solid-phase oxidant NaOH, see Figure 5. In addition, since Na_2SiO_3 was produced when the solid-phase oxidant NaOH was used for polishing, and Na_2SiO_3 was easily removed by dissolution, more small pits with a depth of 1400 nm were produced on the surface, see Figure 9. As a result, the polishing surface roughness was also at a maximum when the solid-phase oxidant NaOH was used, see Figure 6.

5. Conclusions

(1) After dry polishing SiC with all five solid-phase oxidants, oxygen was detected on the surface, but the percentage of oxygen atoms on the surface after polishing varied. The highest percentage of oxygen atoms was observed after dry polishing SiC with Na_2CO_3—1.5 H_2O_2 and the lowest percentage of oxygen atoms was observed on the surface after dry polishing with $KClO_3$.

(2) From the XRD results, it couldbe seen that the appearance of surface oxygen was due to the tribochemical reaction between the five solid-phase oxidants and the SiC in the polishing process. The reaction product was known to be silicon oxides, and the main substance was SiO_2. In addition, under the action of friction, due to the high flash point temperature at the polishing interface, SiC reacted with H_2O and generated a SiO_2 oxide layer on the SiC surface.

(3) The material removal rate was calculated by measuring the mass before and after polishing, and the highest material removal rate couldbe obtained after dry polishing of SiC with NaOH and the lowest material removal rate could be obtained after dry polishing with $KClO_3$.

(4) After polishing SiC with oxidant NaOH, soluble Na_2SiO_3 was generated. Therefore, more obvious scratches and pits appeared on the surface of SiC, and the roughness hada substantial increase.The surface roughness of the remaining four solid-phase oxidants didnot change significantly after polishing.

Author Contributions: Conceptualization, J.S. and W.Q.; methodology, W.Q.; software, W.Q.; validation, W.Q., W.X. and X.C.; formal analysis, Z.W.; investigation, W.Q.; resources, W.Q.; data curation, W.Q.; writing—original draft preparation, W.Q.; writing—review and editing, W.Q., J.S. and Z.W.; visualization, W.Q.; supervision, J.S., Z.W.; project administration, J.S.; funding acquisition, J.S. All authors have read and agreed to the published version of the manuscript.

Funding: This work was supported by the National Natural Science Foundation of China (No. U1804142) and the Science and Technology Research Project of Henan Province (No. 192102210058).

Conflicts of Interest: The authors declare no conflict of interest.

References

1. Lagudu, U.R.; Isono, S.; Krishnan, S.; Babu, S.V. Role of ionic strength in chemical mechanical polishing of silicon carbide using silica slurries. *Colloids Surf. A Physicochem. Eng. Asp.* **2014**, *445*, 119–127. [CrossRef]
2. Zhao, Q.; Sun, Z.; Guo, B. Material removal mechanism in ultrasonic vibration assisted polishing of micro cylindrical surface on SiC. *Int. J. Mach. Tools Manuf.* **2016**, *103*, 28–39. [CrossRef]
3. Li, B. Investigation on Wide Bandgap Semiconductor Material. *Equip. Electron. Prod. Manuf.* **2010**, *39*, 5–10.
4. Yamaoka, H.; Uruga, T.; Arakawa, E.; Matsuoka, M.; Ogasaka, Y.; Yamashita, K.; Ohtomo, N. Development and Surface Evaluation of Large SiC X-ray Mirrors for High-Brilliance Synchrotron Radiation. *Jpn. J. Appl. Phys.* **1994**, *33*, 6718–6726. [CrossRef]
5. Jenny, J.R.; Skowronski, M.; Mitchel, W.C.; Hobgood, H.M.; Glass, R.C.; Augustine, G.; Hopkins, R.H. Vanadium related near-band-edge absorption bands in three SiC polytypes. *J. Appl. Phys.* **1995**, *78*, 3160–3163. [CrossRef]
6. Zhuang, D.; Edgar, J.H. Wet etching of GaN, AlN, and SiC: Areview. *Mater. Sci. Eng. R* **2005**, *48*, 1–46. [CrossRef]
7. Han, R. Research Status of Third generation semiconductor materials SiC planarization technology. *Chem. Intermed.* **2018**, *11*, 73–75.
8. Jiang, P.; Wang, C.; Kang, R.; Guo, D. Design and realization of UML-based CMP bontrol system. *Equip. Electron. Prod. Manuf.* **2008**, *8*, 48–52.
9. Zhang, W.; Yang, X.; Shang, C.; Hu, X.; Hu, Z. High Speed Lapping of SiC Ceramic Material with Solid (Fixed) Abrasives. *J. China Ordnance* **2005**, *1*, 225–228.
10. Zhu, Y.; Wang, J.; Li, J. Research on the polishing of silicon substrate by fixed abrasive pad. *China Mech. Eng.* **2009**, *20*, 723–727.
11. Zhu, Y.; He, J. A model of material removal rate with fixed abrasive pad. *Dimond Abras. Eng.* **2006**, *153*, 39–45.
12. Li, P.; Li, J.; Wang, J. Study on the lapping process of sapphire substrate through fixed abrasive. *J. Synth. Cryst.* **2013**, *42*, 2258–2264.
13. Zhong, X. *Development of Wmocr Polishing Plate Used in Dynamic Friction Polishing of Diamond*; Dalian University of Technology: Dalian, China, 2015.
14. Li, J. Recent advances and trends in tribology. *Lubes Fuels* **2007**, *17*, 10.
15. Muratov, V.A.; Fischer, T.E. Tribochemical Polishing. *Annu. Rev. Mater. Sci.* **2002**, *30*, 27–51. [CrossRef]
16. Zhu, Z.; Muratov, V.; Fischer, T.E. Tribochemical polishing of silicon carbide in oxidant solution. *Wear* **1999**, *225–229*, 848–856. [CrossRef]
17. Kitaoka, S.; Tsuji, T.; Yamaguchi, Y. Tribochemical wear theory of non-oxide ceramics in high temperature and high-pressure water. *Wear* **1997**, *205*, 40–46. [CrossRef]
18. Otani, Y.; Xu, J.; Adachi, K. First-Principles Molecular Dynamics Study of Silicon-Based Ceramics: Different Tribochemical Reaction Mechanisms during the Running-in Period of Silicon Nitride and Silicon Carbide. *J. Phys. Chem. C* **2020**, *124*, 20079–20089. [CrossRef]
19. Zhou, F.; Kato, K.; Adachi, K. Friction and wear properties of CNx/SiC in water lubrication. *Tribol. Lett.* **2005**, *18*, 153–163. [CrossRef]

20. Zhou, S.; Xiao, H. Study of high-temperature friction chemistry of silicon carbide and its complex-phase ceramics. *China Ceram. Ind.* **2002**, *9*, 16–20.

21. Zhou, S.; Xiao, H. Tribo-chemistry and wear map of silicon carbide ceramics. *J. Chin. Ceram. Soc.* **2002**, *30*, 641–644.

22. Liebau, F. Zeolites and clathrasils—Two distinct classes of framework silicates. *Zeolites* **1983**, *3*, 191–193. [CrossRef]

23. Tieu, A.K.; Wan, S.; Kong, N. Excellent melt lubrication of alkali metal polyphosphate glass for high temperature applications. *RSC Adv.* **2014**, *5*, 1796–1800. [CrossRef]

24. Tieu, A.K.; Kong, N.; Wan, S. The Influence of Alkali Metal Polyphosphate on the Tribological Properties of Heavily Loaded Steel on Steel Contacts at Elevated Temperatures. *Adv. Mater. Interfaces* **2015**, *2*, 1500032. [CrossRef]

25. Matsuda, M.; Kato, K.; Hashimoto, A. Friction and Wear Properties of Silicon Carbide in Water from Different Sources. *Tribol. Lett.* **2011**, *43*, 33–41. [CrossRef]

26. Presser, V. *Oxidation and Wet Wear of Silicon Carbide*; University Tübingen: Tübingen, Germany, 2009.

27. Kuhlmann-Wilsdorf, D. Flash temperatures due to friction and Joule heat at asperity contacts. *Wear* **1985**, *105*, 187–198. [CrossRef]

28. Quinn, T. Computational methods applied to oxidational wear. *Wear* **1996**, *199*, 169–180. [CrossRef]

29. Tomizawa, H.; Fischer, T.E. Friction and Wear of Silicon Nitride and Silicon Carbide in Water: Hydrodynamic Lubrication at Low Sliding Speed Obtained by Tribochemical Wear. *ASLE Trans.* **1986**, *30*, 41–46. [CrossRef]

30. Xu, J.; Kato, K. Formation of tribochemical layer of ceramics sliding in water and its role for low friction. *Wear* **2000**, *245*, 61–75. [CrossRef]

31. Gates, R.S.; Hsu, S.M. Tribochemistry Between Water and Si3N4 and SiC: Induction Time Analysis. *Tribol. Lett.* **2004**, *17*, 399–407. [CrossRef]

32. Qin, F.; Gu, D. Thermal Decomposition Reaction Kinetics of Sodium Percarbonate. *J. East China Univ. Sci. Technol. (Nat. Sci.)* **2001**, *27*, 214–216.

33. Wang, H.; Zhao, H.; Qu, G. Research on Isothermal Decomposition Kinetics of Sodium Percarbonate. *J. Henan Norm. Univ. (Nat. Sci.)* **2001**, *29*, 58–61.

34. Galwey, A.K.; Hood, W.J. Thermal decomposition of sodium carbonate perhydrate in the solid state. *Chem. Inf.* **1979**, *10*, 1810–1815. [CrossRef]

35. Wang, Q.; Li, J.; Zhao, Y. Study on the dealkalization of red mud by suspension and carbonation. *Chin. J. Environ. Eng.* **2009**, *3*, 2275–2280.

36. Li, Z. A Study on the Origination of the Laboratory Method of Oxygen Preparation by Potassium Chlorate and Manganese Dioxide. *Univ. Chem.* **2020**, *35*, 268–273.

37. Huang, H.; Li, Y. Research of the Thermal Decomposition Process of Potassium Permanganate. *Chin. J. Chem. Educ.* **2009**, *30*, 65–66.

38. Tan, R.; Chen, G. A Study on the Mechanism of KMnO4 Thermolysis. *J. Guizhou Norm. Univ. (Nat. Sci.)* **1990**, *1*, 20–25.

39. Mao, L. *The Thermal Decomposition of KMnO4, KClO3 and Other Inorganic Solids*; Guangxi Normal University: Guilin, China, 2008.

40. Wang, G.; Zhong, X. Recovery and Utilization of By-products from Decomposition of Potassium Permanganate to Oxygen. *Educ. Chem.* **2000**, *12*, 41–42.

41. Liu, H.; Shen, T.; Cao, W. Stability of the salt iodization agent potassium iodate: A differential thermal analysis. *Chin. J. Endem.* **2012**, *31*, 684–686.

42. Chen, X.; Juan, L.I.; Deying, M.A. Fine Machining of Large-Diameter 6H-SiC Wafers. *J. Mater. Sci. Technol.* **2006**, *22*, 681–684.

43. Ming, J.; Komanduri, R. On the finishing of Si3N4 balls for bearing applications. *Wear* **1998**, *215*, 267–278.

44. Kitaoka, S.; Tsuji, T.; Katoh, T. Tribological Characteristics of SiC Ceramics in High-Temperature and High-Pressure Water. *J. Am. Ceram. Soc.* **1994**, *77*, 1851–1856. [CrossRef]

45. Zhou, L. Chemomechanical polishing of silicon carbide. *Jelectrochemsoc* **1997**, *144*, 161–163. [CrossRef]

46. Li, J.; Huang, J.; Tan, S. Tribological properties of toughened SiC ceramics under water-lubricated sliding. *J. Chin. Ceram. Soc.* **1998**, *26*, 305–312.

47. Hao, L. *Study on Electrical Friction Characteristics of Silicon Carbide Ceramics during Polishing*; Harbin Institute of Technology: Harbin, China, 2008.

48. Jin, L.; Scheerer, H.; Andersohn, G. Experimental study on the tribo-chemical smoothening process between self-mated silicon carbide in a water-lubricated surface contact reciprocating test. *Friction* **2019**, *7*, 181–191. [CrossRef]

49. Jin, L.; Scheerer, H.; Andersohn, G.; Oechsner, M. Tribological behavior of surface contact friction test with water lubrication at elevated temperatures. *Tribol. Und Schmier.* **2017**, *64*, 11–17.

 *micromachines*

Article

Modeling and Simulation of the Surface Generation Mechanism of a Novel Low-Pressure Lapping Technology

Ninghui Yu [1], Lihua Li [2],* and Chea-su Kee [3]

[1] Shenzhen Research Institute, The Hong Kong Polytechnic University, Shenzhen 518057, China; ninghui-a.yu@polyu.edu.hk

[2] Sino-German College of Intelligent Manufacturing, Shenzhen Technology University, Shenzhen 518118, China

[3] School of Optometry, The Hong Kong Polytechnic University, Hong Kong, China; c.kee@polyu.edu.hk

* Correspondence: lilihua@sztu.edu.cn

Abstract: Aluminum alloy (Al6061) is a common material used in the ultraprecision area. It can be machined with a good surface finish by single-point diamond turning (SPDT). Due to the material being relatively soft, it is difficult to apply post-processing techniques such as ultraprecision lapping and ultraprecision polishing, as they may scratch the diamond-turned surface. As a result, a novel low-pressure lapping method was developed by our team to reduce the surface roughness. In this study, a finite element model was developed to simulate the mechanism of this novel lapping technology. The simulation results were compared with the experimental results so as to gain a better understanding of the lapping mechanism.

Keywords: ultraprecision lapping; Al6061; FEM

Citation: Yu, N.; Li, L.; Kee, C.-s. Modeling and Simulation of the Surface Generation Mechanism of a Novel Low-Pressure Lapping Technology. *Micromachines* **2021**, *12*, 1510. https://doi.org/10.3390/mi12121510

Academic Editors: Benny C. F. Cheung and Jiang Guo

Received: 12 November 2021
Accepted: 2 December 2021
Published: 4 December 2021

Publisher's Note: MDPI stays neutral with regard to jurisdictional claims in published maps and institutional affiliations.

1. Introduction

Precision lapping and polishing are difficult for the polishing of softer metals such as aluminum alloys. As shown in Figure 1, many scratches occur on the aluminum alloy mirror surface after ultraprecision bonnet polishing. For aluminum and its alloys, single-point diamond turning (SPDT) is commonly used because of its good machinability. It is important to develop a method to remove the tool marks that are inevitably generated during the SPDT process (Figure 2). Hence, we developed a precision low-pressure lapping method for softer metals [1]. This method uses a brush to drive abrasive particles to roll and slide on the workpiece and has a very a small material removal rate as compared with other conventional precision lapping and polishing methods. The traditional lapping model based on the Preston formula is not suitable for this method. In the general lapping and polishing process, the common mechanisms are microcutting, delamination, slurry erosion, chemical mechanical polishing, etc. [2]. In our study, the brush touches the lapped surface very gently during the lapping process and drives the abrasive particles to move on the surface. The likely mechanism is that the abrasive particles receive large momentum during the lapping process. As a result, the abrasive particles impinge the asperities of the workpiece surface during the sliding and rolling process, causing plastic deformation of the asperities which are worn away, thereby reducing the surface roughness.

Since the asperities of the surface are removed by the impingement of the abrasive particles, the finite element method (FEM) can be used to model and simulate the surface generation in this low-pressure lapping. The simulation results provide a better understanding of the surface generation mechanism and can be used to design the lapping parameters of the actual experiment.

The finite element method (FEM) is an important analytical method for analyzing the deformation process of materials. Komvopoulus et al. used the finite element software Abaqus to simulate an area changing from elastic to plastic deformation under load [3]. This research shows that the material change process depends on strain hardening characteristics and accumulated plastic deformation at a specific load but has little to do with the elastic modulus. Kogut and Etsion studied the deformation process of the contact between a sphere and a rigid plane by the finite element simulation software ANSYS in 2002 [4]. The von Mises yield criterion was introduced to analyze the deformation process of the material, from elastic to plastic. The results showed that the elastic model cannot fully explain the elastic to plastic deformation process.

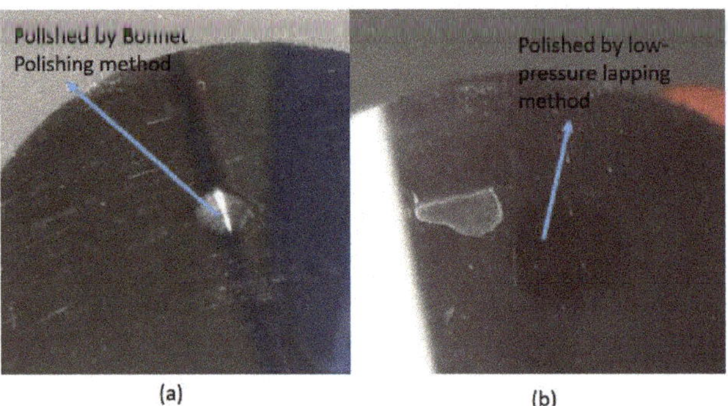

Figure 1. Result of Al6061 mirror surface polished by (**a**) Bonnet polishing method and (**b**) our low-pressure lapping method.

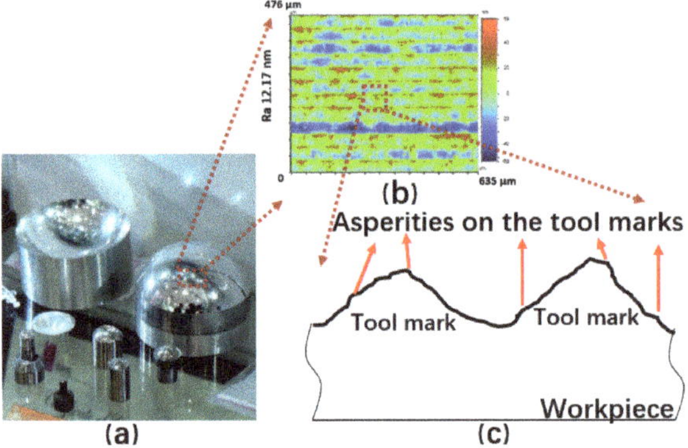

Figure 2. The aluminum alloy workpiece (**a**) looks shiny after SPDT but it actually has nanometer-level tool marks (**b**) on the surface. Asperities also exist on the tool mark (**c**).

The basic contact model proposed by Greenwood and Williamson in 1966 is the basis of many subsequent microcontact models [5]. In this study, it is assumed that the contact surface is rough, the asperity heights on the surface follow a Gaussian distribution with the same curvature, and the asperities do not affect each other when they are under pressure. Contact in this model makes use of the Hertzian approach to calculate the pressure distribution during the contact process. The significance of this model is that it provides a preliminary contact model and makes use of a plasticity index that determines when the material transforms from elastic deformation to plastic deformation under a specific load.

Another classic model was established by Chang et al. based on the Chang, Etsion, and Bogy (CEB) elastic–plastic model by Tabor [6,7] and was the first attempt to determine the boundary state of a material from elastic to plastic. This model and the Greenwood and Williamson model provide good simulation for the contact process in a real contact area at very high or very low plasticity indices [5].

Using the Greenwood and Williamson model, Jackson and Green calculated the yield point of the material on the basis of the asperity contact and the von Mises criterion [8]. Kadin et al. found that the yield point of the material was greatly affected by Poisson's ratio and strain hardening [9]. Peng et al. simulated the elastic–plastic contact process of rough surfaces by the finite element method [10]. The simulation model involves asperity contact with a rigid body. In 2017, Almuramady and Borodich conducted a theoretical and experimental comparison of the plastic behavior of an actual rough surface [11]. Although researchers have been continually studying the elastic–plastic contact of rough surfaces by finite element methods, they are rarely used to simulate the actual lapping process.

Since the core of the low-pressure lapping method involves abrasive particles continuously impinging on the tool marks, which are deformed and worn away gradually. In this study, the surface generation mechanism was simulated using current elastic–plastic contact models.

2. Modeling Processes

2.1. Parameter Setting

The simulation software Abaqus was used in this study. In the simulation, two different heights of tool marks on the Al6061 surfaces were designed to be 1 µm and 20 nm. In order to reduce the calculation time, only the tip of the tool mark was used, which means that the intercepted tool mark heights were 1 nm and 0.02 nm, respectively, in this simulation. Due to the surface with a tool mark height of 20 nm being almost close to the plane, in this simulation, it could be assumed that each tool mark was independent, and the movement of the abrasive particles on the surface was not affected by the adjacent tool mark. In this simulation, the tool mark was a long-ridged shape form; thus, when the abrasive grains collided from the side, it could be simplified into a 2D tool mark model. According to the experiment, the diameter of the abrasive particles in this simulation was 20 nm. In order to make the operation converge, a plane surface was added to the starting side, as shown in Figure 3.

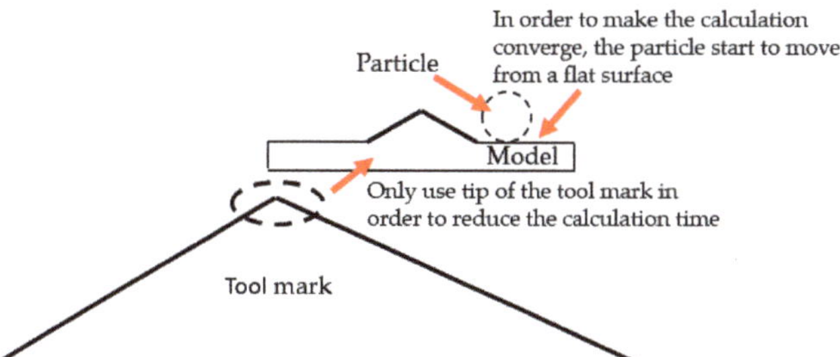

Figure 3. Schematic of the finite element model.

Nanometer-level thermal expansion was identified to be an issue for some technologies such as thermal-assisted machining [12], while others take advantage of it for novel tool-displacement technologies [13]. Since the lapping speed in this research was low and the fiber gap on the lapping head helped to dissipate heat to some degree, this simulation only had a mechanical relationship and no variables such as temperature and magnetism were involved. Accordingly, the only material parameters needed to be set were mass density, elastic behavior, and plastic behavior. In the Abaqus software, there is no unit system, and it is necessary to set the unit according to the content of the simulation. To ensure the correct operation result, the unit setting needs to ensure consistency; otherwise, the results have no actual physical meaning. In Abaqus, units can be divided into two types: fundamental units and derived units. A derived unit is a combination of fundamental units. Since this simulation was in the micrometer range, the common meter-scale units were converted into micrometer units (Table 1) and were used in the simulation. The workpiece material in this simulation was aluminum alloy (Al6061) with a density of 2.7×10^{-15} kg/μm³, Young's modulus of 6.89×10^{4} MPa, and yield stress of 276 MPa. In the assembly process, the initial position of the abrasive particle was on the right side of the tool mark, and the moving direction was from right to the left.

Table 1. Units used in the simulation.

Fundamental Units	
Length	mm
Force	μN
Time	s
Mass	kg
Derived Units	
Pressure	MPa
Velocity	μm/s
Density	kg/μm³
Young's modulus	MPa
Yield stress	MPa

In the simulation, Al6061 alloy was considered as an elastic–plastic material according to the Johnson Cook plasticity model [14]. The model can be expressed as

$$\overline{\sigma}= [A\ +\ B(\overline{\varepsilon}^{pl})^{n}][1+C\ln(\frac{\dot{\overline{\varepsilon}}^{pl}}{\dot{\varepsilon}_0})](1-\hat{\theta}^{m}),\tag{1}$$

where $\overline{\sigma}$ is the yield, $\overline{\varepsilon}^{pl}$ and $\dot{\overline{\varepsilon}}^{pl}$ are the equivalent plastic strain and plastic strain rate, and $A, B, C, n, m,$ and $\dot{\varepsilon}_0$ are material parameters. When the material temperature is lower than the transition temperature, the coefficients are as shown in Table 2.

Table 2. Johnson Cook coefficients.

A	B	C	n	m	$\dot{\varepsilon}_0$	θ_{melt}	$\theta_{transition}$
289.6 MPa	203.4 MPa	0.011	0.35	1.34	1.0 s^{-1}	925.37 K	294.26 K

When the abrasive particles slide and roll on the surface of the workpiece, the polishing time is t, the rotational speed is S, the wool is bent after contact with the workpiece, and the angle of bending is θ. Since the diameter d of the thickest part of the wool is about 100 μm, one fiber sweeps N times on point P, which can be expressed as follows (Figure 4):

$$N = \frac{2\pi(R - \tan\theta\cdot D)}{d}\cdot tS.\tag{2}$$

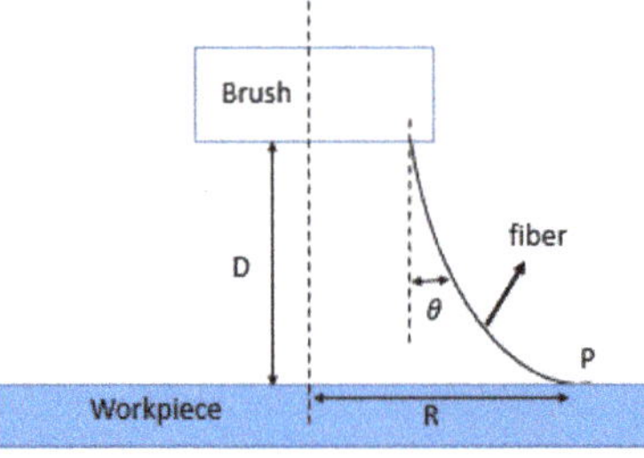

Figure 4. Schematic of the lapping process. One fiber on the brush is swept from the workpiece surface.

The speed at which the abrasive particle passes through the P point is

$$v = 2\pi\kappa \cdot S \cdot R, \tag{3}$$

where κ is a constant that represents how much of the fiber speed is transmitted to the abrasive particles.

The wear model for the nodes in this simulation was built based on the Archard model.

$$q = \frac{kPA\eta}{H}, \tag{4}$$

where q is the wear rate, k is a dimensionless constant obtained from experimental results and depends on several parameters such as surface quality and surface hardness process, P is the normal pressure, η is the interface slip rate, and H is the material hardness. In this simulation, the surface was expressed by the change of nodes on the surface. Each node transmitted force to the next node according to the software settings. The wear rate can be expressed as

$$q(t) = \frac{k}{H} \int P(x,t)\eta(x,t)dA, \tag{5}$$

where x is the node position, and t is the time. Then, as a function of the Eulerian steady-state transport procedure [15], the model can change to a time-independent expression.

$$q(t) = \frac{k}{H} \int P(u)\eta(u)T(u)du, \tag{6}$$

where u is the position along the edge of the grid, and $T(u)$ is the width of the adjacent grid at position u. The wear rate can be expressed as a function of local material change.

$$q(t) = \int w(u)T(u)du, \tag{7}$$

Since the software needs a discrete form expression, Equations (7) and (8) were combined and discretized as

$$\sum_{i=1}^{N} w_i A_i = \frac{k}{H} \sum_{i}^{N} P_i \eta_i A_i, \tag{8}$$

where w is the node wear velocity, and A_i is the node contact area. The expression for w is

$$w = \frac{k\sum_{i=1}^{N} P_i \eta_i A_i}{H\sum_{i=1}^{N} A_i}. \tag{9}$$

In this simulation, the momentum of the abrasive particles depended on the weight and speed of the abrasive particles. Since the individual abrasive particles were very small, the weight of the individual abrasive particles was about 5×10^{-18} kg according to the density of the abrasive material (2.65 g/cm^3) and its size. From Equation (3), the different speeds of the abrasive particles when R is 5 mm were determined, as shown in Table 3.

Table 3. Speed of the particle in the simulation.

1	Rotational speed: 200 rpm	6.28×10^6 μm/s
2	Rotational speed: 400 rpm	1.26×10^7 μm/s
3	Rotational speed: 600 rpm	1.88×10^7 μm/s
4	Rotational speed: 800 rpm	2.51×10^7 μm/s
5	Rotational speed: 1000 rpm	3.14×10^7 μm/s

2.2. Software Setting

In the preprocessing stage, a two-dimensional model was first established in Abaqus, and the model was meshed (Figure 5). The boundary conditions and contact conditions of the contact model were then defined, and the abrasive particles were given an initial velocity. In the process of simulating the elastic–plastic contact between the abrasive particles and the surface of the workpiece, the contact process of the finite element simulation was a discontinuous constraint behavior, allowing the forces of the various nodes to be transmitted to other nodes. The premise of this constraint is that the two surfaces are to make contact. As a result, when analyzing the contact process, it is necessary to confirm that the two surfaces have been in contact and produced constraints.

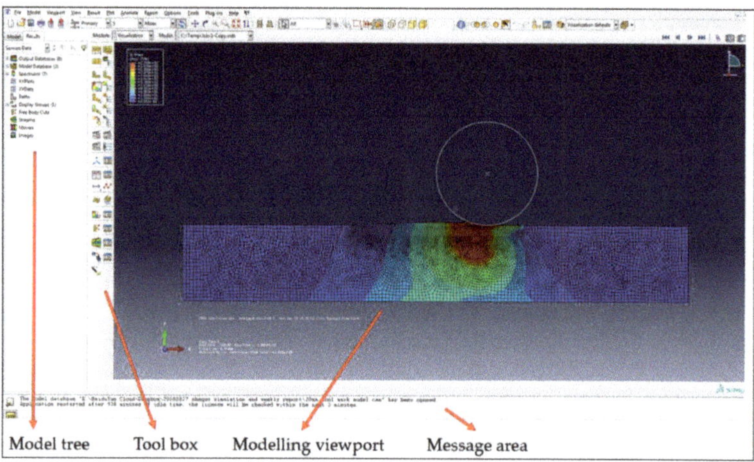

Figure 5. Two-dimensional model in Abaqus.

Since the simulation was an impact process, the simulation made use of the Dynamic Explicit module from Abaqus. The software stores the simulation results as a binary file for the post-processing stage. In the post-processing stage, the simulation process can be visually displayed by the Abaqus software. At the same time, the basic variables can also be calibrated using the simulation results, such as displacement, stresses, and forces.

In the simulation, fatigue failure generally began with microcracks in the material. When the stress sufficiently accumulated in the structures, such as impurities, dislocations, etc., the material began to yield [16]. Generally, the fatigue crack is divided into three stages: initiation, propagation, and final fracture [17]. The early fatigue models were generally based on Hertzian contact to determine the maximum point of force to determine crack initiation. The recent rolling contact fatigue studies consider a rough surface as an important parameter [18,19].

Fatigue life is usually defined as a series of cycles or times that cause fatigue damage and cause a final failure [20]. In this simulation, the fatigue life was the stage at which the tool mark began to plastically deform after the abrasive grains repeatedly impacted on the tool mark.

3. Results and Discussion

Tables 4 and 5 show the stress maps of the two different tool mark height models impacted by particles at different speeds. It can be seen from the two sets of simulation results that, when the abrasive grains were rolling and sliding at high speed on the surface of the workpiece, a greater speed led to a greater impact force on the surface. When an abrasive grain hit the tool mark fast enough, it bounced after the impact and was then pressed back by the fiber, repeatedly following the hit–bounce process. It can be seen from

(d) and (e) in Table 4 and (d) and (e) in Table 5 that, as the momentum of the abrasive particles increased, the stress on the contact surface increased, and the distance between the two high stress areas became wider. This result indicates that a high rotational speed of the lapping pad does not mean a good lapping result for the actual lapping experiment. When the speed of the particle was high enough, the surface started to show pits and scratches because the particle impact on the surface was not even. Some particles may hit the surface to generate pits. Scratches may also be generated due to plowing by particles with large momentum. The experimental results also showed this phenomenon. We used a brush to lap the surface, and the diameter of the particles in the slurry was 15 nm [1]. As shown in Figure 6, after 5 min of lapping, the original surface (Figure 6a) had almost no change (Figure 6b) at a rotational speed of 500 rpm. However, there were pits and scratches generated on the lapped surface (Figure 6d) when the rotational speed was 1500 rpm (lapping time 20 min).

Table 4. The 1000 nm high tool mark workpiece surface under particle impact at different speeds.

Tool Mark Height	Rotational Speed	Results
(a) 1000 nm	200 rpm	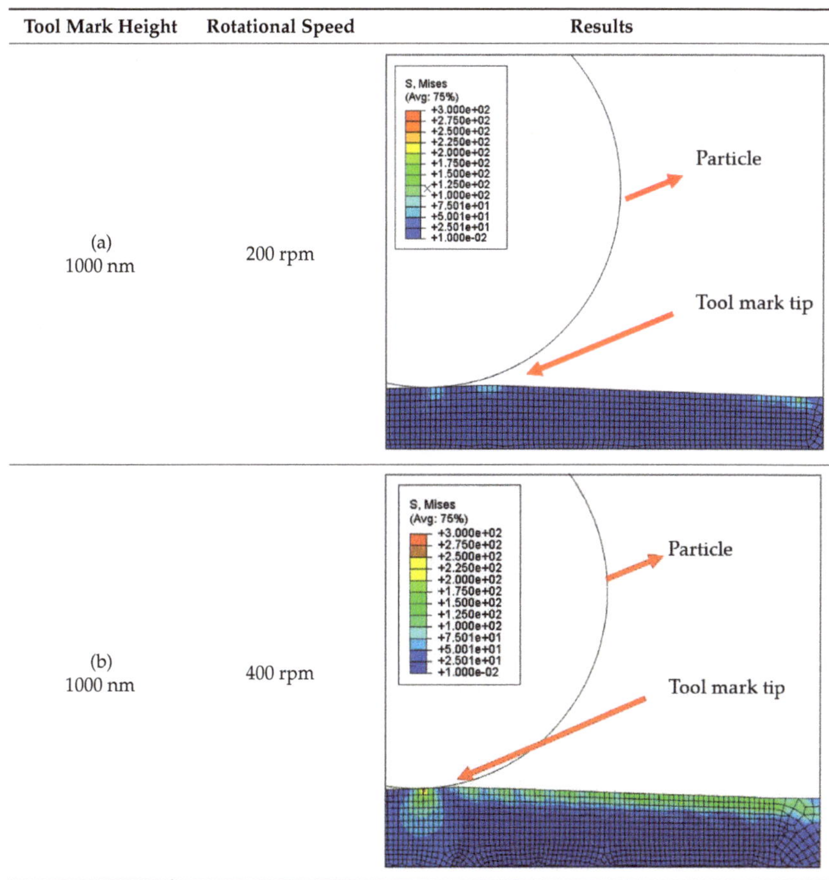
(b) 1000 nm	400 rpm	

Table 4. *Cont.*

Tool Mark Height	Rotational Speed	Results
(c) 1000 nm	600 rpm	
(d) 1000 nm	800 rpm	
(e) 1000 nm	1000 rpm	

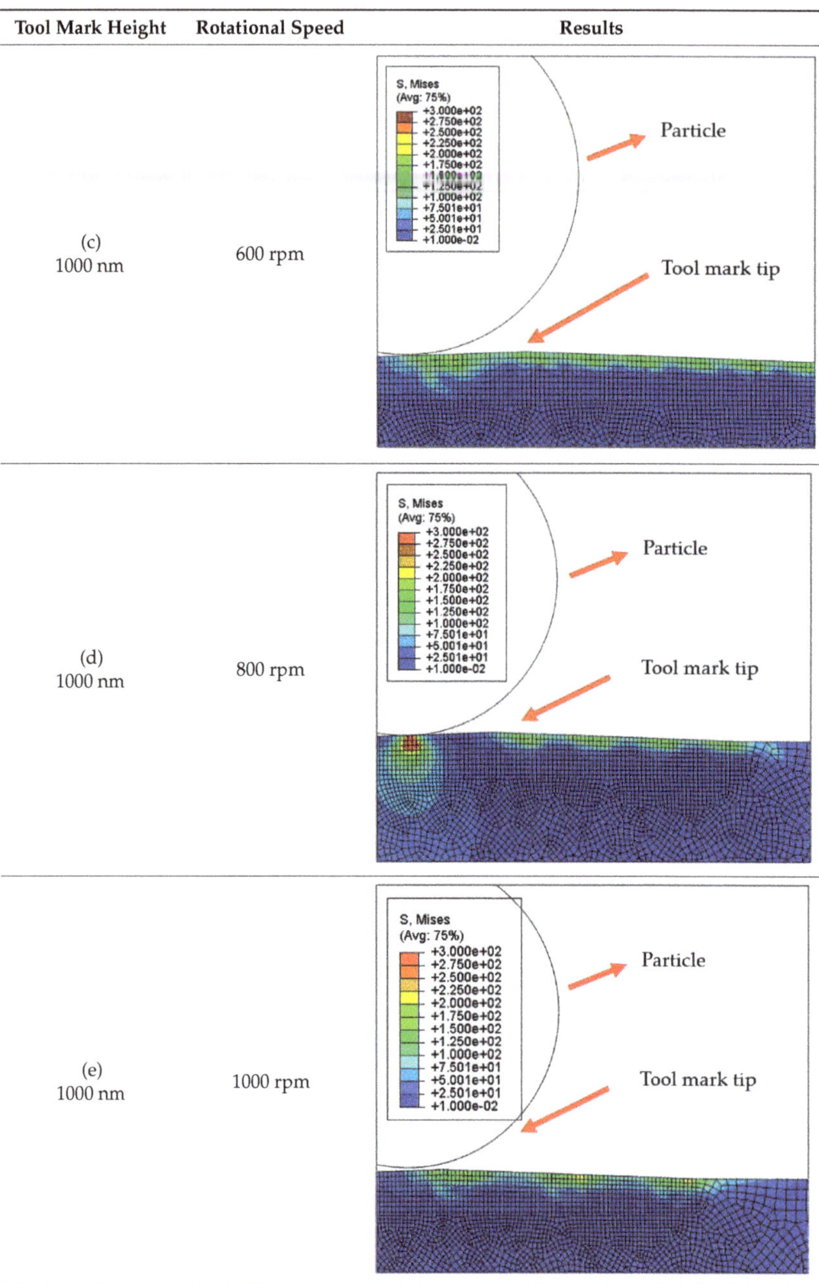

Table 5. The 20 nm high tool mark workpiece surface under particle impact at different speeds.

Tool Mark Height	Rotational Speed	Results
(a) 20 nm	200 rpm	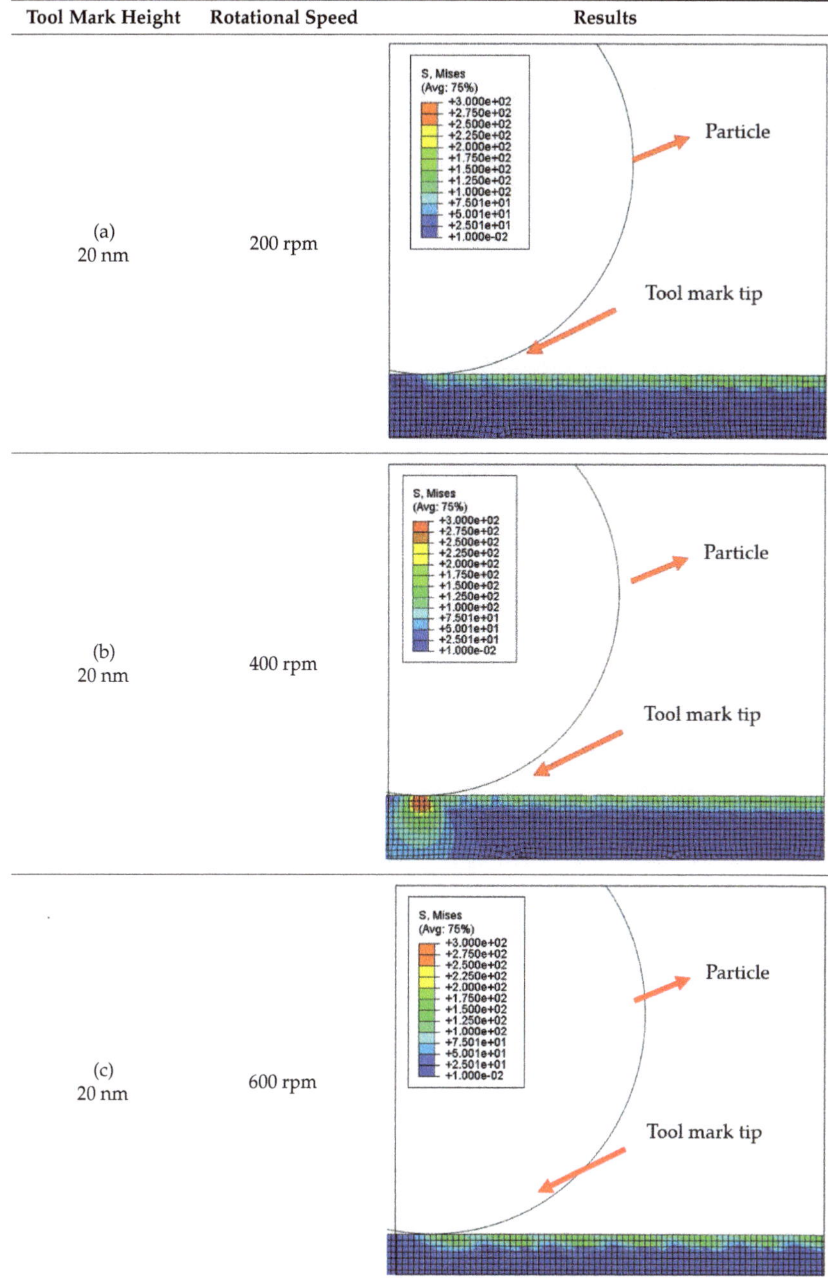
(b) 20 nm	400 rpm	
(c) 20 nm	600 rpm	

Table 5. *Cont.*

Tool Mark Height	Rotational Speed	Results
(d) 20 nm	800 rpm	
(e) 20 nm	1000 rpm	

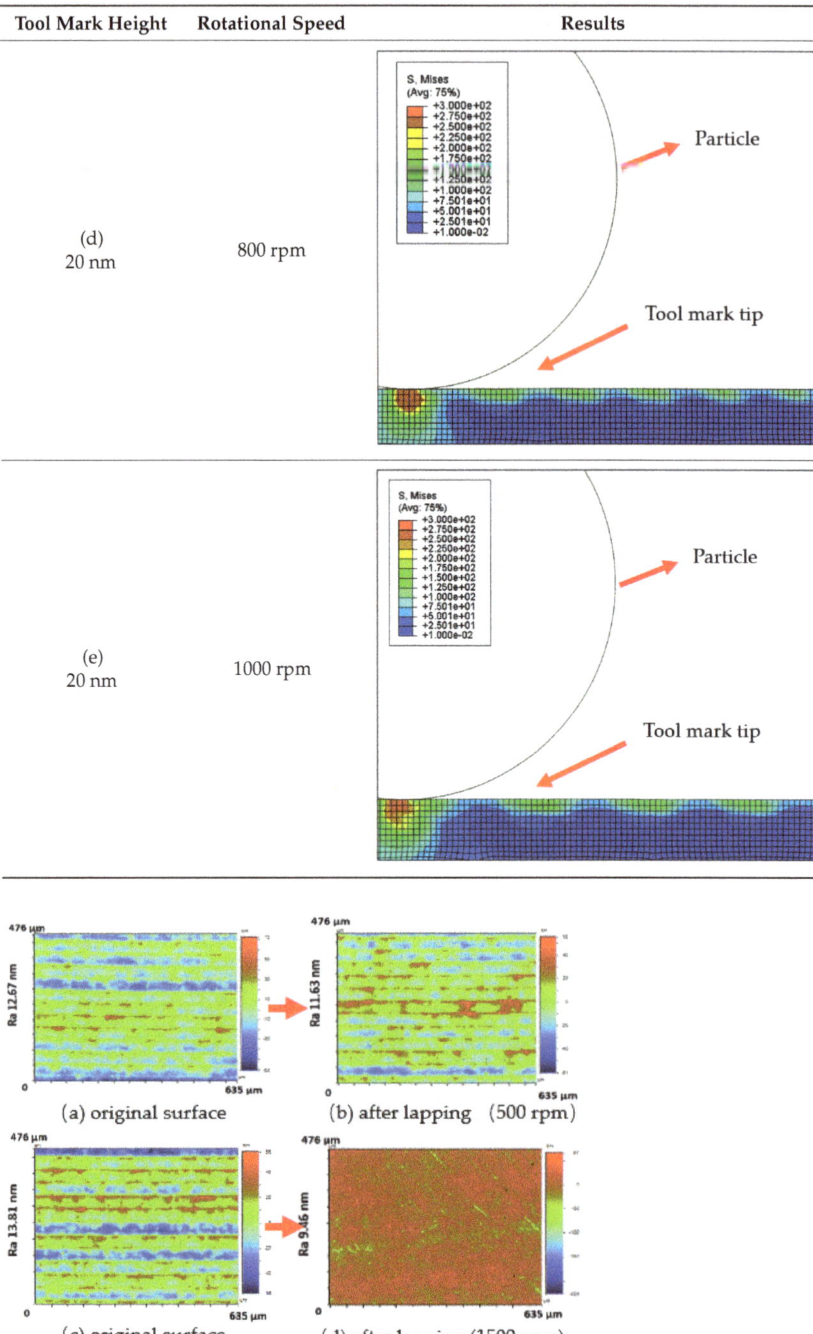

Figure 6. Two surfaces (**a,c**) with tool marks lapped under different rotational speed. The surface (**b**) had almost no change at a rotational speed of 500 rpm. The surface (**d**) started to show pits and scratches at a rotational speed of 1500 rpm [1].

Figures 7 and 8 show the stress map of the tool marks under different impact speeds of the particles. Since the yield strength of the workpiece in this simulation was 276 MPa, it can be seen from Figures 7 and 8 that, when the rotational speed was above 800 rpm, the stress of the abrasive grains in the simulation reached the material yield limit, and the contact area began to change from elastic deformation to plastic deformation. It can seen from Figures 9 and 10 that, when the rotational speed reached 1000 rpm, plastic strain began to occur on the surface of the workpiece.

In the mechanical lapping process, the contact process between the particles and the workpiece determines the surface generation mechanism. Rubbing or plowing are widely accepted as the lapping material removal mechanism [21].

In this research, the diameter of the wool fiber was around 100 nm, and the diameter of the abrasive particle was 15 nm. Studies have shown that SiO_2 particles are generally circular in the slurry [22]. As a result, the possible modes of interaction between a particle and the workpiece surface can be generalized into three categories:

Mode 1: The abrasive particles roll freely on the surface of the workpiece. The efficiency of material removal would be small, close to zero.

Mode 2: The abrasive particles plow a groove on the surface of the workpiece, and the material in the groove is extruded to the front and sides of the groove. In this case, the material removal rate is also close to zero, but the material stacked on the edge of the groove may be taken away by the subsequent abrasive particles.

Mode 3: The abrasive particle cuts a groove while a long strip of chip is removed from the groove. Ideally, the material is completely removed from the groove, and the material removal rate is close to 100%.

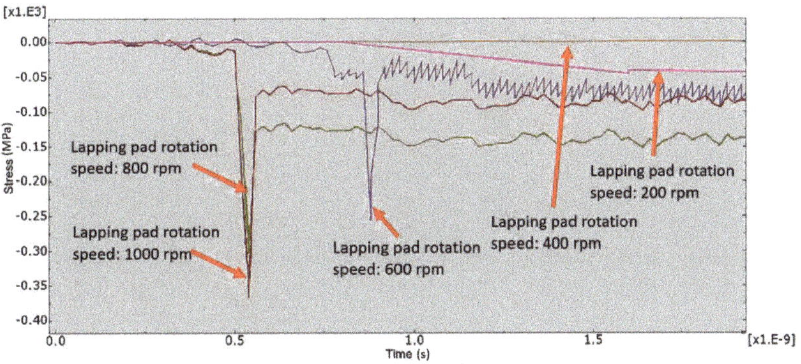

Figure 7. Stress–time curve of 1000 nm tool mark surface under different speeds.

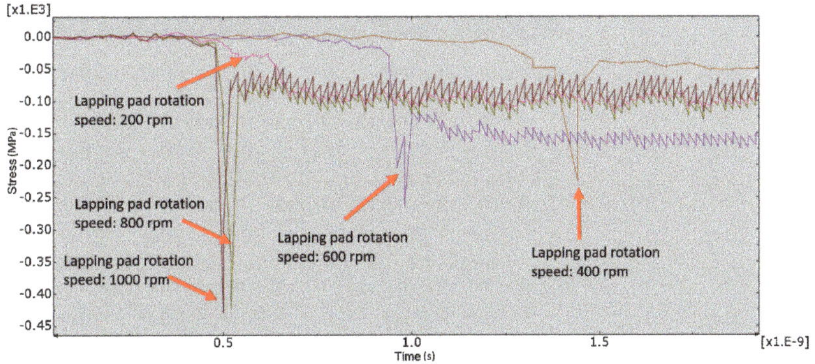

Figure 8. Stress–time curve of 20 nm tool mark under different speeds.

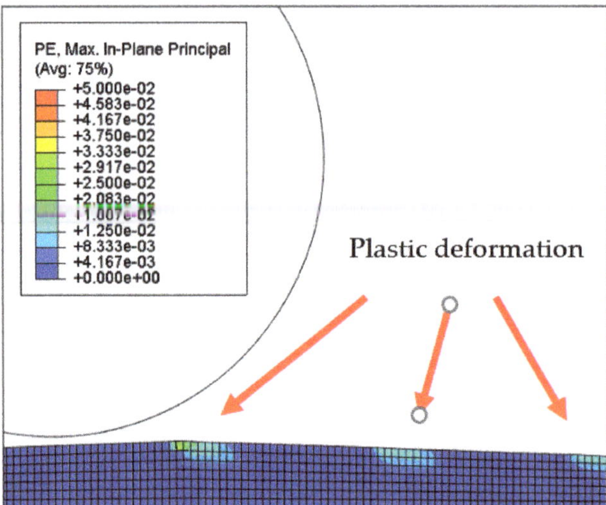

Figure 9. Plastic strain map of the 1000 nm high tool mark surface.

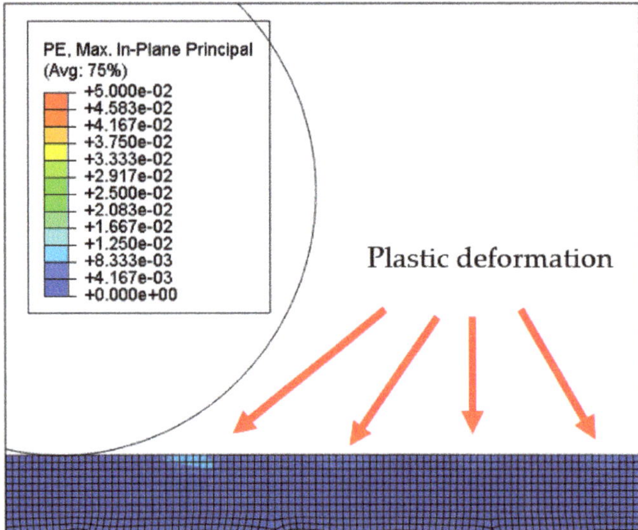

Figure 10. Plastic strain map of the 20 nm high tool mark surface.

Mode 1 is usually called three-body abrasion, and the other two are generally known as two-body abrasion. Since the metal surface is soft, the abrasive particles move at a high speed under the action of the fiber and may be embedded in the workpiece surface, thereby changing from mode 1 to mode 2 or mode 3. However, since the material removal rate of this lapping method is very low, mode 1 plays the main role.

It can be seen from the SEM image that some mottled surfaces were formed on the lapped surface (Figure 11), and some platelet-like materials were taken away during the lapping process (Figure 12). Some researchers believe that, when the abrasive particles are small, to a certain extent, such as less than 0.5 μm, the main mechanism in the lapping process is platelet delamination. However, it is still hard to explain why there was a wide range of scratch-free areas on the lapped surface.

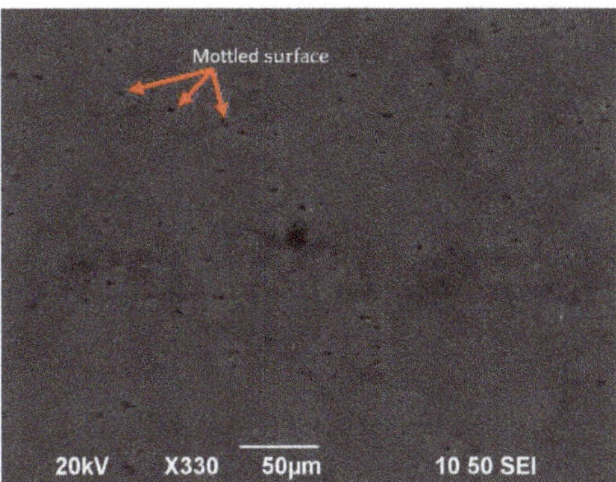

Figure 11. Mottled surface after lapping process (rotational speed 1000 rpm, time 15 min, slurry: SiO$_2$).

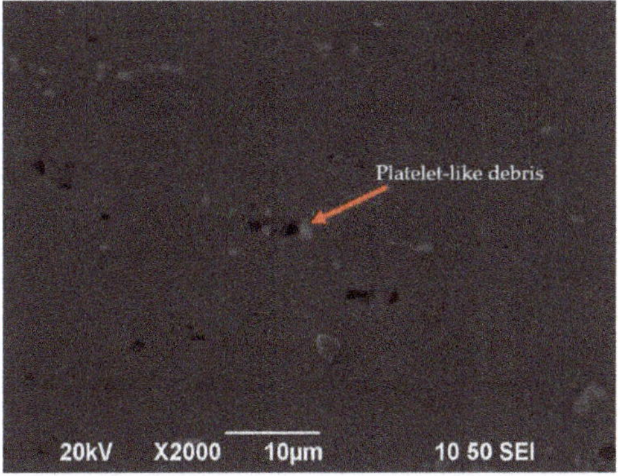

Figure 12. Platelet-like debris on the surface after lapping process (rotational speed 1000 rpm, time 15 min, slurry: SiO$_2$).

For a single abrasive particle, it can microcut or plow the workpiece surface during the lapping process. However, as can be seen from the SEM image in Figure 13, a groove was rarely produced on the surface during the lapping process. In addition, the abrasive particles themselves were relatively round. As a result, the single-particle erosion mechanism only occupied a small proportion of the material removal in the low-pressure lapping process.

For the mechanism in which the abrasive particles are embedded on the fiber to form a cantilever system to establish a removal rate, the surface of the workpiece should have many shallow, long scratches. Although this method makes it is easier to lap a mirror surface, it can be seen from Figure 13 that only one shallow and long scratch existed. As a result, this mechanism is not suitable for this lapping method.

In the metal cutting process, there is a unique type of material removal known as extrusion cutting where there is a physical constraint ahead of the tool or particle [23,24] which may also have been involved in our lapping process. During the lapping process, many abrasive particles pass through the tool marks surface. When two abrasive particles pass through asperities, the front abrasive particles may provide a physical constraint, while the rear abrasive particles do the microcutting. In this scenario, the ductile mode cutting energy dominates [25], and plastic deformation occurs on the asperities, which is a mechanism worthy of further verification by experimentation.

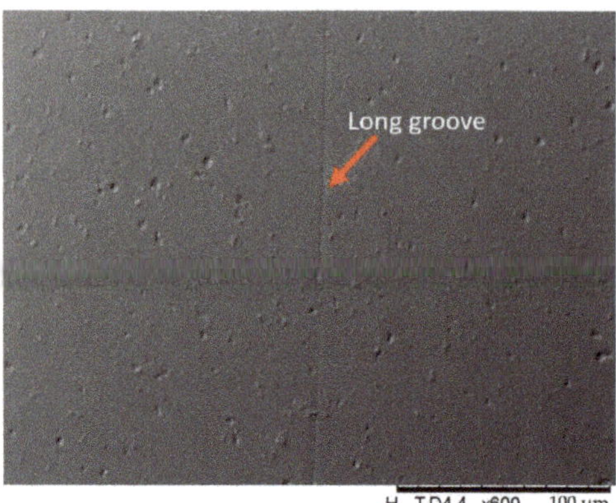

Figure 13. Long scratch on the lapped surface (rotational speed 1000 rpm, time 15 min, slurry particle: Al_2O_3).

Micromachining by abrasives contained in a carrier paste is a mechanism which has no microcutting behavior. This mechanism requires the lapping pad to be softer than the workpiece and the abrasive particles to be sufficiently small. At the same time, the abrasive particles should be easily embedded on the fiber. In this low-pressure lapping process, the fiber of the brush is far softer than the surface of the workpiece, while the abrasive grain is much smaller than the fiber; moreover, because the fiber has a good friction, the abrasive grain should be easily embedded on the fiber or driven by the fiber for high-speed motion. As a result, this mechanism is more suitable for the low-pressure lapping process. The high-speed movement of the abrasive particles means that there is sufficient momentum to laterally impact on the surface. When the momentum is accumulated to a certain level, the asperities on the surface of the workpiece are worn away, thereby reducing the surface roughness.

Through the above discussion, a possible mechanism for the low-pressure lapping method is that the fiber containing the carrier paste is worn away by the tiny asperities from the workpiece as the simulation illustrated in this paper. Although the simulation assumes that the speed of the abrasive particles is equal to the speed of the fiber on the lapping pad, the surface formation mechanism of the polishing method can be successfully verified from the simulation results. That is, the tool marks or asperities on the surface of the workpiece were deformed by the continuous impact of the abrasive grains and finally removed. The number of impacts was calculated according to Equation (3). When lapped for 10 min, the surface was impacted 3×10^6 times. This is far more than the 10^5 times of impact required for high cycle fatigue of typical materials. This cyclic stress is large enough to cause fatigue. As a result, it can be regarded as microscale high-cycle fatigue and can also be seen as impact fatigue for tool marks or asperities. These asperities and tool marks are worn away, and the worn surface continues to be impacted by the abrasive particles, which leads to the small-scale material removal rate.

According to our previous research results, there is almost no change in the lapped surface when the rotational speed is 500 rpm [1]. By finite element simulation, it was found that the rotational speed is the most important lapping parameter in this polishing method. When the rotational speed is lower than a certain level, it is difficult to achieve material removal. This also explains why the asperities on the tool mark surface were removed first, followed by the tool mark itself (Figure 14).

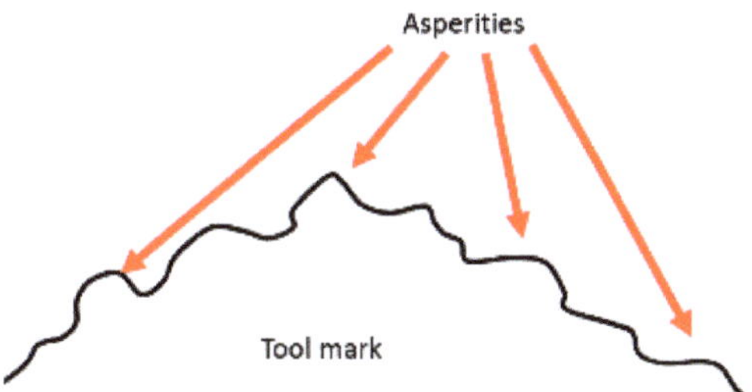

Figure 14. The asperities on the tool mark surface were removed first, followed by the tool mark itself.

4. Conclusions

This research described the modeling and simulation of the surface generation mechanism of a novel low-pressure lapping method by the finite element method. The results indicate that the rotational speed plays a major role in the low-pressure lapping process. When the rotational speed was higher than a certain level, i.e., 1000 rpm in the experiment, the abrasive particles driven by the fiber obtained sufficient momentum to impinge on the surface of the workpiece. This finally wore the asperities and tool marks away. This model not only describes the surface generation mechanism of the low-pressure lapping method, but can also be used to predict the possible lapping parameters when different materials and forms of the workpiece are used. The recommended combination of lapping parameters determined from this simulation is to control the rotational speed at 800–1500 rpm and use particles smaller than the diameter of the fiber (i.e., 20 nm in this experiment).

Author Contributions: Conceptualization, N.Y. and L.L.; methodology, N.Y.; investigation, N.Y.; writing—original draft preparation, N.Y.; writing—review and editing, N.Y. and L.L.; supervision, L.L. and C.-s.K. All authors have read and agreed to the published version of the manuscript.

Funding: This research was funded by Shenzhen Technology University grant number [20211061010009].

Acknowledgments: The authors would like to express their sincere thanks to The Hong Kong Polytechnic University, Shenzhen Technology University (the Natural Science Foundation of Top Talent of SZTU with grant no.20211061010009) and The Hong Kong Polytechnic University Shenzhen Research Institute, also thanks to the Greater Bay Area (GBA) Start-up postdoc Program.

Conflicts of Interest: The authors declare no conflict of interest.

References

1. Yu, N.H.; Chan, C.Y.; Li, L.H.; Lee, W.B. Spectral analysis of surface roughness features of a lapped ultraprecision single-point diamond machined surface. *Int. J. Adv. Manuf. Technol.* **2016**, *88*, 1407–1415. [CrossRef]
2. Bartels, P.H.; Samuels, L.E. Metallographic Polishing by Mechanical Methods. *Trans. Am. Microsc. Soc.* **1972**, *91*, 248. [CrossRef]
3. Kral, E.R.; Komvopoulos, K.; Bogy, D.B. Elastic-Plastic Finite Element Analysis of Repeated Indentation of a Half-Space by a Rigid Sphere. *J. Appl. Mech.* **1993**, *60*, 829–841. [CrossRef]
4. Kogut, L.; Etsion, I. Elastic-Plastic Contact Analysis of a Sphere and a Rigid Flat. *J. Appl. Mech.* **2002**, *69*, 657–662. [CrossRef]
5. Greenwood, J.A.; Williamson, J.B.P. Contact of nominally flat surfaces. *Proc. R. Soc. London. Ser. A Math. Phys. Sci.* **1966**, *295*, 300–319. [CrossRef]
6. Chang, W.-R.; Etsion, I.; Bogy, D.B. An Elastic-Plastic Model for the Contact of Rough Surfaces. *J. Tribol.* **1987**, *109*, 257–263. [CrossRef]
7. Tabor, D. *The Hardness of Metals*; Clarendon Press: Oxford, UK, 1951; pp. 213–219.
8. Jackson, R.L.; Green, I. A Finite Element Study of Elasto-Plastic Hemispherical Contact Against a Rigid Flat. *J. Tribol.* **2005**, *127*, 343–354. [CrossRef]
9. Kadin, Y.; Kligerman, Y.; Etsion, I. Multiple Loading-Unloading of an Elastic-Plastic Spherical Contact. In Proceedings of the World Tribology Congress III, Washington, DC, USA, 12–16 September 2005; pp. 215–216. [CrossRef]
10. Peng, H.; Liu, Z.; Huang, F.; Ma, R. A study of elastic–plastic contact of statistical rough surfaces. *Proc. Inst. Mech. Eng. Part J. Eng. Tribol.* **2013**, *227*, 1076–1089. [CrossRef]

11. Almuramady, N.; Borodich, F. Simulations of sliding adhesive contact between microgear teeth in silicon-based MEMS work in a vacuum environment. In Proceedings of the 25th UKACM Conference on Computational Mechanics, Birmingham, UK, 12–13 April 2017; pp. 38–41.

12. Lee, Y.J.; Chaudhari, A.; Zhang, J.; Wang, H. Thermally Assisted Microcutting of Calcium Fluoride Single Crystals. In *Simulation and Experiments of Material-Oriented Ultra-Precision Machining*; Zhang, J., Guo, B., Zhang, J., Eds.; Springer: Berlin/Heidelberg, Germany, 2019; pp. 77–102.

13. Schönemann, L.; Riemer, O. Thermo-mechanical tool setting mechanism for ultra-precision milling with multiple cutting edges. *Precis. Eng.* **2019**, *55*, 171–178. [CrossRef]

14. Fish, J.; Oskay, C.; Fan, R.; Barsoum, R. *Al 6061-T6-Elastomer Impact Simulations*; Electronic Document; Rensselaer Polytechnic Institute: Troy, NY, USA, 2005.

15. Qi, J.; Herron, J.R.; Sansalone, K.H.; Mars, W.V.; Du, Z.Z.; Snyman, M.; Surendranath, H. Validation of a Steady-State Transport Analysis for Rolling Treaded Tires. *Tire Sci. Technol.* **2007**, *35*, 183–208. [CrossRef]

16. Oila, A.; Bull, S. Assessment of the factors influencing micropitting in rolling/sliding contacts. *Wear* **2005**, *258*, 1510–1524. [CrossRef]

17. Verdu, C.; Adrien, J.; Buffière, J. Three-dimensional shape of the early stages of fatigue cracks nucleated in nodular cast iron. *Mater. Sci. Eng. A* **2008**, *483-484*, 402–405. [CrossRef]

18. Terrin, A.; Dengo, C.; Meneghetti, G. Experimental analysis of contact fatigue damage in case hardened gears for off-highway axles. *Eng. Fail. Anal.* **2017**, *76*, 10–26. [CrossRef]

19. Paulson, N.R.; Sadeghi, F.; Habchi, W. A coupled finite element EHL and continuum damage mechanics model for rolling contact fatigue. *Tribol. Int.* **2017**, *107*, 173–183. [CrossRef]

20. Suresh, S. *Fatigue of Materials*; Cambridge University Press, Massachusetts Institute of Technology:Cambridge, MA USA, 1998.

21. Rahman, M.A.; Amrun, M.R.; Rahman, M.; Kumar, A.S. Variation of surface generation mechanisms in ultra-precision machining due to relative tool sharpness (RTS) and material properties. *Int. J. Mach. Tools. Manuf.* **2017**, *115*, 15–28. [CrossRef]

22. Turley, D.M.; Samuels, L.E. The nature of mechanically polished surfaces of copper. *Metallography* **1981**, *14*, 275–294. [CrossRef]

23. Chiffre, L.D. Extrusion cutting of brass strips. *Int. J. Mach. Tool Des. Res.* **1983**, *23*, 141–151. [CrossRef]

24. Cai, S.L.; Chen, Y.; Ye, G.G.; Jiang, M.Q.; Wang, H.Y.; Dai, L.H. Characterization of the deformation field in large-strain extrusion machining. *J. Mater. Process. Technol.* **2015**, *216*, 48–58. [CrossRef]

25. Lee, Y.J.; Kumar, A.S.; Wang, H. Beneficial stress of a coating on ductile-mode cutting of single-crystal brittle material. *Int. J. Mach. Tools. Manuf.* **2021**, *168*, 03787. [CrossRef]

micromachines

Article

Simulation, Modeling and Experimental Research on the Thermal Effect of the Motion Error of Hydrostatic Guideways

Pengli Lei, Zhenzhong Wang *, Chenchun Shi, Yunfeng Peng and Feng Lu

Department of Mechanical and Electrical Engineering, Xiamen University, Xiamen 361005, China; kenash@mail.ustc.edu.cn (P.L.); shichenchun@nbu.edu.cn (C.S.); pengyf@xmu.edu.cn (Y.P.); 19920191151156@stu.xmu.edu.cn (F.L.)
* Correspondence: wangzhenzhong@xmu.edu.cn

Abstract: Hydrostatic guideways are widely applied in ultra-precision machine tools, and motion errors undermine the machining accuracy. Among all the influence factors, the thermal effect distributes most to motion errors. Based on the kinematic theory and the finite element method, a 3-degrees-of-freedom quasi-static kinematics model for motion errors containing the thermal effect was established. In this model, the initial state of the closed rail as a "black box" is regarded, and a self-consistent setting method for the initial state of the guide rails is proposed. Experiments were carried out to verify the thermal motion errors simulated by the finite element method and our kinematics model. The deviation of the measured thermal vertical straightness error from the theoretical value is less than 1 μm, which ensured the effectiveness of the model we developed.

Keywords: hydrostatic guideways; thermal error; finite element simulation; motion errors

Citation: Lei, P.; Wang, Z.; Shi, C.; Peng, Y.; Lu, F. Simulation, Modeling and Experimental Research on the Thermal Effect of the Motion Error of Hydrostatic Guideways. *Micromachines* **2021**, *12*, 1445. https://doi.org/10.3390/mi12121445

Academic Editors: Benny C. F. Cheung and Jiang Guo

Received: 20 September 2021
Accepted: 22 November 2021
Published: 25 November 2021

Publisher's Note: MDPI stays neutral with regard to jurisdictional claims in published maps and institutional affiliations.

1. Introduction

High-precision and large-diameter aspheric optical components play a key role in advanced optical systems. The typical application areas are inertial confinement nuclear fusion devices and large telescopes. Laser nuclear fusion technology is a cutting-edge technology that uses inertial confinement to achieve controllable nuclear fusion. The National Ignition Facility (NIF) is the largest laser inertial confinement nuclear fusion device in the world today [1,2], whose optical array includes 192 laser beams. More than 7000 large-aperture optical originals are required. In recent years, research on NIF devices has also made breakthrough progress [3,4]. In this context, the research and development of high-quality and high-efficiency processing equipment for large-aperture optical components are imminent.

Ultra-precision manufacturing technology and equipment are important means to achieve mass production of those optical components. The processing order of these large-aperture optical components is usually "grind first, then polish" [5]. First, the aspheric surface directly forms via the ultra-precision grinding processing method. The optical components are then polished by the conformal polishing process. Eventually, the high-precision polishing method is utilized to further minimize the surface error. Therefore, high efficiency and high surface accuracy are realized in the grinding process, which helps to shorten the processing cycle. To this end, our research group has successfully participated in the development of UPG80, UPG60, and other series of ultra-precision grinding machines for optical components [6], all of which adopt hydrostatic guide rails.

Due to their high precision, high rigidity, and high load-bearing characteristics, the hydrostatic guideways are widely applied in the field of ultra-precision machine tools [7,8]. The friction pair interface of the hydrostatic guideway is separated by a layer of oil film, which avoids dry friction and abrasion. The thickness of the oil film generally ranges from 10 μm to 40 μm [9]. The well-known, large-scale optical ultra-precision grinding machine OAGM 2500 adopts hydrostatic guide rails for both the X-axis and Y-axis [10].

There has been abundant research in the field of motion accuracy of hydrostatic guides. Park [11,12] proposed the transfer function method to study the movement error of the worktable of the hydrostatic guideway. Zha [13] established a static analysis model of a four-pad open hydrostatic guideway, by simplifying the oil film at the oil pad into a linear spring unit. Afterwards, Zha [14] took the open hydrostatic guideway of a gantry grinder as an analysis object and proposed an effective method for compensation of motion errors. Xue [15] took the four-pad closed hydrostatic guide rail as the theoretical analysis object and studied the error averaging effect. Shi [16] analyzed the motion straightness theory of hydrostatic guideways with a quasi-static model, concluding that the motion error of the worktable is affected by surface profiles of both the upper and lower guide rails in the closed hydrostatic guideway, and the lower guide rail dominates [17].

All the studies mentioned above focus on the geometric error of the hydrostatic guide rail system. However, sufficient research has shown that the thermal deformation also contributes to the processing error of ultra-precision machining distinctly, up to 30–70% [18]. For the moving parts of the machine tools, the translational axis contains six motion errors [19]. The thermal deformation of the machine tool will cause the spatial movement of the moving parts, which directly results in the positioning error, the angular error, and the straightness error. Therefore, the study on motion accuracy considering the thermal effect is urgently required for the practical application of the hydrostatic guideway. Liu [20] studied the thermal deformation of machine tool components. Zhang [21] established a model of the relationship between spatial error and thermal factors. Liu [22] introduced the research progress of the methods of avoiding thermal errors and the methods of thermal error compensation.

The finite element method has a wide range of applications in the research of static pressure systems. Jeon [23] used the finite element method to characterize the oil film between the moving parts and the surface in the hydrostatic guide rails. Shamoto [9] also conducted static analysis on the relationship between the reaction force of the oil pads and the geometric error of the guide rails based on the finite element method.

Therefore, based on the kinematic model, combined with the thermal deformation of the hydrostatic guideway in the actual process, and supported by the finite element simulation and the experimental data of the hydrostatic platform, a 3-degrees-of-freedom (3-DOF) quasi-static kinematics model for motion errors containing the thermal effect for the hydrostatic guideway is established. Experimental results show that the model has a certain effect on thermal error prediction.

2. Modeling

The ultra-precision grinding machine UPG60 developed by our research group shown in Figure 1a,b, is the structure with the shield removed. The three linear axes of the machine are closed hydrostatic guideways with Progressive Mengen (PM) flow controllers as the restrictors. In this paper, we took the X-axis of UPG60 as the analyzed object and carried out a series of experimental studies by setting up a hydrostatic experimental platform. Figure 1c shows the X-axis structure of the grinding machine UPG60, which contains two closed hydrostatic guide rails and a worktable with six hydrostatic guide shoes (HGSs). Rail A has four rail surfaces: the bottom rail surface 1, the top rail surface 2, and the side rail surfaces 3 and 4, while Rail B has only two rail surfaces, the bottom rail surface 5 and the top rail surface 6, as shown in Figure 2a. The worktable with six HGSs is shown in Figure 3. It should be pointed out that, only HGS1 and HGS2 have four oil pads, and the other four HGSs are all guide shoes with only two oil pads. The hydrostatic guide shoes on the same side of the worktable are equally spaced, and the six HGSs are adjusted repeatedly during assembly until the plane accuracy meets the design requirements. The movement of the hydrostatic worktable can be driven by the rotation of the screw rod, and the six HGSs together with the guide rails are completely separated by a layer of oil film in the working state, so the worktable is suspended in a closed hydrostatic guide system.

The motion error of the hydrostatic worktable is affected by both the parameter of the restrictors and the surface shape of the guide rails.

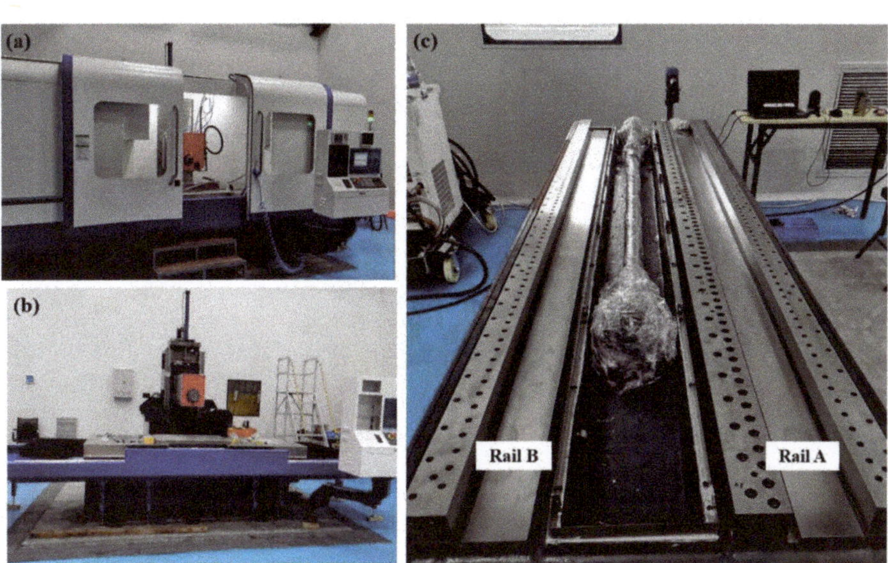

Figure 1. UPG60 ultra-precision grinding machine. (**a**) With the shield. (**b**) Without the shield. (**c**) Structure of the X-axis guideway.

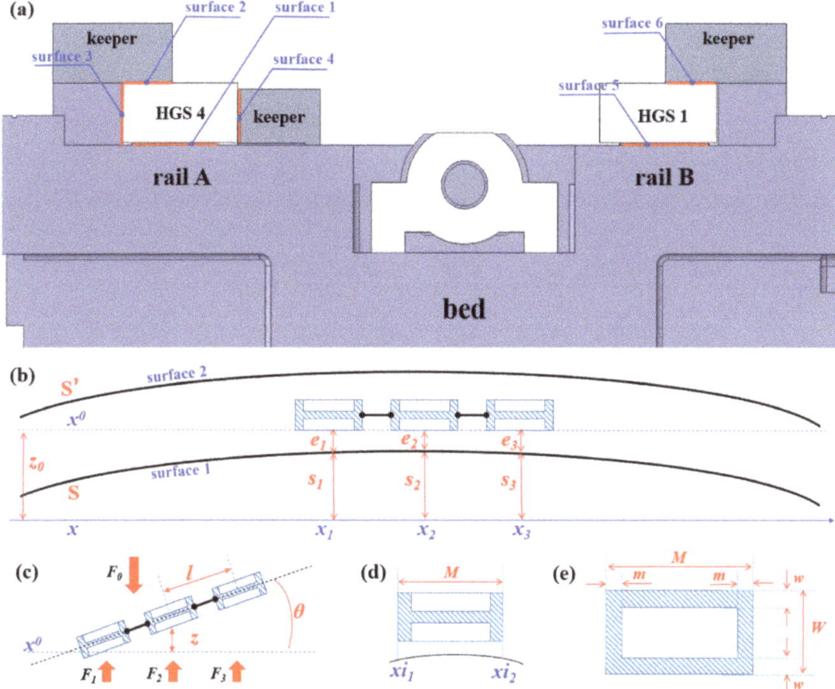

Figure 2. Modeling of the closed hydrostatic guideway with six pads. (**a**) View of the closed hydrostatic guideways along X-axis. (**b**,**c**) Kinematics of the three HGSs on the same side of the worktable. (**d**,**e**) detailed parameters of the HGS.

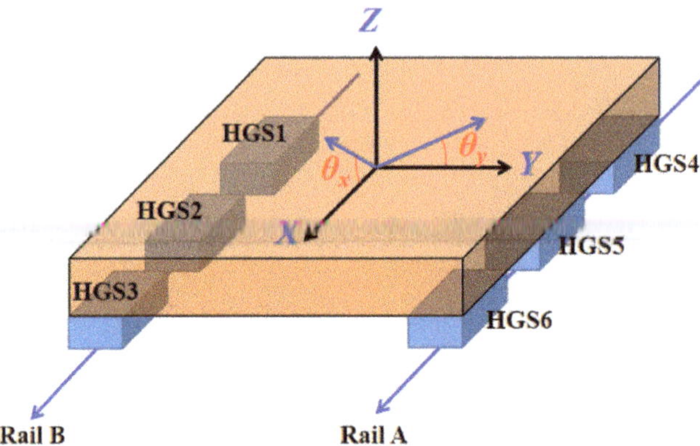

Figure 3. Three-degrees-of-freedom (3-DOF) kinematics model of the worktable.

Figure 2 shows the principle of a single-sided closed hydrostatic guide rail, which contains three hydrostatic guide shoes. When modeling the motion errors, we mainly analyzed the thickness of the oil film of the pads. Among them, S is the profile curve of the bottom guide rail, while S' is the profile curve of the top. x represents the reference coordinate axis of the X-axis, and x_0 is the reference axis parallel to the bottom surface of the three HGSs. Then, z_0 represents the relative value of the vertical motion error of the worktable. At the same time, e_1, e_2, and e_3, respectively, represent the average oil film thickness of the bottom oil pads of the three HGSs, which are:

$$\bar{e}_i = z \pm \frac{l_i}{2} \cdot \sin\theta + \int_{xi_1}^{xi_2} \frac{(z_0 - S)}{M \cdot \cos\theta} dx$$

$$\approx z \pm \frac{l_i}{2} \cdot \theta + \int_{xi_1}^{xi_2} \frac{(z_0 - S)}{M} dx \tag{1}$$

Among them, z_i is the vertical motion straightness of the i-th guide shoes, and θ is the deflection angular error.

M-2 m, W, W-2 w represents the length and width of the oil pad, respectively, as shown in Figure 2e. According to References [15,24], it is believed that the topographic error of the guide rail in the width direction is relatively small, and only the topography error in the length direction of the guide rail is taken into consideration. So, S_A and S_B are used to define the topography error functions of Guide Rail A and Guide Rail B, respectively.

In the quasi-static process, the worktable is in a state of equilibrium with force and moment [16]. Therefore, the overall load F_0 of the worktable and the oil film reaction force F_i of the oil pads satisfy the following equations:

$$\sum_i F_i = F_0$$

$$\sum_i F_i \cdot l_i = 0 \tag{2}$$

The oil film reaction force corresponding to each HGS is the resultant force of the oil film reaction forces of the top and bottom oil pads, namely:

$$F_i = p_{ri}^{bottom} \cdot A_i^{bottom} - p_{ri}^{top} \cdot A_i^{top} \tag{3}$$

Among them, p_{ri}^{bottom} is the reaction pressure of the bottom oil pad of the i-th guide shoe, while p_{ri}^{top} is the reaction pressure of the top oil pad of the i-th guide shoe. A_i^{bottom} is the effective bearing area of the bottom oil pad of the i-th guide shoe, and A_i^{top} represents that of the top pad.

As introduced above, the 6 HGSs are all equipped with PM flow controllers. There is a detailed study on the characteristics of PM flow controllers in References [25,26]. According to References [16,17], the oil film pressure of the hydrostatic guide shoes with PM flow controller is:

$$p_{ri} = \frac{Q_0 \cdot R_{hi}}{1 - \frac{Q_0 \cdot (K_r - 1) \cdot R_{hi}}{P_s}} \tag{4}$$

Among them, Q_0 represents the initial flow corresponding to the pressure of the oil pad is zero; K_r is one of the characteristic parameters of the PM flow controller, which represents the specific flow ratio; P_s is the oil supply pressure of the hydraulic station; and R_{hi} is the flow resistance at the oil pad, whose mathematical expression is as follows:

$$R_h = \frac{3\eta}{h^3 \cdot \left(\frac{M-m}{2w} + \frac{W-w}{2m}\right)} \tag{5}$$

Here, η refers to the dynamic viscosity of the hydrostatic oil, and h can be replaced by the average oil film thickness e_i at the i-th oil pad. Then, we use M_1, m_1, W_1, and w_1 to represent the parameters of the top pads and M_2, m_2, W_2, w_2 to represent the parameters of the bottom pads. Thus, the oil film reaction force can be expressed as:

$$F_i = \frac{Q_0 \cdot 3\eta}{\overline{e}_i^3 \cdot \left(\frac{M_2 - m_2}{2w_2} + \frac{W_2 - w_2}{2m_2}\right) - Q_0 \cdot (K_r - 1) \cdot \frac{3\eta}{P_s}} \cdot (M_2 - m_2) \cdot (W_2 - w_2)$$
$$- \frac{Q_0 \cdot 3\eta}{\overline{e}_i^3 \cdot \left(\frac{M_1 - m_1}{2w_1} + \frac{W_1 - w_1}{2m_1}\right) - Q_0 \cdot (K_r - 1) \cdot \frac{3\eta}{P_s}} \cdot (M_1 - m_1) \cdot (W_1 - w_1) \tag{6}$$

At the same time, it should be pointed out that when the shape of Guide Rail A and Guide Rail B are different (generally), the motion error of the worktable can be represented by the vertical motion error z and the two angular errors θ_x and θ_y, as shown in Figure 3.

Therefore, the expression of the average oil film thickness of the bottom pads of the 6 HGSs are shown as follows:

$$\overline{e}_1 = z - l \cdot \theta_x - l_y \cdot \theta_y + \int_{x1_1}^{x1_2} \frac{(z_0 - S_A)}{M} dx$$
$$\overline{e}_2 = z - l_y \cdot \theta_y + \int_{x2_1}^{x2_2} \frac{(z_0 - S_A)}{M} dx$$
$$\overline{e}_3 = z + l \cdot \theta_x - l_y \cdot \theta_y + \int_{x3_1}^{x3_2} \frac{(z_0 - S_A)}{M} dx$$
$$\overline{e}_4 = z - l \cdot \theta_x + l_y \cdot \theta_y + \int_{x4_1}^{x4_2} \frac{(z_0 - S_B)}{M} dx \tag{7}$$
$$\overline{e}_5 = z + l_y \cdot \theta_y + \int_{x5_1}^{x5_2} \frac{(z_0 - S_B)}{M} dx$$
$$\overline{e}_6 = z + l \cdot \theta_x + l_y \cdot \theta_y + \int_{x6_1}^{x6_2} \frac{(z_0 - S_B)}{M} dx$$

Among them, l_y is the distance between the two guide rails. Combined with the oil pressure expression, by solving the force balance and moment balance equations, a set of parameters $(z, \theta x, \theta y)$ can be obtained, which are the motion errors of the moving parts under the 3DOF model. The detailed parameters of the model are shown in Table 1.

Table 1. Parameters of the model.

Parameters	l				l_y		Rail Length	
value	350 mm				410 mm		2460 mm	
parameters	η				Q_0		K_r	
value	0.095×10^{-6} N·s/mm^2				0.3667×10^{-3} mm^3/s		2.8	
parameters	P_s				F_0		T	
value	3.2 MPa				9535 N		20 °C	
parameters	Top pads/mm				Bottom pads/mm			
	M_1	m_1	W_1	w_1	M_2	m_2	W_2	w_2
value	200	12	43	12	200	12	75	12

3. Experiment

Experimental setup: We designed a single-axis, high-precision, hydrostatic guideway reliability and accuracy retention verification platform, through which we could monitor and measure the temperature, displacement, and motion error of the worktable during the experiment.

The purpose of the experiment is to explore the performance of the accuracy of the static pressure rail under different temperature conditions. Collecting temperature information and motion error data at the same time in the experiment, we could then calculate the corresponding results through the theoretical model. Comparison of theoretical predictions and experimental data can further verify the accuracy of the theoretical model and its scope of applications. We therefore hope to provide theoretical support for the thermal error compensation of the hydrostatic guideway by verifying the accuracy of the model.

The platform is shown in Figure 4. The temperature data of the key distribution points of the hydrostatic rail platform was measured by a group of temperature sensors (PT100 platinum thermal resistance; measuring range: $-70\ °C$–$500\ °C$; sensitivity: $0.1\ °C$). The motion straightness was measured by the Renishaw XK10 alignment laser system, and the angular error was measured by an electronic level meter.

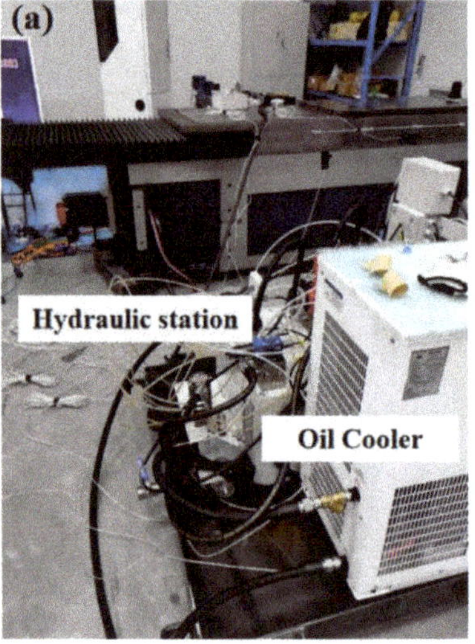

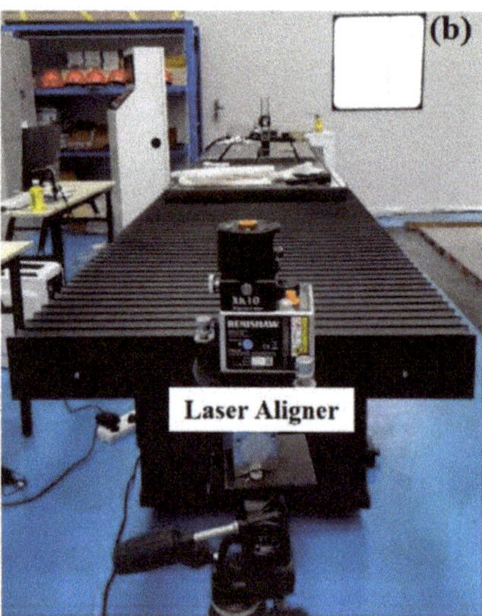

Figure 4. Experimental platform of the hydrostatic guideway. (**a**) Hydraulic station and the oil cooler. (**b**) Hydrostatic guideway.

The experimental platform was equipped with a hydraulic station system and a hydrostatic oil cooling circulation system. Among them, the hydraulic oil converged into the oil storage tank from the oil outlet of the guide rail and then entered the cooler for cooling, and the oil outlet of the cooler also flowed into the oil storage tank (that is, the return oil of the guide rail and the cooling oil are mixed and cooled in the oil storage tank at the same time). Finally, the oil was output to the guide rail through a hydraulic pump. All experiments were performed under isolated constant temperature conditions to avoid environmental temperature fluctuations from affecting the temperature monitoring data of the rail platform. By adjusting the cooling parameters of the oil cooler, we set up two sets of experiments.

Experimental results: In the data collected by the temperature sensor, we only took the temperature of the oil inlet of the guide rail, the temperature of the oil outlet of the guide rail, the temperature of the oil storage tank, and the temperature of the environment as the main analysis objects. It can be seen that the cooling parameters of the oil cooler in the two experiments are different. The cooling temperature of the oil cooler in the first experiment was controlled at about 17 °C, and in the second experiment, it was set to about 19 °C. The purpose was to control and compare the oil temperature of different oil inlets to explore the influence of temperature on the motion error of the worktable.

In Figure 5a,b, the experimental results show that during the whole process, the temperature at the oil inlet of the guide rail changed most obviously, showing a trend of rising first and then stabilizing. While the temperature at the outlet continues to rise slowly, the temperature curves in the figure fluctuate, which is caused by the intermittent working mode of the oil cooler. The stable values of the temperature rising of the oil inlet in the two experiments were 2.5 °C and 3.7 °C, with a difference of about 1 °C.

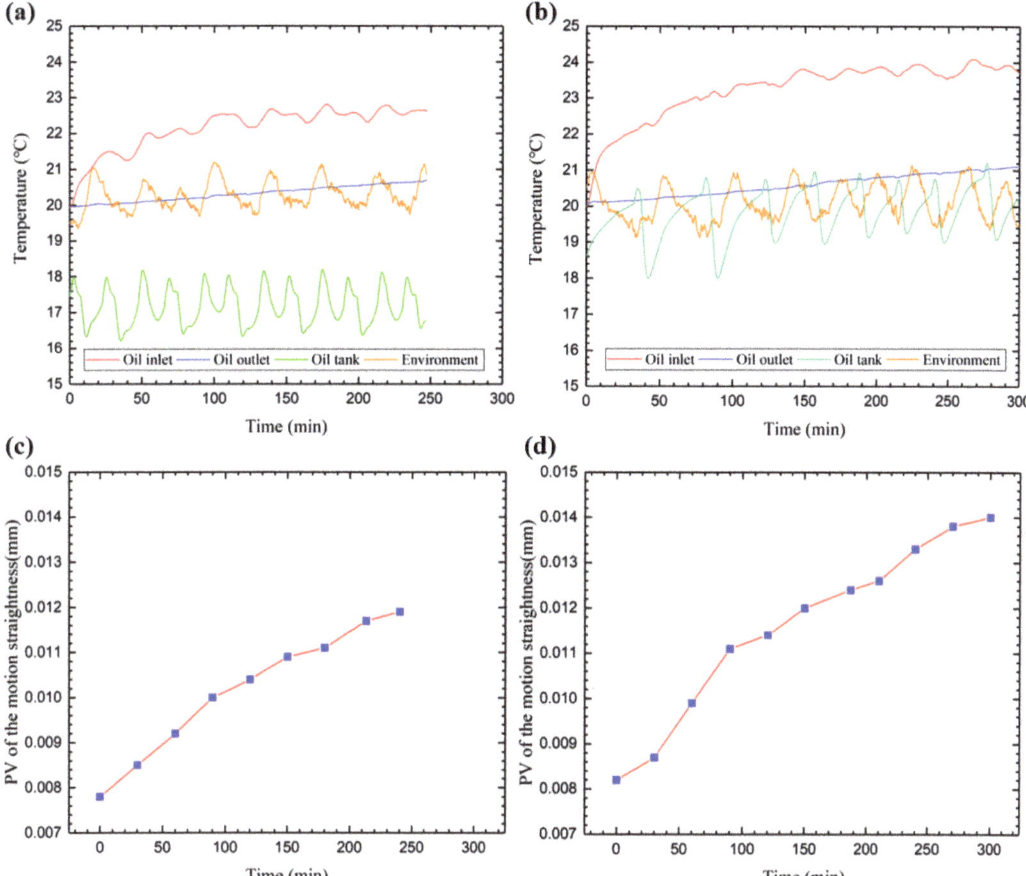

Figure 5. Experimental results. (**a**) Temperature monitoring data in Experiment #1. (**b**) Temperature monitoring data in Experiment #2. (**c**) Motion straightness in Experiment #1. (**d**) Motion straightness in Experiment #2.

Figure 5c,d reveals the trend of the peak–valley (PV) value of the motion straightness of the worktable over time. It can be seen that the vertical straightness error of the workbench gradually increases with the experimental time. At the same time, comparing the two sets of curves, it can be found that under the same initial state, the higher the oil inlet temperature, the greater the vertical straightness error of the final state. It shows that the temperature of the oil inlet is the decisive factor for the accuracy of the moving parts of the hydrostatic guideway.

In Figure 6, we compared the initial vertical motion straightness and the final vertical motion straightness under the two experimental conditions. The measurement results show that the initial motion straightness is a convex shape (which is related to the initial topography of the rail surface of the platform), and as the experiment progresses and the temperature gradually increases, its appearance remains unchanged, but the amplitude gradually increases. Similarly, the higher the oil temperature, the higher the amplitude increase.

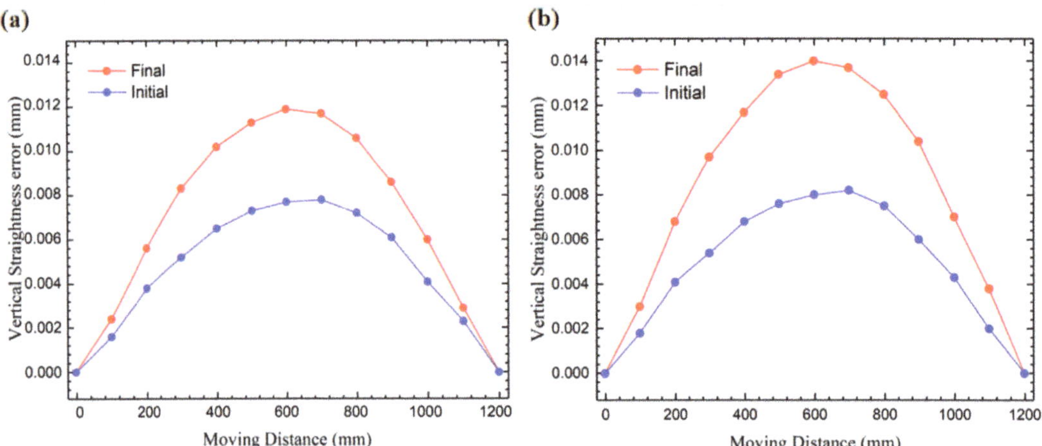

Figure 6. Vertical straightness error under different experimental conditions. (**a**) Experiment #1: oil temperature rises 2.5 °C. (**b**) Experiment #2: oil temperature rises 3.7 °C.

4. Discussion

To explore the characteristics of the straightness of the table motion under the influence of different oil temperatures, in this section, combined with the finite element thermal structure coupling simulation, we will use the kinematics model established in the second section to illustrate the experimental results in detail.

4.1. Self-Consistent Setting of the Initial State of the Guide Rails

Previous research is all based on the kinematics theory, establishing the quasi-static theoretical model of the motion error of the hydrostatic guideway. Based on the proposed model, the posture of the worktable at each position in the guide rails can be solved, and then the motion errors can be obtained. In the research process, the initial data of the surface profile of the hydrostatic guide rails were measured, and then the motion straightness of the table movement was calculated according to the established model. Reference [6] also achieved an ideal prediction result.

However, in fact, due to the hydrostatic guide rail being in a closed state during operation, and under different environmental conditions such as load or a certain operating time, the initial topography of the guide rail tends to change to a certain extent. During the normal operation of the machine tool, it is impossible to disassemble it frequently to measure the data of the guideway surface. Therefore, predicting the motion straightness based on the data measured before packaging the guide rails is still not effective. Therefore, in this paper, we propose a reverse calculation idea, that is, considering the basic accuracy of the kinematics model, combined with the measurement results of the motion straightness, a set of reasonable initial parameters of the rail surface are given. In short, we regard the closed guide rail as a "black box", using the established kinematics model, combined with the "external" measurement results, and infer a set of reasonable "black box" initial parameters. We use that set of initial parameters to predict the motion accuracy under other conditions. We call this process from effect to cause the "self-consistent setting of the initial state of the guide rails".

The surface profile of the guide rails is usually characterized by a chord function. For example, in the calculation in Reference [6], a set of sine functions is used to express the surface parameters of the guide rail before fine grinding. In this paper, we set the following function as the initial rail surface profile:

$$S_{ini} = A_s + B_1 \cdot x + B_2 \cdot x^2 + B_3 \cdot x^3 + B_4 \cdot x^4 \tag{8}$$

Among them, x is the length coordinate of the guide rail. A_s, B_1, B_2, B_3, and B_4 are undetermined parameters. Further, we simplified the model conditions, considering that the parameters of Guide Rail A and Guide Rail B are the same and considering the oil film clearance to be the design value of 50 μm. According to the measurement results of the initial vertical straightness error in the third section (the blue curve in Figure 6), the theoretical results calculated by the model are close to the test results by setting a set of A_s, B_1, B_2, B_3, and B_4 values. In this article, we established a 3DOF kinematics model, so in addition to the vertical motion error z, there are two angular errors θ_x and θ_y (as shown in Figure 3). Here, we take the two main error terms of z and θ_x as the research objects. In the experiment, the angular error θ_x is measured by an electronic level meter. The setting values of the initial parameters of the guide rails are shown in Table 2. Under those parameters, the comparison curve of the vertical straightness error and angle θ_x calculated by the model with the measured data is shown in Figure 7.

Table 2. Parameter setting values of the initial state of the guide rail.

Parameter	A_s	B_1	B_2	B_3	B_5
value	-0.01132	4.15244×10^{-5}	-5.67375×10^{-9}	-7.56398×10^{-12}	1.13377×10^{-15}

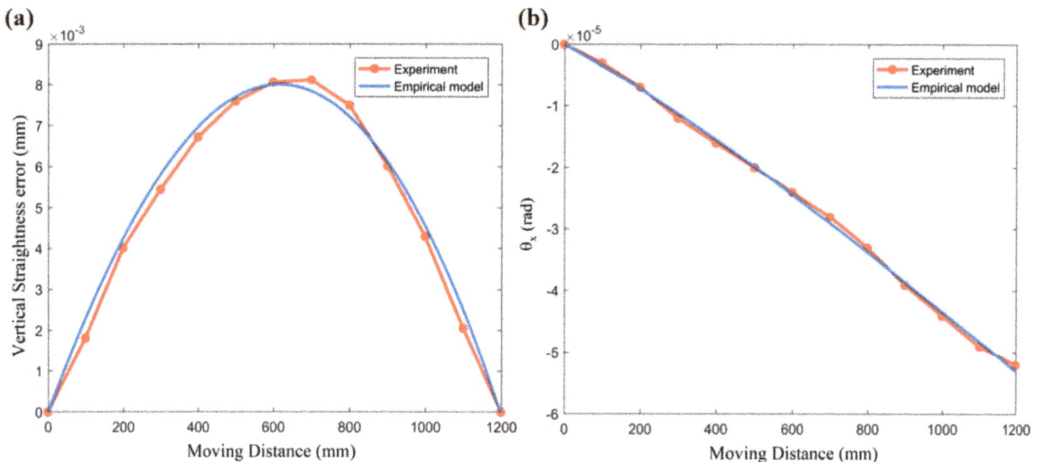

Figure 7. Initial motion errors. (**a**) Initial vertical straightness of the worktable. (**b**) Initial angular motion error of the worktable.

In Figure 7, taking the parameter values in Table 2 as the input data of the "black box", a set of vertical motion straightness error z and angular error θ_x (set the initial value to zero) can be calculated through the kinematics model, which fit well with the experimental results.

Here, we take the coefficient of correlation to denote the prediction accuracy of the curves in Figure 7. The mathematical expression can be given as [6]:

$$R^2 = \left(\frac{\sum (MD - \overline{MD})(TR - \overline{TR})}{\sqrt{\sum (MD - \overline{MD})^2 (TR - \overline{TR})^2}} \right) \qquad (9)$$

Among them, MD and $\overline{MD}$ are the measured data (the red line in Figure 7) and the corresponding mean value, respectively. TR and $\overline{TR}$ are the predicted theoretical result (the blue line in Figure 7) and the corresponding mean value, respectively. So, the coefficient of correlation of the vertical straightness error (R_z^2) and the coefficient of correlation of the angular error θ_x (R_a^2) can be calculated as 0.9785 and 0.9972, respectively. In fact, the optimal fitting parameters are more than the values given in Table 2. However, the following analysis results show that the selection of this set of parameters as the initial rail surface profile is sufficiently accurate in the prediction of thermal errors.

4.2. Finite Element Simulation of Thermal Deformation of Guide Rail Surface

In this section, the finite element analysis method is used to simulate the thermal deformation of the hydrostatic guideway to obtain the topographic parameters of the surface of the guideway after heating, which is used as the inputs of the model.

The material parameters are shown in Table 3. The steady-state thermal analysis module and the statics analysis module are combined to calculate the thermal deformation of the rail surface under the influence of the temperature rising of the oil film.

Table 3. Material parameters of HT300.

Parameters	Volume Density (g/cm^3)	Young's Modulus (GPa)	Poisson's Ratio	Thermal Expansion Coefficient (K^{-1})	Thermal Conductivity (Wm^{-1} K^{-1})	Specific Heat Capacity (Jkg^{-1} K^{-1})
value	6.6–7.4	115–160	0.23–0.27	8.5×10^{-6}	39.2	470

The meshing size is 20 mm, and the basic parts of the machine tool are set as the material of HT300, while the remaining parts are set as structural steel. The main heat sources in the working process of the hydrostatic guide rails are the power consumed during the movement of the moving guide rail and the heat generated by the linear motor. Part of the power consumed by the moving guide is the friction power consumed by the shearing oil film, and the other part is the power consumed when the hydraulic oil flows, that is, the output power of the oil pump. This consumed power will increase the temperature of the motor and the oil film and transfer heat into the moving rail through heat conduction, causing thermal deformation of the rail surface and ultimately leading to motion errors. As the motor is fixed outside of the bed, and the heat is not easy to conduct, it is not considered as the main heat source. Therefore, in the simulation process, the temperature rise of the oil film is used as the internal heat source.

When analyzing the thermal deformation caused by the internal heat source, the external heat dissipation of the machine tool must also be considered. Therefore, the load is the heating oil film of the guide rails, and the boundary condition is set as the air convection on the surface of the platform. The simulated initial temperature is 20 °C, which is a constant temperature of the environment. The convection exchange coefficient of the oil film is 300 W/(m^2 × °C). According to the two experimental conditions, the rising of the oil film temperature during the simulation is set to 2.5 °C and 3.7 °C, respectively.

It can be seen in Figure 8 that the area with the highest temperature rise is inside the guide rail, and the maximum temperature is 22.11 °C and 23.12 °C, respectively. Under the influence of this temperature field, the maximum thermal deformation of the whole experimental platform is 15.7 μm and 23.2 μm, respectively.

In order to explore the influence of thermal deformation on the hydrostatic guide rails, we extracted the deformation data of the two guide rails, namely the deformation of the bottom surface of Guide Rail A and Guide Rail B and the deformation of the corresponding top surface, as shown in Figure 9. The simulation results show that the rail surface becomes convex after being heated. The greater the temperature rise of the oil film, the greater the deformation of the guide rail surface.

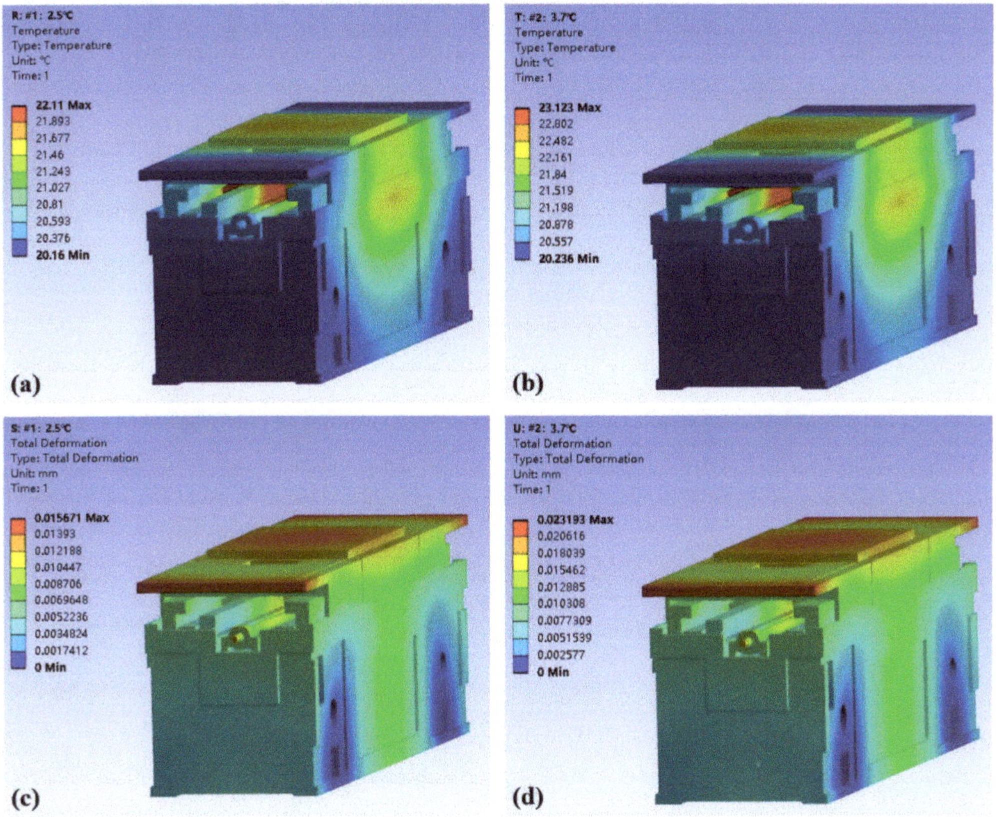

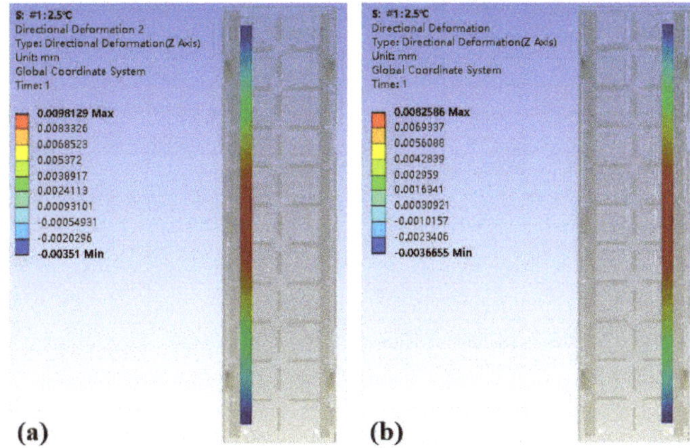

Figure 8. Simulation results. (**a**) Temperature field of Experiment #1, in which the temperature rise of the oil inlet is 2.5 °C. (**b**) Experiment #2, in which the temperature rise of the oil inlet is 3.7 °C. (**c**) Total thermal deformation in Experiment #1. (**d**) Total thermal deformation in Experiment #2.

Figure 9. *Cont.*

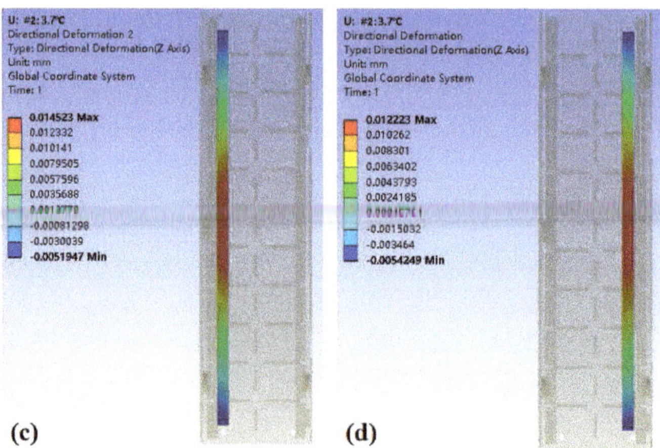

Figure 9. Thermal deformation of the rail surface. (a) Surface deformation of Rail A in Experiment #1; (b) Surface deformation of Rail B in Experiment #1; (c) Surface deformation of Rail A in Experiment #2; (d) Surface deformation of Rail B in Experiment #2.

Further, we made the curve shown in Figure 10 through the polynomial fitting. The thermal deformation fitting curve of the rail surface in Figure 10a can be expressed as follows:

$$
\begin{aligned}
\Delta S_A(2.5\,^\circ\text{C}) &= -0.00328 + 1.42831 \times 10^{-5} \cdot x - 1.86401 \times 10^{-8} \cdot x^2 + 5.15349 \times 10^{-11} \cdot x^3 \\
&\quad -5.31123 \times 10^{-14} \cdot x^4 + 2.11543 \times 10^{-17} \cdot x^5 - 2.93378 \times 10^{-21} \cdot x^6 \\
\Delta S_B(2.5\,^\circ\text{C}) &= -0.0036 + 1.2941 \times 10^{-5} \cdot x - 6.58874 \times 10^{-9} \cdot x^2 + 2.00016 \times 10^{-11} \cdot x^3 \\
&\quad -2.42081 \times 10^{-14} \cdot x^4 + 1.02108 \times 10^{-17} \cdot x^5 - 1.45877 \times 10^{-21} \cdot x^6
\end{aligned}
\tag{10}
$$

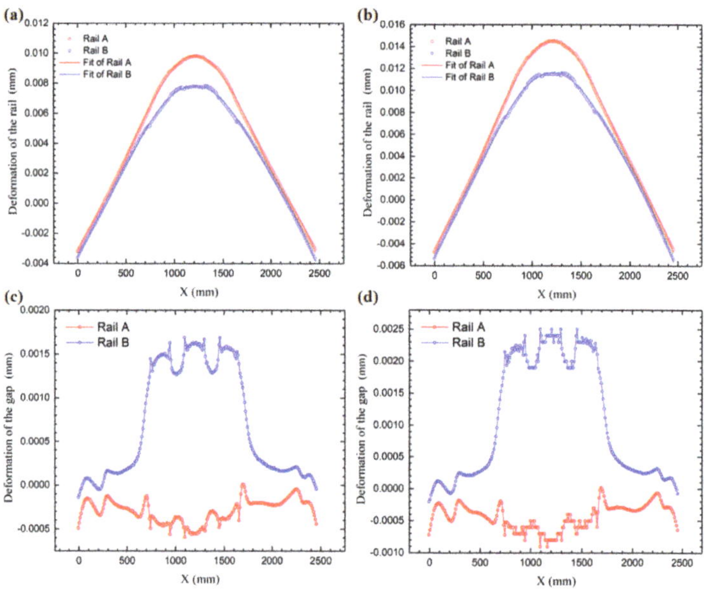

Figure 10. Simulation results and the fitted curve. (a) Fitting curve of the thermal deformation of the guide rail surface in Experiment #1; (b) Fitting curve of the thermal deformation of the guide rail surface in Experiment #2; (c) Deformation of the oil film clearance in Experiment #1; (d) Deformation of the oil film clearance in Experiment #2.

Similarly, the fitting curve of the thermal deformation of the guide rail surface in Figure 10b can be expressed as follows:

$$
\begin{aligned}
\Delta S_A(3.7\,^\circ\mathrm{C}) &= -0.00486 + 2.11799 \times 10^{-5} \cdot x - 2.77544 \times 10^{-8} \cdot x^2 + 7.65134 \times 10^{-11} \cdot x^3 \\
&\quad -7.87634 \times 10^{-14} \cdot x^4 + 3.13557 \times 10^{-17} \cdot x^5 - 4.34737 \times 10^{-21} \cdot x^6 \\
\Delta S_B(3.7\,^\circ\mathrm{C}) &= -0.00532 + 1.90975 \times 10^{-5} \cdot x - 9.45497 \times 10^{-9} \cdot x^2 + 2.90384 \times 10^{-11} \cdot x^3 \\
&\quad -3.5359 \times 10^{-14} \cdot x^4 + 1.49358 \times 10^{-17} \cdot x^5 - 2.13443 \times 10^{-21} \cdot x^6
\end{aligned}
\tag{11}
$$

Figure 10c,d, respectively, corresponds to the difference between the thermal deformation of the top surface of the guide rail and that of the bottom surface of the guide rail, which equals the oil film gap value. It can also be inferred that the higher the oil film temperature, the greater the change in the oil film clearance. However, the overall change in the oil film clearance is relatively small, and it is concentrated in the middle of the guide rail.

4.3. Thermal Model of Motion Straightness

Here, we combine the initial rail surface profile parameters in Section 4.1 and the thermal deformation data of the rail surface in Section 4.2, that is, the expression of the rail surface profile can be shown as follows:

$$
\begin{aligned}
\Delta S_A(2.5\,^\circ\mathrm{C}) &= -0.01132 + 4.15244 \times 10^{-5} \cdot x - 5.67375 \times 10^{-9} \cdot x^2 \\
&\quad -7.56398 \times 10^{-12} \cdot x^3 + 1.13377 \times 10^{-15} \cdot x^4 \\
&\quad -0.00328 + 1.42831 \times 10^{-5} \cdot x - 1.86401 \times 10^{-8} \cdot x^2 + 5.15349 \times 10^{-11} \cdot x^3 \\
&\quad -5.31123 \times 10^{-14} \cdot x^4 + 2.11543 \times 10^{-17} \cdot x^5 - 2.93378 \times 10^{-21} \cdot x^6 \\
\Delta S_B(2.5\,^\circ\mathrm{C}) &= -0.01132 + 4.15244 \times 10^{-5} \cdot x - 5.67375 \times 10^{-9} \cdot x^2 \\
&\quad -7.56398 \times 10^{-12} \cdot x^3 + 1.13377 \times 10^{-15} \cdot x^4 \\
&\quad -0.0036 + 1.2941 \times 10^{-5} \cdot x - 6.58874 \times 10^{-9} \cdot x^2 + 2.00016 \times 10^{-11} \cdot x^3 \\
&\quad -2.42081 \times 10^{-14} \cdot x^4 + 1.02108 \times 10^{-17} \cdot x^5 - 1.45877 \times 10^{-21} \cdot x^6
\end{aligned}
\tag{12}
$$

$$
\begin{aligned}
\Delta S_A(3.7\,^\circ\mathrm{C}) &= -0.01132 + 4.15244 \times 10^{-5} \cdot x - 5.67375 \times 10^{-9} \cdot x^2 \\
&\quad -7.56398 \times 10^{-12} \cdot x^3 + 1.13377 \times 10^{-15} \cdot x^4 \\
&\quad -0.00486 + 2.11799 \times 10^{-5} \cdot x - 2.77544 \times 10^{-8} \cdot x^2 + 7.65134 \times 10^{-11} \cdot x^3 \\
&\quad -7.87634 \times 10^{-14} \cdot x^4 + 3.13557 \times 10^{-17} \cdot x^5 - 4.34737 \times 10^{-21} \cdot x^6 \\
\Delta S_B(3.7\,^\circ\mathrm{C}) &= -0.01132 + 4.15244 \times 10^{-5} \cdot x - 5.67375 \times 10^{-9} \cdot x^2 \\
&\quad -7.56398 \times 10^{-12} \cdot x^3 + 1.13377 \times 10^{-15} \cdot x^4 \\
&\quad -0.00532 + 1.90975 \times 10^{-5} \cdot x - 9.45497 \times 10^{-9} \cdot x^2 + 2.90384 \times 10^{-11} \cdot x^3 \\
&\quad -3.5359 \times 10^{-14} \cdot x^4 + 1.49358 \times 10^{-17} \cdot x^5 - 2.13443 \times 10^{-21} \cdot x^6
\end{aligned}
\tag{13}
$$

Taking the above equations as the input of the model, the thermal error prediction curve of the moving parts of the hydrostatic rail can be obtained. As shown in Figure 11, Figure 11a,b represents the vertical straightness error curve of the oil film temperature rise of 2.5 °C and 3.7 °C, respectively, under the two experimental conditions. The blue curves represent the initial straightness error, and the red curves represent the vertical straightness error affected by heat. Those curves are close to the actual measured vertical straightness error data shown in Figure 12 in both trend and value.

In order to verify the accuracy of the thermal error prediction, Figure 12 shows the comparison between the theoretical results and the measured data of the motion error of the worktable. We choose vertical motion straightness z and angular error θ_x as analysis objects. Figure 12a,c shows the comparison between the predicted results and the experimental results under the condition of the oil temperature rising by 2.5 °C. Figure 12b,d shows the comparison between the predicted results and the experimental results under the condition of the oil temperature rising by 3.7 °C. This result shows that under the same initial parameters of the guide rail, we can obtain relatively accurate predictions for the thermal influence of the motion errors under different temperature conditions.

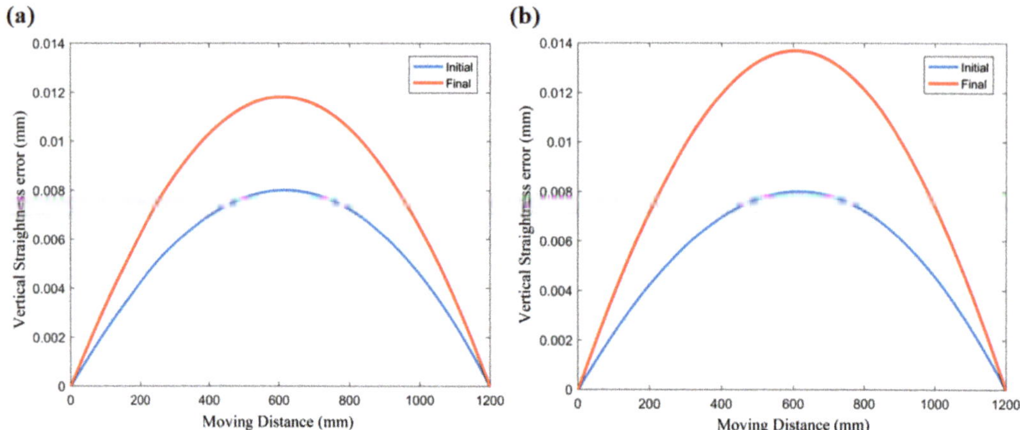

Figure 11. Prediction curve of the vertical straightness. (**a**) Experiment #1: temperature rise of the oil inlet is 2.5 °C. (**b**) Experiment #2: temperature rise of the oil inlet is 3.7 °C.

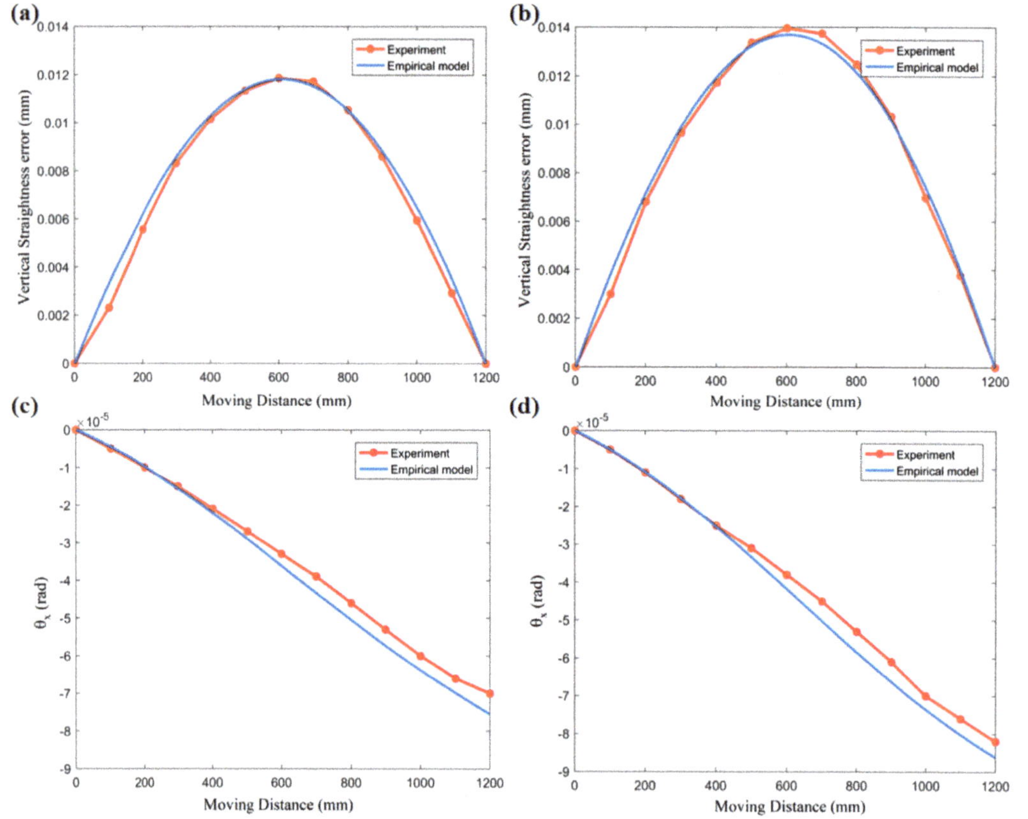

Figure 12. Motion errors prediction curve of comparison with the experimental data. Vertical straightness error under the condition of (**a**) Experiment #1: temperature rise of the oil inlet is 2.5 °C. (**b**) Experiment #2: oil temperature rising by 3.7 °C. Angular motion error under the condition of (**c**) Experiment #1: oil temperature rising by 2.5 °C. (**d**) Experiment #2: oil temperature rising by 3.7 °C.

In fact, through the following analysis, it can be found that the amplitude of the initial rail surface has little effect on the thermal influence of the motion errors. Through Equation (8) in Section 4.1, we obtained that the initial rail surface parameters satisfy:

$$S_{ini\_0} = -0.01132 + 4.15244\times10^{-5} \cdot x - 5.67375\times10^{-9} \cdot x^2 \\ -7.56398\times10^{-12} \cdot x^3 + 1.13377\times10^{-15} \cdot x^4 \tag{14}$$

Then, we set a series of initial rail surface shapes with different PV:

$$S_{ini\_\alpha} = \alpha \times (-0.01132 + 4.15244\times10^{-5} \cdot x - 5.67375\times10^{-9} \cdot x^2 \\ -7.56398\times10^{-12} \cdot x^3 + 1.13377\times10^{-15} \cdot x^4) \tag{15}$$

By changing the value of α, a series of initial rail surface shapes with different amplitudes can be obtained, and a series of corresponding initial straightness errors can be obtained through the model. Then, taking this series of initial rail surface profiles as the input, and combining it with the thermal deformation of the guide rails, we finally obtain the prediction curve of the vertical straightness error. Subtracting the amplitude of the red curve from the PV of the blue curve in Figure 11, we can obtain the difference in vertical straightness error, that is, the thermal error of motion straightness. Here, we set the α value from 0 to 1.2 so that the initial profile of the rail changes from a straight line to a curve with a PV of 34 µm. Thus, the PV range of the initial vertical straightness error is 0 to 15 µm (as shown in the horizontal axis of Figure 13). Based on those different initial PV values of the guide rail surface, combined with the thermal deformation, we can obtain a set of PV values of the initial vertical straightness error (AS_{ini}) and the final vertical straightness error (AS_{heated}) so that AS_{heated} minus AS_{ini} represents the thermal error of the vertical straightness. Then, Figure 13a can be obtained, which shows that the PV of the initial rail surface has no significant effect on the thermal error of the moving parts, and the oil temperature is the decisive factor for the thermal error. This is also the reason why the selection of a set of appropriate parameters as the initial rail surface profile is sufficiently accurate in the prediction of thermal error as mentioned in Section 4.1.

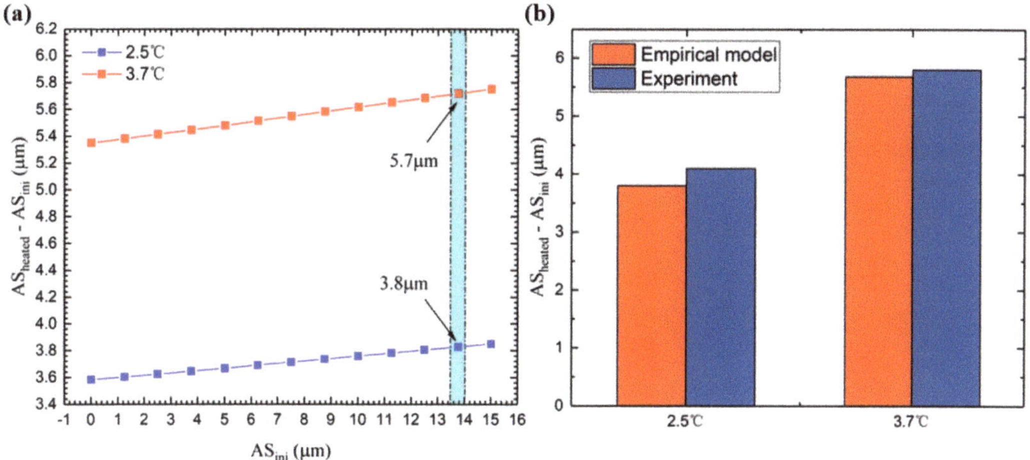

Figure 13. Thermal error analysis results. (**a**) Effect of the initial vertical straightness error on the thermal error. (**b**) Comparison between theoretical results and the measured data.

In the actual experimental measurement process, the value of the initial vertical straightness error fluctuates within a certain range, as shown in the blue area of Figure 13a. However, since the PV of the initial rail surface does not have a significant effect on the thermal error of the moving parts, we still obtained a sufficiently accurate value of the final motion error, as shown by the arrows in Figure 13a. Under the condition that the oil temperature rises 2.5 °C, the PV of the vertical straightness error of the hydrostatic worktable increased by 4.1 μm. Correspondingly, its theoretical value is 3.8 μm. Similarly, under the condition of an oil temperature rise of 3.7 °C, the PV of the vertical straightness error of the hydrostatic worktable increased by 5.8 μm, and the theoretical value is 5.7 μm. The comparison result is shown in Figure 13b.

5. Conclusions

In the research of this paper, firstly, a 3-DOF quasi-static kinematics model was established based on the flow characteristics of the PM flow controller and the parameters of the guide rails and guide shoes. At the same time, by setting up an experimental platform, we carried out a series of experimental studies on the thermal effect of the motion error. Comparing the measurement data of the vertical straightness error under the two experimental conditions (oil temperature rising 2.5 °C and 3.7 °C, respectively), it is shown that the initial vertical straightness error of the experimental platform is a convex shape. As the experiment progressed, the temperature of the oil increased, as did the PV value of the vertical straightness error, but the shape of the vertical straightness error remained unchanged. Similarly, the higher the oil temperature, the higher the PV increase.

On the basis of the established 3-DOF quasi-static kinematics model, we proposed a self-consistent setting method of the initial state of the guide rails so that the parameters of the initial shape of the guide rails in a closed state can be given. Furthermore, with finite element simulation, we obtained the profile of the thermal deformation of the rail surface. The motion error with the effect of heat was predicted by the 3-DOF quasi-static model, and the error between the theoretical results and the measured results on the vertical straightness error was less than 1 μm. The results show that the PV value of the initial rail surface has no significant effect on the thermal error of the moving parts, while the oil temperature is the decisive factor for the thermal error. Through the self-consistent setting method of the initial state of the guide rails, the prediction of the thermal effect of the motion error has a certain accuracy.

Author Contributions: Conceptualization, P.L. and Z.W.; methodology, P.L. and C.S.; software, P.L. and F.L.; validation, P.L. and C.S.; formal analysis, Z.W.; investigation, P.L.; resources, Y.P.; data curation, P.L.; writing—original draft preparation, P.L.; writing—review and editing, C.S.; supervision, Z.W.; project administration, Y.P. All authors have read and agreed to the published version of the manuscript.

Funding: This research received no external funding.

Institutional Review Board Statement: Not applicable.

Informed Consent Statement: Not applicable.

Data Availability Statement: Not applicable.

Conflicts of Interest: The authors declare no conflict of interest.

References

1. Stolz, C.J. The National Ignition Facility: The world's largest optics and laser system. In Proceedings of the Optical Design and Testing III(Part One of Two Parts), Beijing, China, 12–15 November 2007.
2. Wisoff, P.; Bowers, M.W.; Erbert, G.V.; Browning, D.F.; Jedlovec, D.R. NIF injection laser system. In Proceedings of the SPIE-The International Society for Optical Engineering, San Jose, CA, USA, 25 January 2004; Volume 5341, pp. 146–155.
3. Cheng, B.L.; Bradley, P.A.; Finnegan, S.M.; Thomas, C.A. Fundamental factors affecting thermonuclear ignition. *Nucl. Fusion* **2021**, *61*, 096010. [CrossRef]
4. Thomas, C.A.; Campbell, E.M.; Baker, K.L.; Casey, D.T.; Hohenberger, M.; Kritcher, A.L.; Spears, B.K.; Khan, S.F.; Nora, R.; Woods, D.T.; et al. Principal factors in performance of indirect-drive laser fusion experiments. *Phys. Plasmas* **2020**, *27*, 112712. [CrossRef]
5. Guo, Y. Technology and Application of Ultra-precision Machining for Large Size Optic. *Mech. Eng.* **2013**, *49*, 171–178. [CrossRef]
6. Shi, C.; Wang, Z.; Peng, Y.; Lei, P.; Li, C. Quasi-static kinematics model for motion errors of closed hydrostatic guideways in ultra-precision machining. *Precis. Eng.* **2021**, *71*, 90–102. [CrossRef]
7. Liu, C.; Hu, J.; Qian, H. Preview Control of Hydrostatic Guideway for Ultraprecision CNC Machine Tools. *Iran. Sci. Technol. Trans. Mech. Eng.* **2019**, *43*, 749–759. [CrossRef]

8. Wang, Z.; Liu, Y.; Wang, F. Rapid calculation method for estimating static and dynamic performances of closed hydrostatic guideways. *Ind. Lubr. Tribol.* **2017**, *69*, 1040–1048. [CrossRef]
9. Shamoto, E.; Park, C.H.; Moriwaki, T. Analysis and Improvement of Motion Accuracy of Hydrostatic Feed Table. *CIRP Ann.-Manuf. Technol.* **2001**, *50*, 285–290. [CrossRef]
10. Leadbeater, P.B.; Clarke, M.; Wills-Moren, W.J.; Wilson, T.J. A unique machine for grinding large, off-axis optical components: The OAGM 2500. *Precis. Eng.* **1989**, *11*, 191–196. [CrossRef]
11. Park, C.-H.; Oh, Y.-J.; Lee, C.-H.; Hong, J.-H. Theoretical Verification on the Motion Error Analysis Method of Hydrostatic Bearing Tables Using a Transfer Function. *Int. J. Precis. Eng. Manuf.* **2003**, *4*, 64–70.
12. Oh, Y.-J.; Park, C.-H.; Lee, C.-H.; Hong, J.-H. Experimental Verification on Motion Error Analysis Method of Hydrostatic Tables Using Transfer Function. *J. Korean Soc. Precis. Eng.* **2002**, *19*, 64–71.
13. Zha, J.; DLv Jia, Q.; Chen, Y. Motion straightness of hydrostatic guideways considering the ratio of pad center spacing to guide rail profile error wavelength. *Int. J. Adv. Manuf. Technol.* **2016**, *82*, 2065–2073. [CrossRef]
14. Zha, J.; Xue, F.; Chen, Y. Straightness error modeling and compensation for gantry type open hydrostatic guideways in grinding machine. *Int. J. Mach. Tools Manuf.* **2017**, *112*, 1–6. [CrossRef]
15. Xue, F.; Zhao, W.; Chen, Y.; Wang, Z. Research on error averaging effect of hydrostatic guideways. *Precis. Eng.* **2012**, *36*, 84–90. [CrossRef]
16. Shi, C.; Wang, Z.; Peng, Y.; Li, C.; Kong, L. Influence of PM Controller Parameters on Motion Accuracy of Hydrostatic Guideways. *J. Mech. Eng.* **2020**, *56*, 157.
17. Chenchun, S.; Zhenzhong, W.; Yunfeng, P.; Pengli, L.; Chenlei, L. Influence of relative difference between paired guide rails on motion accuracy in closed hydrostatic guideways. *J. Mech. Sci. Technol.* **2020**, *34*, 631–648.
18. Yang, J.; Shi, H.; Feng, B.; Zhao, L.; Ma, C.; Mei, X. Thermal error modeling and compensation for a high-speed motorized spindle. *Int. J. Adv. Manuf. Technol.* **2015**, *77*, 1005–1017. [CrossRef]
19. Chen, D.; Wang, H.; Pan, R.; Fan, J.; Cheng, Q. An accurate characterization method to tracing the geometric defect of the machined surface for complex five-axis machine tools. *Int. J. Adv. Manuf. Technol.* **2017**, *93*, 3395–3408. [CrossRef]
20. Liu, M.; Xiong, Z.; Han, S.; Zhou, S.; Zhou, Z.; Zhou, W. Inverse Finite Element Method for Reconstruction of Deformation in the Gantry Structure of Heavy-Duty Machine Tool Using FBG Sensors. *Sensors* **2018**, *18*, 2173. [CrossRef]
21. Zhang, Y.; Yang, J.; Xiang, S.; Xiao, H. Volumetric error modeling and compensation considering thermal effect on five-axis machine tools. *Proc. Inst. Mech. Eng. Part C J. Mech. Eng. Sci.* **2013**, *227*, 1102–1115. [CrossRef]
22. Liu, K.; Han, W.; Wang, Y.; Liu, H.; Song, L. Review on Thermal Error Compensation for Feed Axes of CNC Machine Tools. *J. Mech. Eng.* **2021**, *57*, 156–173.
23. Jeon, S.Y.; Kim, K.H. A fluid film model for finite element analysis of structures with linear hydrostatic bearings. *Proc. Inst. Mech. Eng. Part C J. Mech. Eng. Sci.* **2004**, *218*, 309–316. [CrossRef]
24. Zhang, P.; Chen, Y.; Zhang, C.; Zha, J.; Wang, T. Influence of geometric errors of guide rails and table on motion errors of hydrostatic guideways under quasi-static condition. *Int. J. Mach. Tools Manuf.* **2018**, *125*, 55–67. [CrossRef]
25. Gao, D.; Zhao, J.; Zhang, Z.; Zheng, D. Research on the Influence of PM Controller Parameters on the Performance of Hydrostatic Slide for NC Machine Tool. *J. Mech. Eng.* **2011**, *47*, 186–194. [CrossRef]
26. Gao, D.; Zhang, D.; Zhang, Z. Theoretical analysis and numerical simulation of the static and dynamic characteristics of hydrostatic guides based on progressive mengen flow controller. *Chin. J. Mech. Eng.* **2010**, *23*, 709–716. [CrossRef]

micromachines

Article

Knowledge-Driven Manufacturing Process Innovation: A Case Study on Problem Solving in Micro-Turbine Machining

Dong Zhang [1], Gangfeng Wang [1,*], Yupeng Xin [2], Xiaolin Shi [3], Richard Evans [4], Biao Guo [5] and Pu Huang [1]

[1] Key Laboratory of Road Construction Technology and Equipment of MOE, School of Construction Machinery, Chang'an University, Xi'an 710064, China; 2019125089@chd.edu.cn (D.Z.); 2020225023@chd.edu.cn (P.H.)
[2] College of Mechanical and Vehicle Engineering, Taiyuan University of Technology, Taiyuan 030024, China; xinyupeng@tyut.edu.cn
[3] School of Mechanical Engineering, Northwestern Polytechnical University, Xi'an 710072, China; sxl86@mail.nwpu.edu.cn
[4] Faculty of Computer Science, Dalhousie University, Halifax, NS B3H 4R2, Canada; R.Evans@dal.ca
[5] China Huayin Ordnance Test Center, Huayin 714200, China; guo_biao1234@126.com
* Correspondence: wanggf@chd.edu.cn

Abstract: Micromachining techniques have been applied widely to many industrial sectors, including aerospace, automotive, and precision instruments. However, due to their high-precision machining requirements, and the knowledge-intensive characteristics of miniaturized parts, complex manufacturing process problems often hinder production. To solve these problems, a systematic scheme for structured micromachining process problem solving and an innovation support system is required. This paper presents a knowledge-based holistic framework that enables process planners to achieve micromachining innovation design. By analyzing innovation design procedures and available knowledge sources, an open multi-source Machining Process Innovation Knowledge (MPIK) acquisition paradigm is presented, including knowledge units and a knowledge network. Further, a MPIK network-driven structured process problem-solving and heuristic innovation design method was explored. Subsequently, a knowledge-driven heuristic design system for machining process innovation was integrated in the Computer-Aided Process Innovation (CAPI) platform. Finally, a case study involving specific process problem-solving and innovation scheme design for micro-turbine machining was studied to validate the proposed approach.

Keywords: micromachining; micro-turbine; computer-aided innovation; innovation design; smart manufacturing; knowledge-based engineering; problem solving

Citation: Zhang, D.; Wang, G.; Xin, Y.; Shi, X.; Evans, R.; Guo, B.; Huang, P. Knowledge-Driven Manufacturing Process Innovation: A Case Study on Problem Solving in Micro-Turbine Machining. *Micromachines* **2021**, *12*, 1357. https://doi.org/10.3390/mi12111357

Academic Editors: Benny C. F. Cheung and Jiang Guo

Received: 15 October 2021
Accepted: 1 November 2021
Published: 3 November 2021

1. Introduction

Advancements in artificial intelligence, advanced manufacturing, automation control, and optics and microelectronics technologies, have evolved the way in which industrial products are designed, with emphasis now being placed on miniaturization and structural complexity [1–5]. Product miniaturization reduces overall weight and material and energy consumption, and improves space utilization, but places greater requirements on parts micromachining technology [6,7]. In the field of modern manufacturing technologies, micromachining refers to the application of precision or ultra-precision cutting tools to remove metal, composite, alloy, ceramics, and other engineering materials, to obtain microscale parts or structures using mechanical force. The geometric size of such micro-parts or structures is often at the centimeter level or smaller, creating difficulties in the conventional fixing and clamping of parts. Due to high-precision requirements and the small size of parts, they can easily deform, especially during the machining of thin-walled structures, micro-holes, and slender shafts [6,8].

Extant research into micromachining technology has predominantly focused on micromachining tools, cutting tools, and the influence of micromachining parameters on machining quality and efficiency. Takacs et al. [9] used fine-grained carbide end mills to conduct experimental research into micromilling for metallic materials. To overcome the nonlinearity of the conventional servomechanism, Wang et al. [10] developed a new control system for three-axis ultra-precision micromilling using coreless linear motors and air bearing slides. Scholars have also optimized the microcutting process parameters and process routes of specific materials or parts. To improve the micro-surface roughness of aluminum alloy, Cardoso and Davim [11] explored the optimization of cutting parameters such as feed rate and machining strategies. In the machining of micro-parts and structures, it is necessary to optimize process planning, not only to ensure machining accuracy and efficiency, but also to control the feed rate and chip thickness to avoid deformation. Son et al. [12] studied the relationship between the friction of a tool-workpiece and the minimum cutting thickness in ultra-precision diamond micro-cutting, and developed a corresponding ultra-precision cutting model. In light of the machining process problems associated with complex parts, several process innovation technologies, and methods for microparts with specific structures have been developed. To avoid tool breakage and to ensure the geometrical accuracy of the machined feature in micromachining, Ba et al. [13] proposed a cutting force prediction method that combined sensitivity analysis and finite element simulation. To solve the contradiction between minimum chip thickness and relatively large tool deformations, under the condition of micromachining, Balazs et al. [14] produced a dynamic milling tool path strategy and performed systematic experiments.

However, the influencing factors and formation mechanism of micromachining process problems are relatively complicated, and effective process innovation and optimization schemes are difficult to design structurally. Most existing innovations are achieved through trial-and-error or unstructured brainstorming sessions, which lead to micromachining process innovations relying on the inspiration of only a few designers, leading to instability in innovations and resource wastage [15–17]. The structured approach of heuristic innovation has distinct characteristics from other traditional problem-solving methods, such as trial-and-error and brainstorming, which usually attempt to directly find specific factual solutions for factual problems.

Computer-Aided Innovation (CAI) technology has provided an effective means to assist designers in obtaining innovation inspiration and in improving efficiency during technological innovation. CAI adopts computer technology, combined with modern design methodologies, innovation thinking theory, knowledge management, information and communication technology, cognitive psychology, and other multidisciplinary fields [18]. To further shorten the product development lifecycle, against the background of CAI, Cugini et al. [19] proposed an integrated CAI methodology to ensure the interoperability of multiple systems by adopting optimization systems as a bridge between CAI and Product Lifecycle Management (PLM) systems. Xu et al. [20] studied knowledge management and its application in product innovation design, and proposed an integrated knowledge modeling and management approach for continuous innovation. Esterhuizen et al. [21] studied the role of knowledge management in enhancing innovation design and identified a knowledge creation path as a critical enabler for innovation capability maturity. To enhance computer-aided problem solving, Duran-Novoa et al. [22] presented a strategy that incorporated dialectical negation operators in evolutionary algorithms and TRIZ (The Theory of Inventive Problem Solving) principles [23]. Similarly, Cakir and Cilsal [24] built a TRIZ-like contradiction matrix-based access system, with a corresponding knowledge base that can provide design recommendations for turning, milling, and drilling processes. Delgado-Maciel et al. [25] proposed a methodology that combines TRIZ tools with system dynamics simulations to speed up new products development and to identify key inventive problems to ensure technical feasibility. By analyzing technological developments in open innovation and Web 2.0, Hüsig and Kohn [26] proposed an Open CAI 2.0 paradigm, based on Closed CAI 1.0. Flores et al. [27] introduced a conceptual framework that combined collective intelligence and a logical TRIZ approach for the inventive problem resolution of Open CAI 2.0.

The aforementioned studies have completed valuable exploratory work into aspects of innovation design methods, theoretical model construction, and system tool development, but have mainly focused on product innovation design rather than manufacturing process innovation. In addition, existing computer-aided tools, such as Computer-Aided Process Planning (CAPP) and PLM, have mainly been used to improve efficiency and the standardization of process planning and management [28,29] rather than to solve manufacturing process problems structurally and create or improve process methods; therefore, they cannot systematically improve the level of manufacturing process research and the development of enterprises [15,30,31]. In considering the intentions and application objects of existing studies, and the fact that the proposed design systems and methods cannot be directly applied to the manufacturing process innovation of conceptual designs, a Computer-Aided Process Innovation (CAPI) methodology and knowledge accumulation framework was proposed in previous research [32]. Whilst CAI is mostly employed in the field of product innovation design, some specific achievements demonstrate that knowledge-driven computer-supported process innovation tools can effectively inspire and guide designers to implement structured process problem solving [24,33–35]. At present, studies into CAPI are at the early stages of development and, therefore, the theory and method of manufacturing process innovation design, as well as knowledge management and applications supporting the innovation design process, need to be systematically explored [32,36,37].

This paper examines how to effectively acquire and organize machining innovation knowledge and how to apply this knowledge to facilitate heuristic process problem-solving and innovation scheme design. By analyzing the innovation design procedure and knowledge sources, several types of process innovation knowledge units are formally represented. Then, an open multi-source Machining Process Innovation Knowledge (MPIK) acquisition paradigm is proposed, and a heuristic machining process problem-solving method, based on innovation knowledge, is explored. Further, a Knowledge-driven Heuristic Design System for Machining Process Innovation (MPI-KHDS) is constructed to support knowledge-driven micro-turbine machining process innovation.

The rest of the paper is organized as follows: In Section 2, a holistic framework of knowledge-based manufacturing process innovation design is introduced. Section 3 presents the innovation knowledge acquisition and management of machining processes, including knowledge units and knowledge networks. Section 4 describes the knowledge network-driven machining process problem-solving and innovation design method. Then, the specific process of problem solving and the innovation design of micro-turbine machining is implemented in Section 5 with the support of the MPI-KHDS prototype system. Finally, conclusions of this study are summarized in Section 6.

2. A Holistic Framework of Knowledge-Based Manufacturing Process Innovation Design

Computer-aided manufacturing process innovation is a knowledge-based design procedure whereby formal process knowledge is derived from multiple manufacturing sources. The knowledge of these sources must be effectively acquired, organized, and formalized into manufacturing process innovation knowledge that can support the corresponding design stages of CAPI. As shown in Figure 1, this paper proposes a holistic framework of a knowledge-based manufacturing process innovation design, which includes two stages of innovation knowledge acquisition and formal expression, and a knowledge-based heuristic manufacturing process innovation design. In considering that MPIK has the characteristics of wide-range dispersion, strong fuzziness, and high experience, an open innovation knowledge acquisition method for the machining process is required in Stage I, which can effectively manage both explicit and implicit knowledge sources. Thus, MPIK can be organized according to the whole procedure of machining process innovation design. In Stage II, on the basis of formal innovation knowledge representation, different types of innovation knowledge units need to be further associated and organized, and a dynamic innovation knowledge network should be constructed to inspire process designers to effectively carry out structured innovation design in complex manufacturing process problem-solving scenarios.

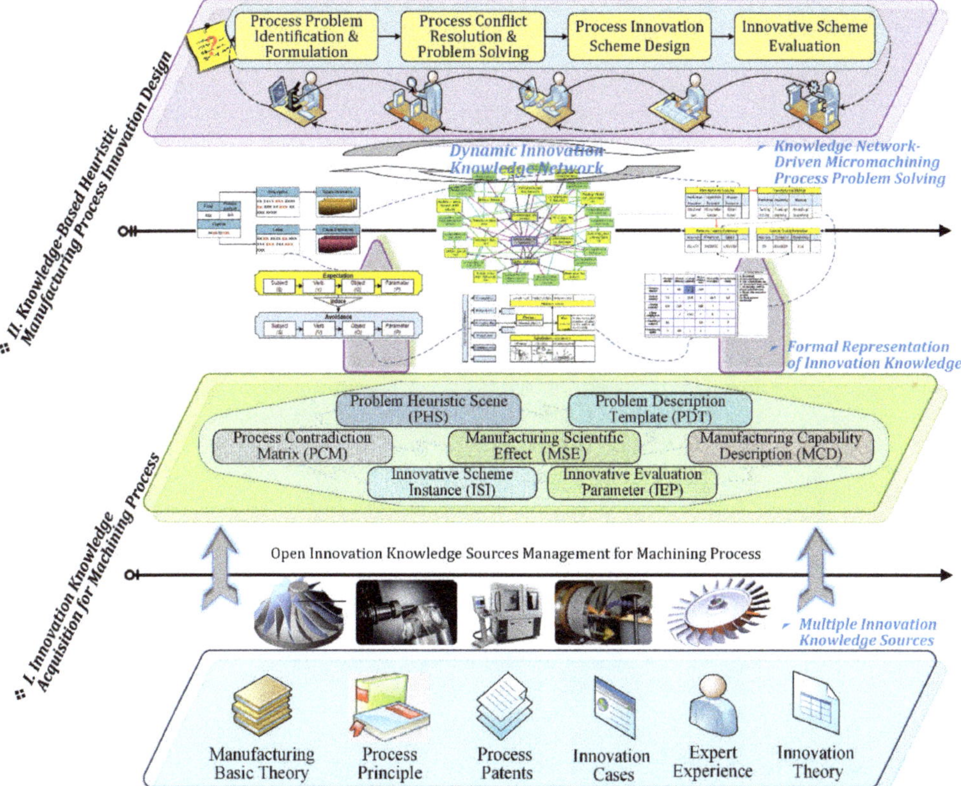

Figure 1. A holistic framework of knowledge-based heuristic manufacturing process innovation.

3. Innovation Knowledge Acquisition and Management for Machining Process

3.1. Sources and Contents of Machining Process Innovation Knowledge

MPIK is the basis of machining process problem solving in the context of computer-aided innovation. High-quality solutions depend on massive process innovation knowledge. Through the analysis of the computer-aided machining innovation design process and the required knowledge content, this paper divides the sources of process innovation knowledge as follows:

(1) Basic manufacturing theories. Basic manufacturing theories refer to the general laws of natural science that can provide guidance for applied research. They are the basic scientific principles that manufacturing process innovation activities should follow, such as the theories and principles of physics, chemistry, geometry, materials, and biology.

(2) Technical process principles. Technical process principles relate to the manufacturing methods and mechanisms that guide the production of parts and entire products in the manufacturing environment, such as cutting and measuring.

(3) Process patents and innovation cases. Process patents are an important knowledge source for manufacturing innovation that reflect the latest advances in multi-disciplinary fields involved in new technologies, new processes, new methods, etc.

Innovation cases can reflect the manufacturing ability and characteristics of an organization and can be used as a powerful reference for solving similar process problems and innovative scheme design.

(4) Expert experience. Expert experience refers to machining know-how, and experience and methods in the field of manufacturing that exist in the minds of manufacturing professionals and operators. Expert experience is a valuable asset accumulated through project experience, so explicit expert experience can be regarded as an important innovation knowledge source for the machining process.

(5) Innovation theories. Innovation theories refer to methodologies that use available resources to create something new, such as TRIZ and OTSM (General Theory of Powerful Thinking) [38,39]. Such theories can provide guidance and assistance for the implementation of manufacturing process innovation.

According to the analysis in Section 2, the procedure of process innovation design can be split into four steps that include: process problem identification and formulation, process conflict resolution and problem solving, process innovation scheme design, and innovative scheme evaluation. The purpose of the innovation process is to identify contradictions in problems and resolve them with dialectical thinking. The Theory of Inventive Problem Solving (TRIZ) follows this approach and receives widespread attention due to its structured procedure of problem solving. In the process problem identification step, typical problem scenarios have heuristic features and can help technicians to improve the efficiency and quality of problem recognition. In the innovative scheme evaluation step, the feasibility of innovative solutions require evaluation according to the evolution law of technology systems and the manufacturing capacity of the corresponding enterprise.

Different types of MPIK must be applied to the corresponding innovation design stage. Thus, MPIK can be divided into the following types to support the entire process of innovation design: Problem Heuristic Scene (PHS), Problem Description Template (PDT), Process Contradiction Matrix (PCM), Manufacturing Scientific Effect (MSE), Innovative Scheme Instance (ISI), Innovative Evaluation Parameter (IEP), and Manufacturing Capability Description (MCD). Table 1 shows the concept and function of each type of MPIK.

Table 1. Concept and function of each type of MPIK.

Type	Concept	Function
Problem Heuristic Scene (PHS)	Abstract and formalized phenomenon description of typical process problems to facilitate the cause analysis of problem scenes	To stimulate technicians to associate their tacit knowledge in order to recognize and analyze problems effectively and correctly
Problem Description Template (PDT)	Formalized representation framework according to the essential structure of machining process problem	Providing constraints for the formal description of process problems to ensure the clarity of problem expression and the understandability
Process Contradiction Matrix (PCM)	The relationships between technical parameters, contradictions, and process innovation principles	Providing the solving direction and principal solution for the structural conflict resolution of process problems
Manufacturing Scientific Effect (MSE)	Manufacturing-oriented and multidisciplinary basic scientific effects	Providing the basic technical framework and scientific effects for preliminary innovative solution design
Innovative Scheme Instance (ISI)	Formalized, practical, and successful schemes for typical process problems	Providing technical implementation reference solutions to support the detailed innovative scheme design
Innovative Evaluation Parameter (IEP)	Evolution path parameters and lifecycle phase parameters of all technical systems	Providing the quantitative evaluation parameters to identify the innovativeness grade of innovative solution
Manufacturing Capability Description (MCD)	Formalized description of manufacturing capability for the specific enterprise	Evaluating the manufacturing feasibility of the innovative scheme in the firm-specific manufacturing environment

3.2. Open Multi-Source Knowledge Acquisition for Machining Process Innovation

Manufacturing experts and technicians possess a powerful ability to solve machining problems, although the discrete and unstructured knowledge cannot be used directly in machining process innovation. Hence, the process of MPIK acquisition should recognize that knowledge can be turned from tacit to explicit, from discrete to associative, and from rough to refined. Considering dynamic knowledge forms, the organizing process of MPIK can be divided into three phases: discrete knowledge, Process Innovation Knowledge Unit (PIKU), and process innovation knowledge network.

The process innovation knowledge network is similar to a biological neural network. Neural networks consist of large numbers of neurons, accept external stimuli, and output a control action through the interaction between neurons. The process innovation knowledge network contains a large number of PIKUs, accepts the stimulus of process problems, and outputs innovative solutions through the interaction between PIKUs. Due to their similarity, we propose a construction method of the process innovation knowledge network by imitating a neural network. Thus, the PIKU, which should have specific interfaces and accurate parameters, can be considered a process innovation knowledge neuron in the knowledge network, and an adequate number of PIKUs constitute a Process Innovation Knowledge Neural Network (PIKNN) in the environment of machining process problem solving.

The PIKU has a knowledge parameter as an input interface and a knowledge result as an output interface, which corresponds to the dendrite and axon of a neuron, respectively. The encapsulation space of a PIKU is mapped to the cell membrane of a neuron. The knowledge attribute is mapped to the cytoplasm of a neuron. The core handling process is mapped to the nucleus of a neuron. The hierarchical structure of the machining process innovation-oriented knowledge network is shown in Figure 2 and formally defined as follows.

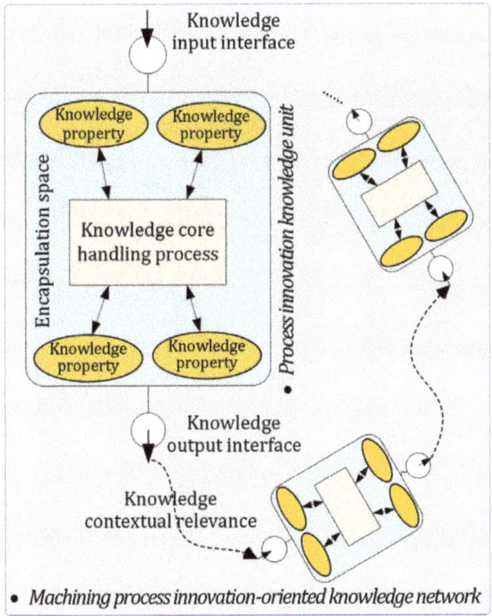

Figure 2. Illustration of the machining process innovation-oriented knowledge network.

Definition 1. *A manufacturing process innovation-oriented knowledge network is a set of spatial knowledge structures, formally represented as*

$$MPIK_\Omega = \{KU, CTR, U\} \tag{1}$$

where KU is a set of multi-type machining process innovation knowledge units, CTR is a set of knowledge contextual relevance for specific machining process innovation scenarios, and U is a set of process innovation participants.

Definition 2. *A MPIK unit has the ability to solve certain types of machining process problems and deliver information. It is denoted as*

$$KU = \langle P,\ I_I,\ I_O,\ E,\ U \rangle \tag{2}$$

where P is a set of knowledge properties, I_I and I_O are the sets of knowledge input interfaces and knowledge output interfaces, respectively; E represents an encapsulation space for the whole knowledge unit. Several types of machining process innovation knowledge, Π_{KU} = PHS,PDT,PCM,MSE, ISI,IEP,MCD, are mainly applied in the machining process innovation design.

Definition 3. *The knowledge contextual relevance of machining process innovation is further denoted by*

$$CTR = \left\{ \langle ku,k,r,k',u \rangle \mid ku \in KU, k,r,k' \in \mathbb{O}, u \in U \right\} \tag{3}$$

where k, r, and k' are ontological entities defined in the machining process innovation domain ontology $\mathbb{O}$, and r stands for a contextual relationship between k and k'.

Definition 4. *Ontology $\mathbb{O}$ consists of a series of concepts and relationships that represent domain knowledge models. It can be represented as*

$$\mathbb{O} := (C, R, \mathcal{E}_R, I_C) \tag{4}$$

where C and R represent a set of classes and a set of relations, respectively; $\mathcal{E}_R \subseteq C \times C$ is a set of relationships between classes, which can be denoted as a set of triples $\{ \langle c,r,c' \rangle \mid c,c' \in C, r \in R \}$; and I_C is a power set of instance sets of a class $c \in C$.

The PIKUs can be divided into several categories, which can be built as knowledge ontologies in accordance with the corresponding knowledge classifications. Meanwhile, the process data ontologies must be built to provide input parameters and output results for the knowledge interfaces. Table 2 describes the basic process data ontologies and the knowledge ontologies in the machining process innovation domain. The semantic relationships between process data ontologies and knowledge ontologies characterize the knowledge interfaces.

Social network technologies, based on relationship networks and interested topics, can provide an exchanging, sharing, and manifesting knowledge platform beyond background and specialty. Wiki technologies can provide a knowledge refining and associating platform through page locking and collaborative editing [32]. Therefore, for the knowledge building stages of PIKU and PIKNN, combining the characteristics of social networks with wikis, an open MPIK acquisition paradigm, based on a social-wiki network, was constructed and is presented in Figure 3.

Table 2. The basic ontologies for machining process innovation.

Type	Ontology Contents
Process Data Ontology	Initial Problem Description (IPD)
	Problem Description with Cause (PDC)
	Formalized Problem Description (FPD)
	Principle Solution (PSU)
	Initial Solution (ISU)
	Detailed Solution (DSU)
	Solution Innovativeness (SIV)
	Solution Manufacturability (SMF)
Knowledge Ontology	Heuristic Neuron (HN)
	Template Neuron (TN)
	Contradiction Neuron (CN)
	Effect Neuron (EN)
	Solution Neuron (SN)
	Innovativeness Neuron (IN)
	Manufacturability Neuron (MN)

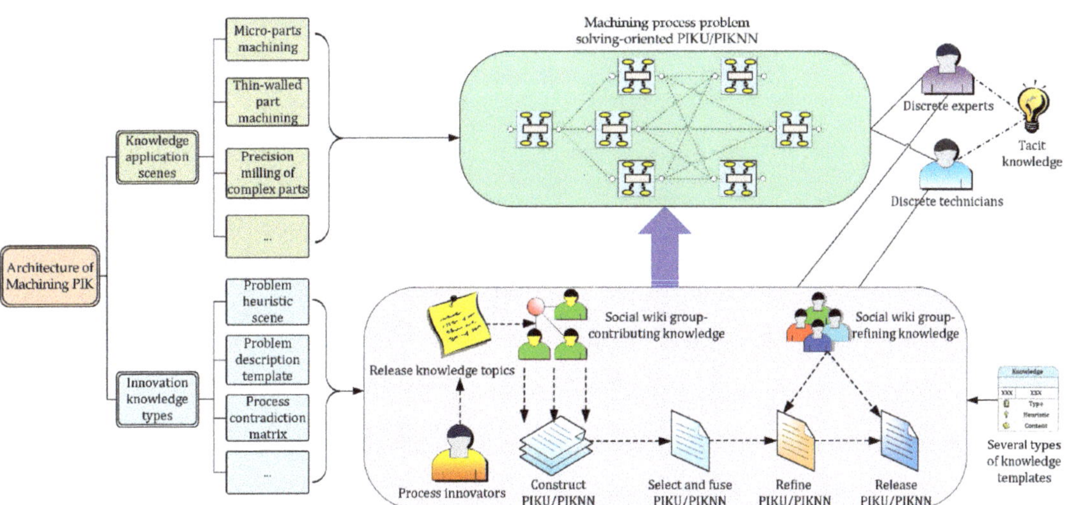

Figure 3. Open knowledge acquisition paradigm for machining process innovation.

From the perspective of knowledge contributors, social wikis can quickly establish innovative communities and seek appropriate participants for knowledge acquisition. From the perspective of MPIK, social wikis can make knowledge contributors participate in the collaborative editing of various innovation knowledge and guarantee the refinement of the knowledge obtained. Meanwhile, the PIKU social wiki network can collect personal discrete knowledge and refine and store it in a public knowledge space, while the PIKNN social wiki network can make technicians collaboratively add knowledge relevance for PIKUs in order to form a dynamic, self-organizing machining process problem-solving-oriented PIKNN.

4. Knowledge-Driven Problem-Solving and Innovation Design Method for Micromachining Process

4.1. Structured Problem-Solving and Innovation Design Procedure of Machining Process

The structured problem solving of machining process innovation can be regarded as a dialectical process from concrete to abstract to concrete again. First, a specific and factual process problem is reduced to its essential state and recorded in a conceptual format. In its conceptual form, the problem can be matched with one or more conceptual solutions. Then, the identified conceptual solution can be transformed into a specific, factual solution that can answer the original factual problem. In this way, heuristic process innovation, using computer-aided technology, can reduce inherent mindset and knowledge limitations [22,40], essentially identifying problems and providing possible heuristic principles or new solutions for the machining process innovators.

The basic procedure for structured heuristic process problem solving is shown in Figure 4 and described as follows.

(1) Process problem identification. The machining process innovator manually inputs the IPD according to the identified machining problem. Then, the HNs that meet the threshold of semantic similarity with the description of IPD are automatically listed for heuristic innovation. The innovator can then interactively input the cause of the process problem and generate the PDC, based on the selected HNs. Finally, the selected HNs transfer the PDC to all TNs.

(2) Problem formal description. After the TNs accept the PDC, they will determine if they can be applied to the PDC according to the problem domain. Then, the innovator can retrieve the objective and obstacles of the problem, based on the PDC and recommended TNs, and interactively input them to form the FPD, which will be transferred to all CNs. Finally, the FPD formation process is constrained and standardized using the domain ontology database.

(3) Innovation principle acquisition. Each CN represents a unit of the machining contradiction matrix. After accepting the FPD, the CNs automatically parse the contradiction parameters in the FPD. If the parsed contradiction parameters are matched with the CNs, the solving principles in the CNs will be extracted as PSU under human–computer interaction and output as the function requirement format to all ENs.

(4) Initial solution generation. Each EN is a triple combination of effect, function, and typical structure. After accepting the PSU, each EN will automatically build a comparison between the function requirement of PSU and its realizable function. If they are matched, the corresponding effect and typical structure will be output as ISU to the SNs.

(5) Detailed solution design. When the SN has accepted the ISU, it calculates the similarity between itself and the ISU, referencing the previous TN, CN, and EN. Then, the system automatically lists the SNs that meet the threshold of similarity for the machining process innovator to refer to. The innovator can then reuse the SNs or interactively input information to generate the DSU inspired by the SNs.

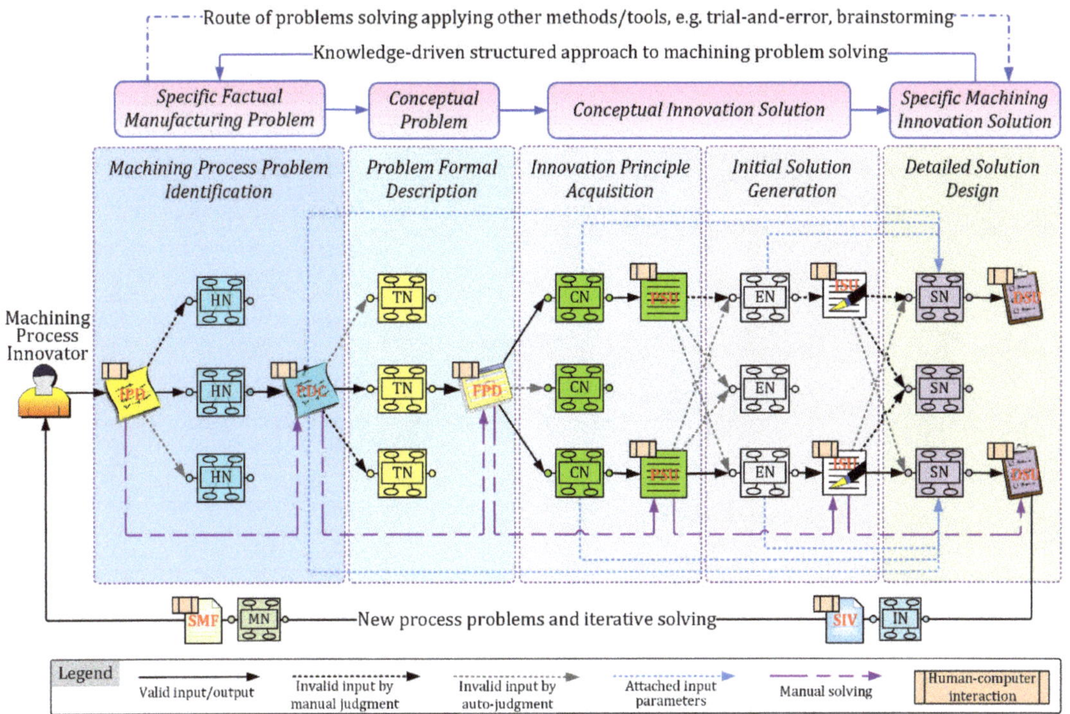

Figure 4. Structured problem-solving and innovation design procedure of machining process.

It should be noted that, after the detailed solution design of machining process innovation is completed, the innovativeness and manufacturability of the scheme need to be further evaluated with the INs and MNs in order to meet the actual manufacturing conditions of the specific organization. In addition, the specific problem solving of machining process innovation design may result in new process problems; thus, iterative problem solving is often required during the entire innovation process.

4.2. Heuristic Innovation Knowledge Processing Architecture of Problem Solving

Many discrete PIKUs with semantic relationships can constitute a static knowledge network. After accepting the input stimulation of machining process problems, the static knowledge network selects the appropriate PIKUs to form a dynamic temporary knowledge network. Then, the knowledge signal will be transmitted to the semantic knowledge network, and the process problem will be solved gradually. The basic rules for machining process problem solving with PIKUs are listed in Table 3.

For complex process problems, there may exist multiple contradictions. In such situations, the problem usually needs to be split into multiple single and simple problems. If the innovative solution leads to new machining process problems, the solving process must be iteratively executed until the contradictions are completely resolved or a compromise is reached. The continuous processing of PIKUs for process problems constitute the process of problem solving and innovation design. Thus, the core of machining process problem solving lies in the knowledge processing of each PIKU as input data. In fact, the knowledge processing of innovation design is a cognitive and reasoning process of PIKUs, and can be regarded as an Integrated Neuro-Cognitive Architecture (INCA) [41]. If the problem is solved through the INCA, it will be clearer and more reasonable, and complies with the innovative thinking process of the innovators. The knowledge processing architecture of the PIKU simulating the INCA for the input problem is shown in Figure 5.

Table 3. The basic rules of knowledge-based heuristic machining process problem solving.

Rule	Description
Interface Matching	Considering that the semantic relations and processing rules among various types of machining PIKUs are explicit, the forward propagation mode is adopted to transfer the parameters without feedback learning. Through the output interface, the current PIKU will select all subsequent PIKUs that can accept their output parameter types and transfer the output results to the input interface of subsequent PIKUs.
Parameter Judgment	There are two methods for judging the effectiveness of input parameters. The first is to determine automatically whether the parameter is effective, according to the processing rules, and the second is to manually intervene. In both ways, as long as the parameters are ineffective, the current machining PIKU will interrupt its knowledge transfer process.
Knowledge Processing	There are two ways to process the input parameters. One way is to automatically process by PIKUs, based on the knowledge properties and processing rules, while the other way requires manual intervention and to output the processing results.
Manual Solving	If there are no suitable machining PIKUs to use at a solving step of process innovation, then the solution process will be transferred to manual solving. Subsequently, the result of manual solving will be transferred to the next-level PIKUs to continue the solving process.

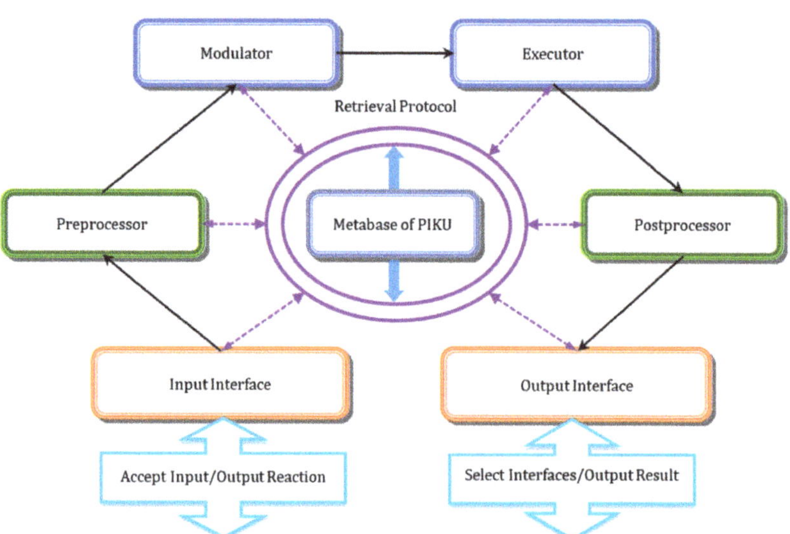

Figure 5. Knowledge processing architecture of machining PIKU.

The architecture of knowledge processing for machining process innovation consists of seven parts. The input interface that simulates the function of the thalamus accepts the input parameters and feeds-back the decision on whether to process or not. The output interface that simulates the function of the motor cortex selects all suitable subsequent PIKUs and transfers the processing result to them. The preprocessor simulating the function of the hippocampus makes pretreatments for the input parameters, for example, extracting contradiction units from the FPD of a machining process problem. The postprocessor that simulates the function of the cerebellum ganglia creates the posttreatments for the processing result in order to meet the formal requirements of the output information, for example, generating the DSU through the typical machining case of the SN and the input information of human–computer interaction. Modulator, simulating the function of the amygdala, provides a control directive, according to the preprocessing result, such as the mode of human interaction and the way of interrupt handling. Executor, which simulates the function of the frontal cortex, controls and finishes the knowledge-based processing, for example, matching the function requirements of the PSU and extracting manufacturing effects and typical structures from the EN. Metabase, which simulates the function of the posterior cortex, stores the meta knowledge of the PIKU, such as process contradiction units and manufacturing effects. The retrieval protocol can support the corresponding components to retrieve the knowledge properties of Metabase and to further process problems. In this way, according to the process problem-solving rules, solving process, and knowledge processing architecture, the machining process problems are expected to be solved heuristically and structured with the semantic knowledge network.

5. Knowledge-Driven Process Problem Solving for Micro-Turbine Machining

Based on the proposed approach, we have designed and implemented a prototype system for knowledge-driven heuristic machining process innovation, named MPI-KHDS, which is integrated as a subsystem on the general CAPI platform containing the basic tools of innovation knowledge management and application.

The MPI-KHDS presents a four-layer-framework, as shown in Figure 6. The knowledge and data layer stores the basic source data of machining process innovation, and the formalized machining process innovation knowledge. The service layer provides access to the knowledge and data layer and supports various system background services of innovation knowledge acquisition and structured process problem solving. The functional layer provides the functions of system configuration and management, innovation knowledge acquisition and management, machining process problem solving and others. The user interaction layer provides a visual human–computer interaction interface for machining process innovation users, e.g., domain expert, machining process innovator, process designer, innovation approach researcher, and machining technician.

To support machining process problem solving effectively, we invited 9 domain experts, 20 machining technicians, and 30 graduate students to participate in the open knowledge acquisition of machining process innovation. After nearly 3 months of knowledge accumulation, the MPI-KHDS stored about 5000 pieces of refined process innovative knowledge in the micromachining field.

The KJ-66 micro-turbine is the core part of high-precision turbojet engines and is a typical micro-part with a complex thin-walled structure. However, there is no through hole in the turbine center. The positioning and clamping methods commonly used cannot be effectively applied to the machining process of this turbine; thus, the process problem relates to how to creatively implement the high-precision machining of the turbine using the existing manufacturing resources and process environment. In the following, we illustrate the application of knowledge-driven heuristic process problem solving for micro-turbine machining.

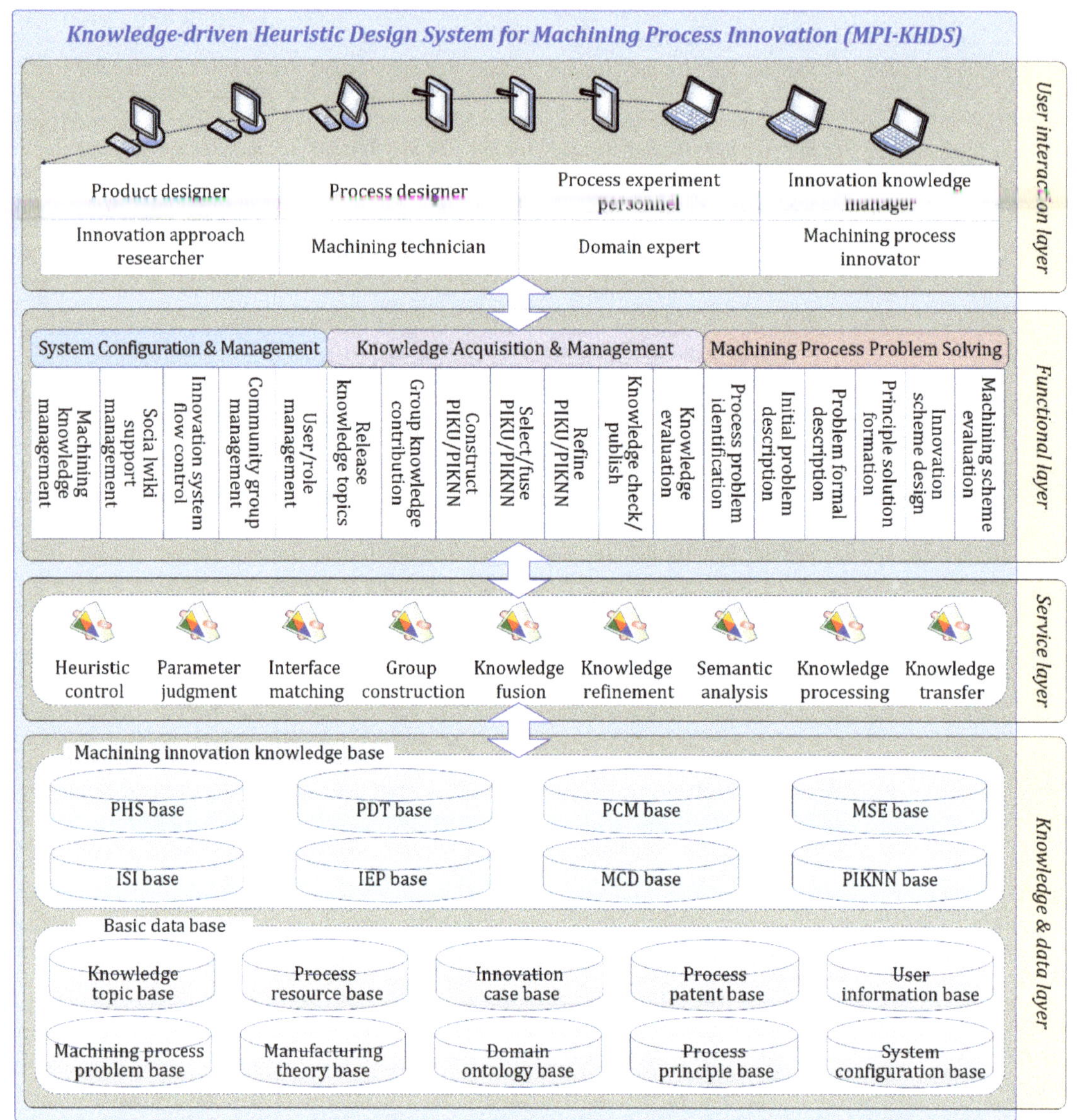

Figure 6. The system framework of MPI-KHDS.

5.1. Process Problems Description of Micro-Turbine Machining

The KJ-66 micro-turbine is composed of a hub, several sets of blades evenly distributed along the circumference of the hub, and hub grooves on both ends. Its structure is shown in Figure 7. The micro-turbine is smaller in size than general turbines, and it requires higher machining accuracy. Its technological characteristics are as follows:

(1) The blade is a thin-walled part with poor process rigidity, more serious deformation, and higher processing accuracy requirements.
(2) As the size of the turbine is reduced, the diameter of the tool is also reduced, and so the rigidity of the tool becomes worse and is easier to break.

(3) The flow path between the blades is narrower and deeper, and the processing space is smaller. The relative swing of the tool during processing can easily cause cutting interference to the adjacent blades.

(4) The turbine center has no through-hole structure, which brings difficulties in turbine clamping.

(5) The thickness of the blade is variable. The thickness at the root of the blade is 1.1 mm, and the thickness at the tip of the blade is only 0.8 mm. The blade is thin and easily deformed. It is required to be free of burrs after machining.

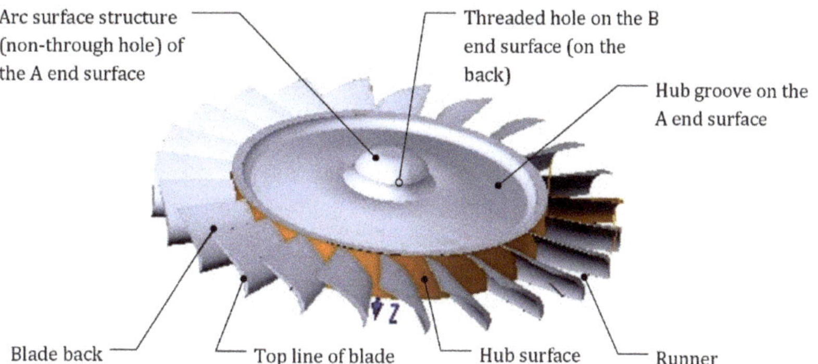

Figure 7. Structure diagram of a micro-turbine.

The two ends of the turbine have bosses and arc surface structures with a height of approximately 4 mm. When machining the turbine blades, the four-jaw chuck of the CNC machine tool cannot directly and effectively clamp the two structures. In the process of micro-turbine blade machining, the chuck and core shaft are often used to achieve positioning and clamping, i.e., after the two end faces are machined, an axial through hole is drilled in the center of the micro-turbine, and the core shaft is clamped through the through-hole; thus, the turbine can be pressed on the chuck through the tightening force of the nut and the thread at one end of the core shaft. This method is suitable for the machining of a through-hole structure turbine. For a turbine without a through-hole, this method will destroy the structural integrity of the part itself. In addition, the positioning of the core shaft adopts a clearance fit, and the vibration generated during the machining process makes it difficult to ensure the accuracy of the blade. Blade machining is key to the manufacturing of the micro-turbine, so reliable clamping of the workpiece ensures accuracy in its size and shape.

According to conventional thinking, one method to solve the above problems is to change the structure of the turbine itself (e.g., drilling a central hole) to make it more convenient for clamping; however, this method will damage the structural integrity of the part. The alternative option is to replace the four-jaw chuck of the universal fixture and use more complex tooling and fixtures. In order to solve this process problem, under the existing processing conditions and without damaging the part structure or introducing too complex toolings and fixtures, we applied the acquired knowledge of the MPI-KHDS system to carry out process problem solving and innovation scheme design.

5.2. Knowledge-Inspired Process Innovation Scheme Design and Machining Experiment

By using the knowledge-inspired innovation guidance of the MPI-KHDS system, the process designer can describe the process problem of turbine machining in detail, and gradually obtain innovation principles and solutions, as shown in Figure 8.

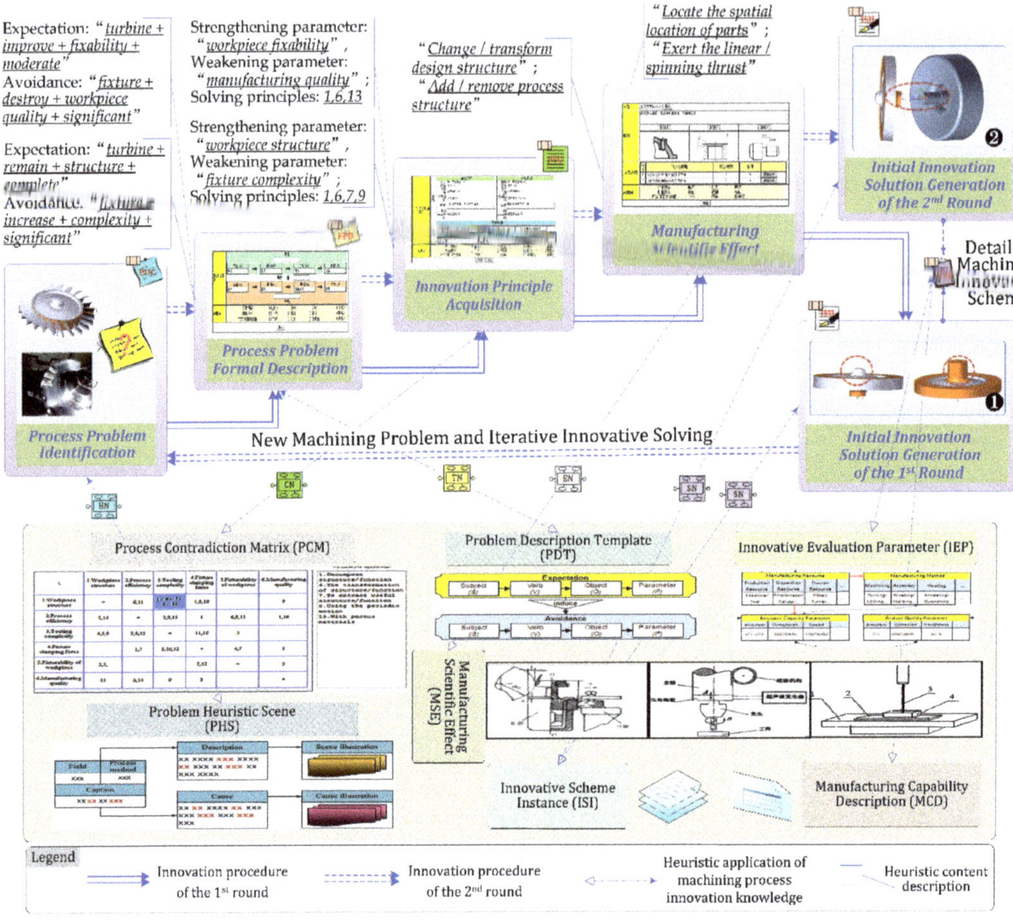

Figure 8. Knowledge-driven heuristic innovation scheme design for micro-turbine machining.

Through process problem identification, with the support of HNs, we define the first process problem to be solved as "how to clamp the turbine so that the blade can be machined with high quality." The designer can formalize the process problem with the support of TNs, and output the FPD in an SVOP format (subject + verb + object + parameter) as: expectation: "turbine + remain + structure + complete"; avoidance: "fixture + increase + complexity + significant." Subsequently, the system will automatically match the corresponding knowledge units in the process contradiction matrix, according to the formal input content and output the corresponding innovation principles. In this problem-solving process, we obtain the following PSU with help from CNs: strengthening parameter: "workpiece structure," weakening parameter: "fixture complexity"; solving principles: 1, 6, 7, 9. The corresponding solution principles are: 1. split structure/function; 6. convert structure form/function; 7. extract useful structure/function; 9. use periodic actions. Through analysis from designers, it is found that the two innovation principles (6, 7) have a greater heuristic effect on solving process problems. The heuristic result that can be obtained is that, in addition to the blades of the micro-turbine, the available parts are the arc bosses on the A end face and the threaded hole structure on the B end face, which can be transformed into useful structures or functions during machining. Thus, the designer can check the heuristic solution principle (6, 7) and select the corresponding industry field, manufacturing technology, and product parts and other keywords. Further, the system will provide the corresponding manufacturing scientific effect (ENs in this step are "change/transform design structure"; "add/remove process structure") to help the designer clarify their innovation ideas. Subsequently, the process innovation knowledge network of the system can be generally matched to the corresponding innovative scheme instances. As there is no ISI matched for this problem, the system will turn to the manual solution here. Based on the solving principles and manufacturing scientific effects obtained, we retrieve the initial innovation solution as follows: the arc surface structure of the micro-turbine A end face can be transformed into a form that can provide clamping, i.e., change its form and function in the machining environment to facilitate the clamping of the four-jaw chuck. After the turbine blade is machined, the process boss can be machined into a circular arc surface structure.

When the turbine blades are processed, there is a reserved process boss on the A end face, which needs to be machined into a circular arc surface structure. Considering that the finishing machining of the blades has been completed, and the turbine blades are thin-walled structures, it is easy to crush the finished blade and even cause blade deformation, if the turbine is clamped using the pressing plate. However, if the turbine blade is not clamped, it will not be possible to accurately cut off the process boss and machine the arc surface structure. Thus, the new process problem is how to clamp the micro-turbine to mill the process boss when the blades have finished being machined. The machining quality of the turbine blades and the clamping capability of the turbines constitute a contradiction that requires iterative innovative problem solving.

According to the above process problem analysis procedure, the second-round innovation principles of problem solving are obtained: strengthening parameter: "workpiece fixability," weakening parameter: "manufacturing quality"; solving principles: 1, 6, 13. Through analysis, it is considered that principle 6 (convert structure form/function) and principle 13 (use its own structure/function) provide significant contributions to this process problem solving. In the first round of process problem analysis, we are inspired by principle 6 by which the structure or function of the arc boss or threaded hole on the two ends can be transformed. However, the process boss on the A end face improves the machining quality of the blade, but causes the current problem. Therefore, the use of the form or function of the threaded hole structure on the B end face may become a problem-solving direction. In combination with the inspiration of using its own structure/function of principle 13, a solution to this problem can be that the threaded hole structure on the B end surface for the final product assembly can be used in workpiece clamping and positioning. Process innovators can consider making full use of the counterbore and thread of the workpiece itself and design a fixture with threads to connect with it, so that the process boss can be milled into the final shape without affecting the surface quality and dimensional precision of the micro-turbine blades.

On the basis of the solving principles for the above two rounds of machining process problems, a new fixture was manufactured and a detailed micro-turbine clamping and machining scheme was designed. After NC tool path planning for the key machining position of the turbine was simulated using ES-Surfmill6.0 software, we carried out the micro-turbine machining experiment using a Smart CNC500 machine tool. In Figure 9, it can be seen that, compared with the original mandrel positioning scheme, the clamping and positioning of the process innovation scheme, with the new fixture, are more reliable, while vibration in the machining process is less, the surface quality of the blade is obviously improved, and there is no tool interference during the machining process.

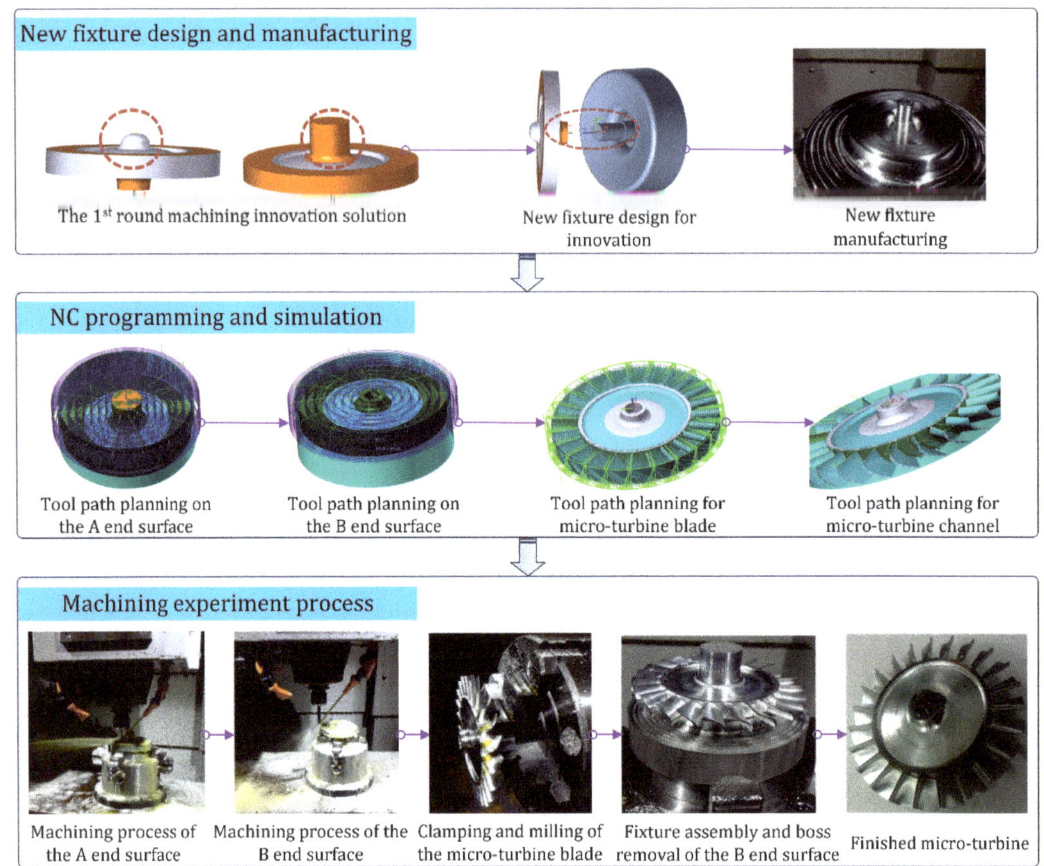

Figure 9. Machining process innovation experiment of the micro-turbine.

6. Conclusions

Process problem solving in micromachining, such as unreliable clamping, thin-walled deformation, and low surface quality, is vital to ensure quality assurance of micropart production. Knowledge plays an intrinsic key role in the procedure of manufacturing process innovation. In this paper, a knowledge network-driven systematic scheme for micromachining process innovation is presented. The main contributions of this study can be summarized as:

- By analyzing the knowledge requirements of computer-aided machining process innovation, several types of MPIK units and the corresponding knowledge network are formally represented. An open multi-source MPIK acquisition and management approach based on collective intelligence is introduced.
- In considering the specific role of formal knowledge in human–computer interaction innovation, a knowledge network-driven structured problem-solving and heuristic innovation design procedure for the machining process is presented that can support process planners in reducing inherent mindsets and individual knowledge limitations and facilitate knowledge-based heuristic innovation.
- The specific micromachining process problem-solving and innovation design for a micro-turbine, without a through-hole, is completed using the innovation support

prototype system, MPI-KHDS. The machining experiment shows that the machining quality of the micro-turbine, with the innovation scheme, is significantly improved.

Author Contributions: Conceptualization, G.W.; methodology, G.W. and D.Z.; software, D.Z., Y.X., X.S. and P.H.; formal analysis, Y.X., X.S., B.G. and R.E.; validation, B.G. and P.H.; writing—original draft preparation, G.W. and D.Z.; writing—review and editing, R.E. and G.W.; funding acquisition, G.W. All authors have read and agreed to the published version of the manuscript.

Funding: This work was supported in part by the Natural Science Basic Research Project of Shaanxi Province, China (Grant No. 2019JM-073) and the National Natural Science Foundation of China (Grant No. 51805043).

Acknowledgments: The authors would also like to express thanks to the collaborators from the Institute of Intelligent Manufacturing of NWPU for their involvements in the CNC machining experiment.

Conflicts of Interest: The authors declare no conflict of interest.

References

1. Moslemipour, G.; Lee, T.; Rilling, D. A review of intelligent approaches for designing dynamic and robust layouts in flexible manufacturing systems. *Int. J. Adv. Manuf. Technol.* **2012**, *60*, 11–27. [CrossRef]
2. Li, B.H.; Hou, B.C.; Yu, W.T.; Lu, X.B.; Yang, C.W. Applications of artificial intelligence in intelligent manufacturing: A review. *Front. Inf. Technol. Electron. Eng.* **2017**, *18*, 86–96. [CrossRef]
3. Zhang, Q.; Lu, J.; Jin, Y. Artificial intelligence in recommender systems. *Complex Intell. Syst.* **2021**, *7*, 439–457. [CrossRef]
4. Feng, Y.X.; Zhao, Y.L.; Zheng, H.; Li, Z.W.; Tan, J.R. Data-driven product design toward intelligent manufacturing: A review. *Int. J. Adv. Rob. Syst.* **2020**, *17*, 1729881420911257. [CrossRef]
5. Wu, L.; Leng, J.; Ju, B. Digital twins-based smart design and control of ultra-precision machining: A review. *Symmetry* **2021**, *13*, 1717. [CrossRef]
6. Lee, Y.J.; Wang, H. Current understanding of surface effects in microcutting. *Mater. Des.* **2020**, *192*, 108688. [CrossRef]
7. Luo, X.; Cheng, K.; Webb, D.; Wardle, F. Design of ultraprecision machine tools with applications to manufacture of miniature and micro components. *J. Mater. Process. Technol.* **2005**, *167*, 515–528. [CrossRef]
8. Friedrich, C.R.; Coane, P.J.; Vasile, M.J. Micromilling development and applications for microfabrication. *Microelectron. Eng.* **1997**, *35*, 367–372. [CrossRef]
9. Takacs, M.; Vero, B.; Meszaros, I. Micromilling of metallic materials. *J. Mater. Process. Technol.* **2003**, *138*, 152–155. [CrossRef]
10. Wang, B.; Zhang, P.; Liang, Y.C.; Dong, S. Development of the control system for three-axis ultraprecision compact micromilling machine. *J. Vac. Sci. Technol. B* **2009**, *27*, 1285–1287. [CrossRef]
11. Cardoso, P.; Davim, J.P. Optimization of surface roughness in micromilling. *Mater. Manuf. Process.* **2010**, *25*, 1115–1119. [CrossRef]
12. Son, S.M.; Lim, H.S.; Ahn, J.H. Effects of the friction coefficient on the minimum cutting thickness in micro cutting. *Int. J. Mach. Tools Manuf.* **2005**, *45*, 529–535. [CrossRef]
13. Ba, S.; Jain, N.; Joseph, V.R.; Singh, R. Integrating analytical models with finite-element models: An application in micromachining. *J. Qual. Technol.* **2013**, *45*, 200–212. [CrossRef]
14. Balazs, B.Z.; Jacso, A.; Takacs, M. Micromachining of hardened hot-work tool steel: Effects of milling strategies. *Int. J. Adv. Manuf. Technol.* **2020**, *108*, 2839–2854. [CrossRef]
15. Ayhan, M.B.; Öztemel, E.; Aydin, M.E.; Yue, Y. A quantitative approach for measuring process innovation: A case study in a manufacturing company. *Int. J. Prod. Res.* **2013**, *51*, 3463–3475. [CrossRef]
16. Yamamoto, Y.; Bellgran, M. Four types of manufacturing process innovation and their managerial concerns. *Proc. CIRP* **2013**, *7*, 479–484. [CrossRef]
17. Bonnardel, N.; Didier, J. Brainstorming variants to favor creative design. *Appl. Ergon.* **2020**, *83*, 102987. [CrossRef] [PubMed]
18. Leon, N. The future of computer-aided innovation. *Comput. Ind.* **2009**, *60*, 539–550. [CrossRef]
19. Cugini, U.; Cascini, G.; Muzzupappa, M.; Nigrelli, V. Integrated computer-aided innovation: The prosit approach. *Comput. Ind.* **2009**, *60*, 629–641. [CrossRef]
20. Xu, J.; Houssin, R.; Caillaud, E.; Gardoni, M. Fostering continuous innovation in design with an integrated knowledge management approach. *Comput. Ind.* **2011**, *62*, 423–436. [CrossRef]
21. Esterhuizen, D.; Schutte, C.S.L.; du Toit, A.S.A. Knowledge creation processes as critical enablers for innovation. *Int. J. Inf. Manag.* **2012**, *32*, 354–364. [CrossRef]
22. Duran-Novoa, R.; Leon-Rovira, N.; Aguayo-Tellez, H.; Said, D. Inventive problem solving based on dialectical negation, using evolutionary algorithms and triz heuristics. *Comput. Ind.* **2011**, *62*, 437–445. [CrossRef]
23. Fiorineschi, L.; Frillici, F.S.; Rotini, F.; Conti, L.; Rossi, G. Adapted use of the triz system operator. *Appl. Sci.* **2021**, *11*, 6476. [CrossRef]

24. Cakir, M.C.; Cilsal, O.O. Implementation of a contradiction-based approach to dfm. *Int. J. Comput. Integr. Manuf.* **2008**, *21*, 839–847. [CrossRef]
25. Delgado-Maciel, J.; Cortes-Robles, G.; Sanchez-Ramirez, C.; Garcia-Alcaraz, J.; Mendez-Contreras, J.M. The evaluation of conceptual design through dynamic simulation: A proposal based on triz and system dynamics. *Comput. Ind. Eng.* **2020**, *149*, 106785. [CrossRef]
26. Hüsig, S.; Kohn, S. "Open cai 2.0"—Computer aided innovation in the era of open innovation and web 2.0. *Comput. Ind.* **2011**, *62*, 407–413. [CrossRef]
27. Lopez Flores, R.; Belaud, J.-P.; Le Lann, J.-M.; Negny, S. Using the collective intelligence for inventive problem solving: A contribution for open computer aided innovation. *Expert. Syst. Appl.* **2015**, *42*, 9340–9352. [CrossRef]
28. Yusof, Y.; Latif, K. Survey on computer-aided process planning. *Int. J. Adv. Manuf. Technol.* **2014**, *75*, 77–89. [CrossRef]
29. El Kadiri, S.; Kiritsis, D. Ontologies in the context of product lifecycle management: State of the art literature review. *Int. J. Prod. Res.* **2015**, *53*, 5657–5668. [CrossRef]
30. García-Alcaraz, J.L.; Alvarado I, A. Problems in the implementation process of advanced manufacturing technologies. *Int. J. Adv. Manuf. Technol.* **2013**, *64*, 123–131. [CrossRef]
31. Al-wswasi, M.; Ivanov, A.; Makatsoris, H. A survey on smart automated computer-aided process planning (acapp) techniques. *Int. J. Adv. Manuf. Technol.* **2018**, *97*, 809–832. [CrossRef]
32. Wang, G.; Tian, X.; Geng, J.; Guo, B. A knowledge accumulation approach based on bilayer social wiki network for computer-aided process innovation. *Int. J. Prod. Res.* **2015**, *53*, 2365–2382. [CrossRef]
33. Duflou, J.R.; D'hondt, J. Applying triz for systematic manufacturing process innovation: The single point incremental forming case. *Proc. Eng.* **2011**, *9*, 528–537. [CrossRef]
34. Cavallucci, D.; Leon, N. Computer-supported innovation pipelines: Current research and trends. *Comput. Ind.* **2011**, *62*, 375–376. [CrossRef]
35. Wang, G.; Tian, X.; Geng, J.; Evans, R.; Che, S. Extraction of principle knowledge from process patents for manufacturing process innovation. *Proc. CIRP* **2016**, *56*, 193–198. [CrossRef]
36. Gao, J.; Nee, A.Y. An overview of manufacturing knowledge sharing in the product development process. *Proc. IMechE Part B J. Eng. Manuf.* **2018**, *232*, 2253–2263. [CrossRef]
37. Beydoun, G.; Hoffmann, A.; Garcia, R.V.; Shen, J.; Gill, A. Towards an assessment framework of reuse: A knowledge-level analysis approach. *Complex Intell. Syst.* **2020**, *6*, 87–95. [CrossRef]
38. Cavallucci, D.; Khomenko, N. From triz to otsm-triz: Addressing complexity challenges in inventive design. *Int. J. Prod. Dev.* **2007**, *4*, 4–21. [CrossRef]
39. Yan, W.; Liu, H.; Zanni-Merk, C.; Cavallucci, D. Ingenioustriz: An automatic ontology-based system for solving inventive problems. *Knowl.-Based Syst.* **2015**, *75*, 52–65. [CrossRef]
40. Chechurin, L.; Borgianni, Y. Understanding triz through the review of top cited publications. *Comput. Ind.* **2016**, *82*, 119–134. [CrossRef]
41. Oentaryo, R.J.; Pasquier, M. Knowledge consolidation and inference in the integrated neuro-cognitive architecture. *IEEE Intell. Syst.* **2011**, *26*, 62–71. [CrossRef]

 *micromachines*

 MDPI

Article

Theoretical Modeling and Experimental Analysis of Single-Particle Erosion Mechanism of Optical Glass

Zhongchen Cao [1,2,*], Shengqin Yan [1], Shipeng Li [1] and Yang Zhang [1]

[1]   Key Laboratory of Advanced Ceramics and Machining Technology, Ministry of Education, Tianjin University, Tianjin 300072, China; ysq@tju.edu.cn (S.Y.); 14221278@bjtu.edu.cn (S.L.); summer0724zy@icloud.com (Y.Z.)
[2]   Key Laboratory of Mechanism Theory and Equipment Design of Ministry of Education, Tianjin University, Tianjin 300072, China
*   Correspondence: charles.cao@connect.polyu.hk; Tel.: +86-22-2740-4915

Abstract: The study of the single-particle erosion mechanism is essential to understand the material removal mechanism in the non-contact polishing process and ultimately ensure the high-efficiency, non-damage, and ultra-smooth processing of optical glass. In this study, the theoretical model of smoothed particle hydrodynamics (SPH) is established to reveal the dynamic removal process of a single particle impacting the optical glass. The single-particle erosion mechanisms, which include ductile–brittle transition, crack initiation, and propagation, are discussed in detail through theoretical simulation. A series of particle impact experiments are designed to validate the correctness of the SPH model. The experimental data show good agreement with the simulation results in terms of the depth and width of the eroded craters. Thereafter, the SPH simulation is conducted by studying the effect of various impact parameters, such as impact speed, impact angle, and abrasive diameter, on the material removal process. With the gradual increase of impact velocity and particle size, the material removal mode changes from plastic removal to brittle removal. Although the large impact velocity and particle size increase the material removal rate, they lead to the occurrence of brittle removal and reduce the surface and sub-surface quality. When the impact angle is between $45°$ and $75°$, the material removal rate is the largest, and the increase of the material removal rate does not cause damage to the subsurface layer of the material.

Keywords: single-particle erosion; smoothed particle hydrodynamics; mechanism; optical glass; subsurface damage

---

Citation: Cao, Z.; Yan, S.; Li, S.; Zhang, Y. Theoretical Modeling and Experimental Analysis of Single-Particle Erosion Mechanism of Optical Glass. *Micromachines* **2021**, *12*, 1221. https://doi.org/10.3390/mi12101221

Academic Editor: Benny C. F. Cheung

Received: 20 July 2021
Accepted: 3 October 2021
Published: 6 October 2021

Publisher's Note: MDPI stays neutral with regard to jurisdictional claims in published maps and institutional affiliations.

1. Introduction

Optical glass has become extensively used in aerospace, defense energy, and microelectronics due to its excellent material properties, such as high-temperature resistance, corrosion resistance, high strength, and good wear resistance [1,2]. To satisfy the stringent precision requirements of modern optical applications, the optical glass should be efficiently fabricated into ultra-smooth surface and low subsurface damage [3]. However, due to its high brittleness and low fracture toughness, brittle fracture is inevitably introduced during the processing of optical glass, which affects its ultimate performance [4,5]. Ultra-precision polishing technology such as fluid jet polishing, abrasive air jet polishing, and disc hydrodynamic polishing can be successfully used to suppress the occurrence of brittle fractures of optical glasses and obtain an ultra-smooth surface [6–8]. During these polishing processes, optical materials are removed by repeated impacts of abrasive particles in the polishing slurry [9,10]. Therefore, studying the single-particle erosion mechanism is necessary to improve our understanding of the material removal and damage control mechanism in the non-contact polishing process, as well as achieve the high-efficiency, damage-free, and ultra-smooth processing of optical glasses.

Many experimental investigations of particle impact tests under static and dynamic contact loads have been conducted to understand the erosion mechanism of brittle ma-

---

terials [11–14]. Evans et al. [15] found that the abrasive erosion mechanism of optical glass is similar to the static indentation mechanism. Lawn et al. [16] further analyzed the mechanism of crack initiation and propagation based on a static indentation experiment. Zhao et al. [17] identified three abrasive–workpiece interaction stages, namely, elastic, plastic, and brittle stages, during the abrasive erosion process. Qi et al. [18] conducted a series of microparticle impact tests and found three categories of the eroded impressions, including craters, scratches, and microdents. In their study, the influence of different impact speeds on the material removal rate was also studied. Hadavi et al. [19] analyzed the effect of impact parameters including impact speed, abrasive size, and incident angle on the brittle fracture and rebound kinematics. They found more larger abrasive fractures than smaller ones at the same impact speed. Although related scholars have conducted experimental research on the erosion mechanism of hard and brittle materials, the erosion theory cannot explain the erosion of micro- and nano-abrasive particles in ultra-precision polishing technology due to the large size of the abrasive particles used in the study. In addition, the experimental study can only be performed by observing the surface and subsurface characteristics of the hard and brittle materials after particle impact, and cannot achieve real-time monitoring of various dynamic information during the impact process, such as crack initiation and growth, stress, phase change, and temperature [18].

To observe the dynamic removal process of abrasive particle erosion, researchers have used numerical simulation methods to study the material removal mechanism of single-particle impact [20,21]. The finite element method (FEM) has been widely used to simulate the erosion removal of ductile and brittle materials [22,23]. Aquaro and Fontani [24] established a FEM simulation model for particle erosion of the ductile and brittle material to study the erosion mechanism of abrasive particles. Yand and Wang [25] developed a finite element model of the erosion by using Johnson–Cook and Johnson–Holmquist material models so as to study the effect of angle and speed on the erosion process. The FEM simulation model was effectively verified by comparing the experimental results with the simulation results [25]. However, the FEM model may result in mesh distortion of large deformation in simulating brittle materials. To effectively simulate the erosion removal process of brittle materials, smooth particle hydrodynamics (SPH), a meshless method, was proposed [26,27]. This method is particularly suitable for solving high-speed collisions and dynamic large deformation problems [28]. Du et al. [29] established a coupled FEM/SPH model to simulate the impact process of angular particles on the float glass. Crack propagation during the erosion process has been thoroughly characterized. This study shows the advantage of SPH in simulating brittle fracture behaviors of glass during the abrasive erosion process. Nishikawa et al. [30] studied the influence of various particle shapes, including cubic and spherical particles, on the erosion behavior of glass based on a coupled FEM/SPH model. The material removal impacted by spherical particles was greater than that of cubic particles. Hao et al. [31] analyzed the mechanism of crack initiation and propagation during the impact process of angular particles on float glass by establishing a coupled FEM/SPH model, and found that incident orientation plays a key role in the erosion process. Dong et al. [32,33] established the SPH constitutive relationship formula describing the plastic behavior and ductile fracture process, and used the improved SPH model to simulate the erosion process of particles with sharp corners. At present, research on the erosion removal mechanism of optical glasses is insufficient, and the influence mechanism of different impact parameters on the material removal process remains to be clarified. Therefore, this study develops the SPH model for single-particle erosion of optical glass, and conducts a theoretical study of various impact parameters, such as impact speed, impact angle, and abrasive diameter, on the material removal process. The single-particle erosion mechanisms, including ductile–brittle transition, crack initiation, and propagation, are discussed in detail through theoretical simulations. A series of particle impact experiments are designed to validate the correctness of the SPH model by comparing the depth and width of the eroded craters between the simulation results and the experimental data.

2. Smooth Particle Hydrodynamics (SPH) Simulation Model for Single-Particle Erosion of Optical Glass

2.1. Basic Theory of SPH Simulation

Due to its adaptability, meshlessness, Lagrangian, and particle properties, SPH provides great advantages in solving large deformations, free surface flows, and complex interface motions. As a relatively mature meshless numerical calculation method, SPH is extensively used in many engineering and scientific fields, including astrophysics, impact explosion, and hydrodynamics [34]. In SPH, the state of the system is expressed by a limited number of particles with a certain space and mass. The continuous function, $f(u)$, that controls the range in which particles interact with one another is expressed as:

$$f(u) = \int_{\Omega} f(u')\delta(u - u')du' \tag{1}$$

where $\delta(u - u')$ is the Dirac delta function, Ω is the integral space, and u is the vector of the particle position. Discrete numerical models cannot be constructed by the Dirac delta function, only with the support of "points" in the calculation process. Thus, the kernel function $T(u - u', h)$ is used to replace the delta function, and the continuous function is modified as follows:

$$\langle f(u) \rangle = \int_{\Omega} f(u')T(u - u', h)du' \tag{2}$$

where h is the smoothing length that determines the influence area of the kernel function, which should meet the requirements of the normalization condition, Delta function property, and compact condition. As the spline kernel function has good accuracy for the three cases, it is universally used in SPH and can be expressed as:

$$T(u - u', h) = \frac{1}{h^d}\theta\left(\frac{\|u - u'\|}{h}\right) \tag{3}$$

where d is the spatial dimension and θ is the spline function. The cubic B-spline function is often used as a representative of the spline function because of its good regularity [35]. This function can be expressed as:

$$\theta(y) = C \times \begin{cases} 1 - 1.5y^2 + 0.75y^3 & |y| \leq 1 \\ 0.25(2 - y^3) & 1 < |y| < 2 \\ 0 & |y| > 2 \end{cases} \tag{4}$$

where C is the normalization constant that can be calculated by the spatial dimension (that is, $y = x/h$) and x is the distance between the two particles.

Through numerical calculation, the expression of the continuous integral of the function $f(u)$ can be transformed into a cumulative summation process of SPH particles. After discretization, the approximate expression of the particle is as follows:

$$\langle f(u_a) \rangle = \sum_{b}^{N} \frac{m_b}{\rho_b} f(u_b)U(u_a - u'_b, h) \tag{5}$$

where m_b, ρ_b, and m_b/ρ_b denote the mass, density, and volume of particle b, respectively. In the initial state, the smooth length, h, can be calculated by giving the position, velocity, and density of particle a. Particle b with a position within $2h$ is then found. The strain and strain rate between two particles can be calculated based on the spatial derivatives of position and velocity. Stress can help obtain the acceleration, and the velocity and position can be updated using Equation (6). The density and energy can be calculated in real time according to the position of the particle by Equations (7) and (8), respectively.

$$\frac{dv_a^{\alpha}}{dt} = \sum_{b}^{N} m_b\left(\frac{\sigma_a^{\alpha\beta}}{\rho_a^2} - \frac{\sigma_b^{\alpha\beta}}{\rho_b^2}\right)\frac{\partial W_{ab}}{\partial u_a^{\beta}} \tag{6}$$

$$\frac{d\rho_a}{dt} = \rho_a \sum_b^N \frac{m_b}{\rho_b}\left(v_b^\beta - v_a^\beta\right)\frac{\partial W_{ab}}{\partial u_a^\beta} \tag{7}$$

$$\frac{dE_a}{dt} = -\frac{\sigma_a^{\alpha\beta}}{\rho_a^2}\sum_b^N m_b\left(v_\alpha^b - v_\alpha^a\right)\frac{\partial W_{ab}}{\partial u_a^\beta} \tag{8}$$

where α and β are space indices.

2.2. Johnson–Holmquist-2 (JH-2) Material Model

The JH-2 material model can accurately describe the fracture and fragmentation of brittle materials under the dynamic impact of abrasive grains [36]. This material model is mainly composed of a strength model, damage model, and state equation. The state equation is expressed by a cubic polynomial for materials with high strain rate, such as fused quartz glass. When $D = 0$, the material is undamaged, and the state equation is as follows:

$$p = K_1\mu + K_2\mu^2 + K_3\mu^3 \tag{9}$$

where p is the hydrostatic pressure, K_1 is the bulk modulus, and K_2 and K_3 are the volumetric strain coefficients. Conversely, when $D > 0$, the material is damaged, with elastic energy gradually reducing and potential energy of the material gradually increasing. The conversion from elastic energy into potential energy needs a correction, and is written as:

$$(\Delta p_{t+\Delta t} - \Delta p_t)\mu_{t+\Delta t} + \frac{1}{2}\left(\Delta p_{t+\Delta t}^2 - \Delta p_t^3\right)/K_1 = \beta\Delta U \tag{10}$$

where β is the energy conversion coefficient and ΔU is the internal energy change.

A new strength model that uses the equivalent stress of the material as a power function of hydrostatic pressure, which is in connection with the strain rate and damage factor, is adopted in JH-2. The new strength model is expressed as:

$$\sigma^* = \sigma_i^* - D\left(\sigma_i^* - \sigma_f^*\right) \tag{11}$$

where σ^* is the equivalent stress, and σ_i^* and σ_f^* are the equivalent stresses of the undamaged and damaged materials, respectively. When $D = 0$, σ_i^* can be expressed as follows:

$$\sigma_i^* = A(p/p_{HEL} + \sigma_{t,m}/p_{HEL})^N\left[1 + C\ln\left(\overset{*}{\varepsilon}/\overset{*}{\varepsilon}_0\right)\right] \tag{12}$$

where A, C, and N are model parameters, $\overset{*}{\varepsilon}$ is the true strain rate, $\overset{*}{\varepsilon}_0$ is the reference strain rate, p_{HEL} is the pressure component of the Hugoniot elastic limit (HEL), p is the hydrostatic pressure, and $\sigma_{t,m}$ is the maximum hydrostatic pressure. The equivalent stress for complete fracture when $D = 1$ is:

$$\sigma_f^* = B(p/p_{HEL})^M\left[1 + C\ln\left(\overset{*}{\varepsilon}/\overset{*}{\varepsilon}_0\right)\right] \tag{13}$$

where B is the model parameter. The damage model of JH-2 can be expressed by:

$$D = \sum\frac{\Delta\varepsilon_p}{D_1(p/p_{HEL} + T/p_{HEL})^{D_2}} \tag{14}$$

where $\Delta\varepsilon_p$ is the integral of the effective plastic strain in a single cycle, and D_1 and D_2 are the damage constants of the material. We need to assess the brittle fracture of fused quartz glass. Due to the relatively low fracture toughness of fused quartz glass, the tensile strength limit is remarkably smaller than the compressive strength limit and the tensile strength is less than the yield limit of the material. Therefore, tensile strength is used to measure the material fracture.

2.3. SPH Model for Single-Particle Erosion of Optical Glass

In the practical polishing process, the alumina particle is usually a polyhedral structure with a multiple-edge radius. To simplify the SPH simulation model of a single alumina particle impacting optical glass, we adopted spherical particles with diameters of 6, 9, and 12.8 μm. A single alumina particle is meshed using the FEM to reduce the simulation time. To balance simulation accuracy and efficiency, we designed the optical glass as a $50 \times 18 \times 0.3$ μm cube that was divided into 60,000 SPH particles by the SPH method. The SPH geometric model of the single alumina particle impacting the optical glass is shown in Figure 1. As described in Section 2.2, the JH-2 model is applied to the optical glass material in the simulation, and relevant parameters in the JH-2 material model of the optical glass are presented in Table 1. The density, elastic modulus, and Poisson's ratio of the single-particle alumina materials are 3.95 g/cm^3, 350 GPa, and 0.22, respectively. The impact parameters of the simulation are shown in Table 2. The impact particles and impacted workpiece are in CONTACT-AUTOMATIC-NODES-TO-SURFACE contact mode, with the single-particle alumina being the main surface and the optical glass being the subordinate particle. BOUNDARY-SPH-SYMMETRY-PLANE is selected to fix the optical glass.

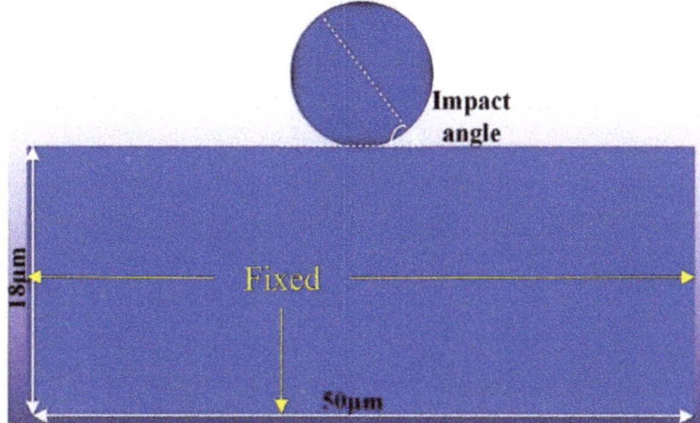

Figure 1. Smooth particle hydrodynamics (SPH) model of a single particle impacting the optical glass.

Table 1. JH-2 parameters of optical glass [37].

Description	Symbol	Value
Density (kg/m^3)	ρ	2530
Shear modulus (Gpa)	G	30.4
Intact normalized strength coefficient	A	0.93
Intact strength exponent	N	0.77
Fractured normalized strength coefficient	B	0.088
Fractured strength exponent	M	0.35
Strain rate coefficient	C	0.003
Net compressive stress at HEL (GPa)	HEL	5.95
Pressure component at HEL	P_{HEL}	2.92
Maximum tensile strength (GPa)	T	0.15
Reference strain rate	ε_0	1
Maximum fracture strength	SFMAX	0.5
Damage coefficient	D1	0.053
Damage exponent	D2	0.85
Bulk modulus (GPa)	K1	45.4
Pressure coefficient (GPa)	K2	−138
Pressure coefficient (GPa)	K3	290

Table 2. Simulation parameters of a single particle impacting the optical glass.

Impact Parameters	Value
Impact speed (m/s)	30, 35, 40, 45, 50, 55
Impact angle (°)	30, 45, 60, 75, 90
Abrasive diameter (μm)	6, 9, 12.8

3. Experiments and Model Validation

3.1. Experimental Design

A specific slurry supply system was used so that alumina slurry with an extremely low concentration is impacted on the surface of the optical glass in a very short time and at a specific injection pressure. Moreover, fixed-point impact on the K9 optical glass is achieved by controlling the motion control platform, and a large nozzle with a diameter of 1 mm is used to effectively identify the impact morphology of a single alumina particle. The material properties of K9 glass are density 2.5 g/cm^3, elastic modulus 81 GPa, Vickers hardness 6.9 GPa, and fracture toughness 0.8 MPa·m$^{1/2}$. The distance between the nozzle and optical glass is 5 mm, which is short enough for us to reasonably assume that the impact angle of a single alumina particle is equal to the incident angle. Additionally, the short impact distance and impact angle caused the velocity fluctuation of the alumina particle to become extremely small. Different impact speeds are obtained by altering the impact pressure, and calculations suggest that the corresponding impact speeds under impact pressures of 0.8, 1.0, 1.2, and 1.5 Mpa are 35, 40, 45, and 50 m/s, respectively. In order to ensure the accuracy of the experiment, the repeatability experiment was carried out by repeating each pressure condition 4 times. The distributions of impact spots under various impact pressures are shown in Figure 2.

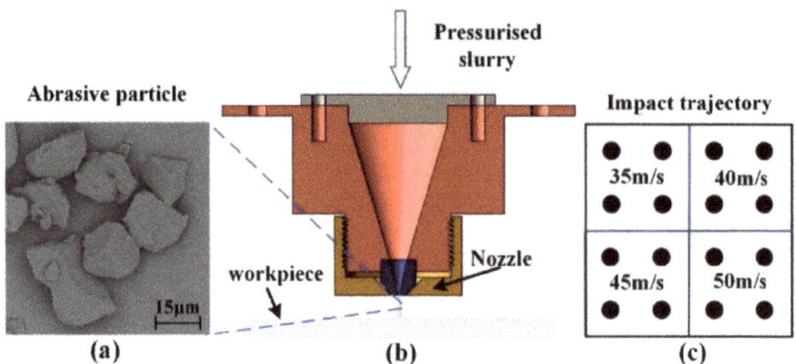

Figure 2. Schematic of experimental setup. (**a**) alumina particles (**b**) polishing nozzle and (**c**) impact test design.

Alumina particles (FUJIMI WA#1000) with an average diameter of 12.8 μm are used in the impact removal experiments, and the morphology of the single alumina particle is obtained through scanning electron microscopy. As shown in Figure 2a, these particles are not exactly spherical. Consequently, we can expect a certain deviation between the simulation of the SPH model, in which the alumina particles are assumed to be spherical, and the experimental data are within an allowable range. A cube workpiece (40 × 40 × 1 mm) made of optical glass was used in the experiments. A scanning electron microscope (Phenom XL, Eindhoven, The Netherlands) and a CSPM5500 scanning probe microscope (Being, Beijing, China) were used to measure the surface morphology and depth information of the craters.

3.2. Model Validation

Figure 3 shows the surface morphology of the optical glass material at different impact speeds. We can observe the removal characteristics (craters, scratches, and microcraters) on the surface of three types of materials, among which craters account for the largest proportion. According to the morphology, when the impact speed is 35 or 40 m/s, the surface is mainly composed of plastic craters, and the material removal mode is mainly plastic removal. However, brittle characteristics begin to form and become more dominant as the impact speed increases, indicating that brittle removal replaces plastic removal as the main removal mode. This result reveals that a lower impact speed can achieve plastic removal of the optical glass and improve the surface morphology, but the material removal rate is reduced.

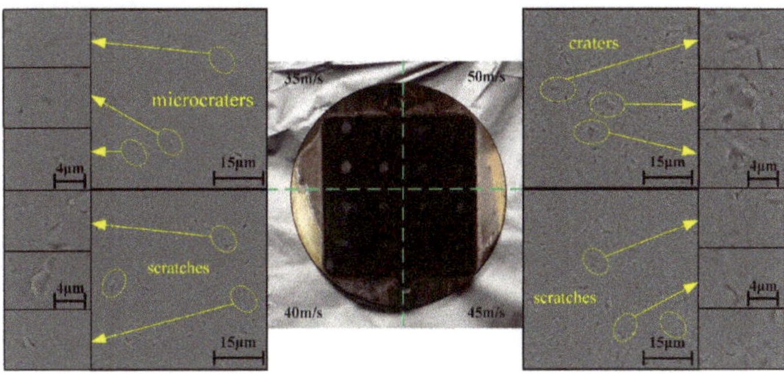

Figure 3. Surface morphology of impacted optical glass under different impact speeds.

Figure 4 displays the crater morphology of the single-particle alumina impact at speeds ranging from 35 to 50 m/s. First, due to the irregularity of the shape of particles, the crater shape is not only spherical, and with the increase of the impact speed, the width of the crater also increases. At a speed no higher than 40 m/s, the removal is mainly realized by plastic extrusion, with no microcracks being generated around the craters. However, when the impact speed reaches 45 m/s or higher, microcracks begin to form around the crater. This result demonstrates that because of the increase in the impact speed, the impact energy of a single particle is greater than the critical energy of the plastic–brittle transition, thereby resulting in the appearance of microcracks. Therefore, by analyzing the crater morphology, we can conclude that the increase of the impact speed leads to the increase of the width of the craters and the formation of transverse cracks on the crater, thereby causing material breakage and affecting the surface quality of the material.

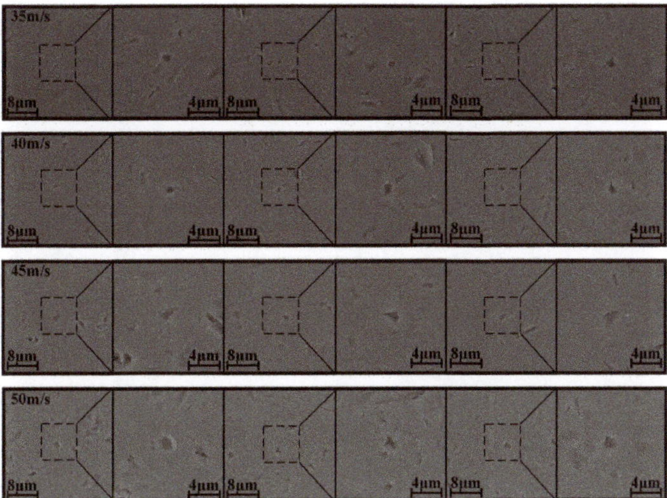

Figure 4. Surface morphology of a single crater under different impact speeds.

Figure 5 illustrates the depth information of the craters at various impact speeds. The irregularities of the shape of the impact particles cause the section of the crater to become a non-standard partial circle. With impact speed increasing from 35 to 50 m/s, we can observe that the impact depth of a single crater ranges from 0.21 to 0.41 µm. The reason is that as the impact speed increases, the impact velocity of a single particle also increases, and consequently, the increased impact kinetic energy of that particle leads to an increase in the impact depth. Thus, the increase in impact speed can increase the depth of the material removal and enhance the material removal rate.

Figure 6 demonstrates the SPH simulation of impact under different impact speeds, with the particle diameter being 12.8 µm and the impact angle being 90°. When the surface is impacted at the impact speed of 30 or 40 m/s, the material removal process depends on the plastic deformation of the material surface caused by extrusion between the impact particles and the surface. Nevertheless, when the impact speed reaches 45 m/s, short transverse cracks can be found, revealing that brittle removal begins to occur at this point. Moreover, as the impact velocity continues to increase, obvious brittle removal occurs: the length of the transverse cracks increases and extends to the surface of the material and radial cracks form on the subsurface. Accordingly, we can reasonably conclude that with the increase of impact speed, the material removal mode changes from plastic to brittle removal. In addition, the critical impact speed of the conversion of the removal mode is approximately 40 m/s, which is basically consistent with the experimental results.

Figure 7 presents a comparison between the SPH simulation and the experiment of the single-particle alumina impacting the optical glass with regard to the depth and width of the removal craters. As the impact speed increases, the maximum crater depth of the simulation ranges from 0.28 to 0.48 µm, while the experimental results are between 0.21 and 0.41 µm, which show good agreement with the simulation in terms of trends and values. For the maximum crater width, the experiment is basically in accordance with the simulation: with the impact speed increasing from 35 to 50 m/s, and the width increasing from 1.6 to 2.9 µm in the experiment and from 2.7 to 4.0 µm in the simulation. Owing to the instability of nozzle pressure and the irregularity of alumina particles in the removal experiment, the data cannot fully fit the simulation but are within the allowable error range. Thus, the correctness of the SPH simulation model is verified by comparing the experimental data with the simulation.

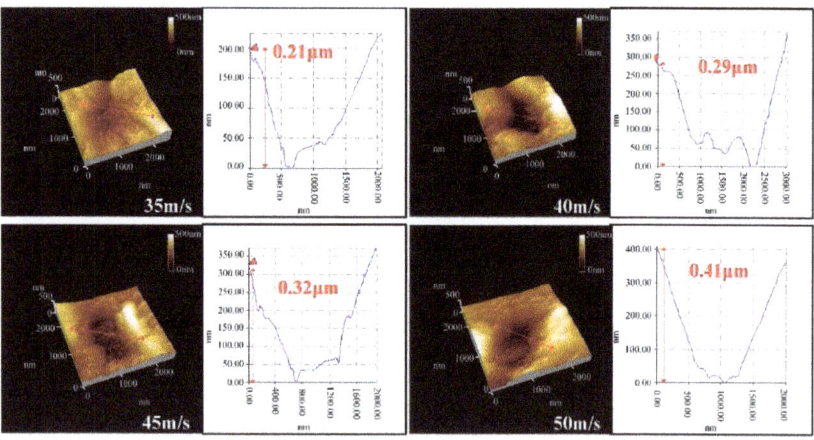

Figure 5. Depth information of a single crater under various impact speeds.

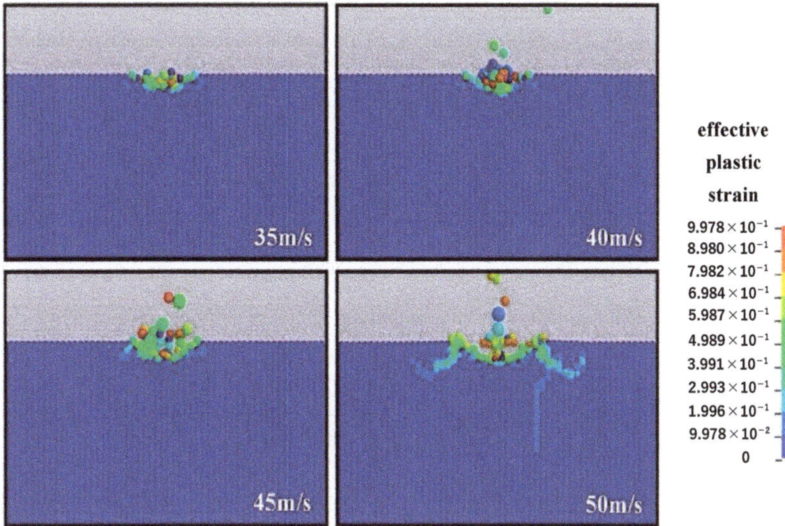

Figure 6. Surface morphology of impacted optical glass under different impact speeds.

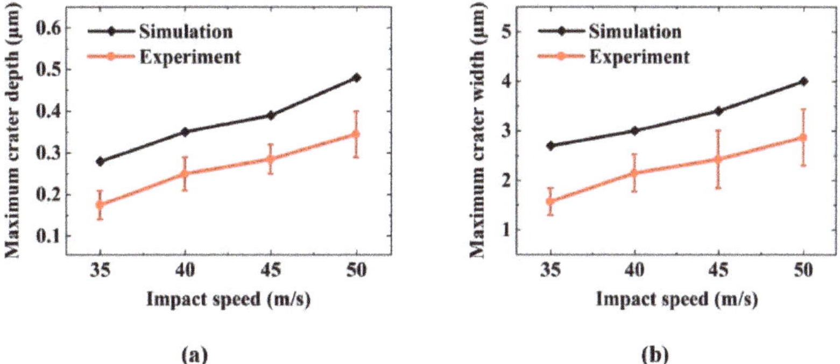

(a)   (b)

Figure 7. Comparison of experimental results and simulation: (**a**) change in maximum depth of crater and (**b**) change in maximum width of crater.

4. Simulation Results and Discussion

4.1. Material Removal Mechanism

Figure 8 illustrates the material removal process of the single alumina particle with a diameter of 12.8 µm impacting the optical glass at the impact speed of 50 m/s and the impact angle of 90°. When the impacting alumina particle starts contacting the optical glass, the material removal is mainly realized by plastic mode, with no brittle crack forming on the surface or subsurface layers of the material, as shown in Figure 8a. However, as the impact process continues, transverse cracks appear around the impact craters and continue to expand, eventually stopping on the surface of the material and resulting in the generation of massive chips, as presented in Figure 8b. When the impact depth continues to increase, a radial crack perpendicular to the surface of the material can be clearly observed in the subsurface layer, which affects the subsequent use of the material, as shown in Figure 8c. Thus, the material removal exhibits significant brittleness removal characteristics when the single-particle alumina with a diameter of 12.8 µm impacts the optical glass at an impact velocity of 50 m/s and an impact angle of 90°. In detail, transverse cracks first occur, followed by radial cracks. The formation of transverse and radial cracks sharply reduces the quality of the material surface and subsurface layers. Based on the result of the SPH simulation, the brittle removal process of the single-particle alumina impacting the optical glass materials is summarized as follows:

1. The particles first cause plastic deformation of the material.
2. Transverse cracks form around the deformation zone and gradually expand to the surface of the material, which eventually leads to the generation of brittle chips, thereby causing material removal.
3. Radial cracks perpendicular to the surface of the material form in the subsurface layer and gradually expand to the deeper subsurface layer.

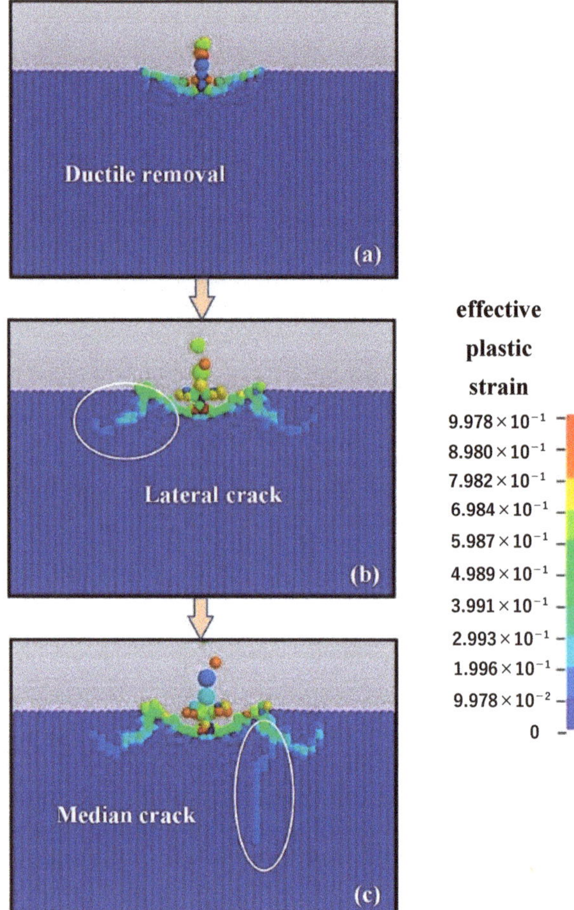

Figure 8. Erosion removal process of optical glass: (**a**) ductile removal, (**b**) lateral crack, and (**c**) radial crack.

4.2. Effect of Impact Angle on Material Removal

Figure 9 shows the surface morphology of the single-particle alumina with a diameter of 12.8 μm impacting the optical glass at an impact speed of 50 m/s and different impact angles (15°, 30°, 45°, 60°, 75°, and 90°). We can observe that the impact angle plays a crucial role in the material removal mode. When the impact angle is 15°, the contact area between the single-particle alumina and material surface is the smallest, which leads to a slight plastic deformation on the material surface. When the impact angle is 30°, microcracks are generated in the subsurface layer of the material. As the impact angle increases and settles between 45° and 75°, transverse cracks are produced due to the extrusion between the impact particles and material, thereby resulting in the generation of massive chips and leading to material removal. Nonetheless, when the impact angle is 45° or 60°, no radial cracks are observed in the subsurface layer of the material, whereas when the angle is 75°, shorter radial cracks are observed in the subsurface layer. The material removal process for the impact angle of 90° is described in Section 4.1. By comparing the surface morphology of the optical glass under different impact angles, we can observe that with the increase of the impact angle, transverse cracks that cause a large amount of material removal are first formed and no subsurface damage is caused. Conversely, when the impact angle is greater than 60°, radial cracks that affect the quality of the material are generated in the subsurface layer of the material.

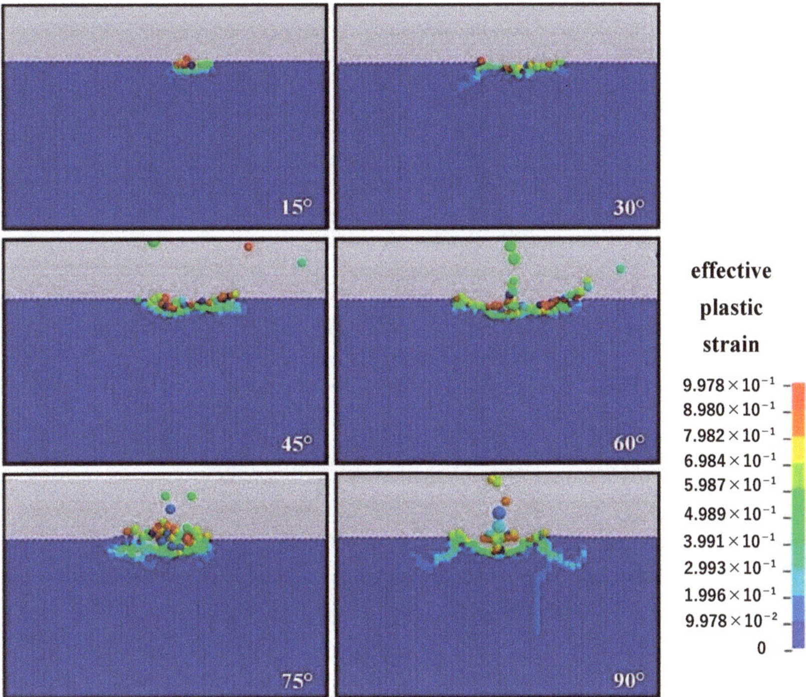

Figure 9. Surface morphology of impacted optical glass under different impact angles.

Figure 10 quantitatively demonstrates the relationship between the maximum depth and maximum width of the craters on the material surface as the impact angle increases. Under the impact angles of 15°, 30°, 45°, 60°, 75°, and 90°, the corresponding maximum depths of the craters are 0.15, 0.29, 0.50, 0.46, 0.62, and 0.48 µm respectively, and the corresponding maximum widths of the craters are 1.8, 3.4, 4.6, 4.9, 5.4, and 4.0 µm, respectively. When the impact angle is 60°, the reason for the decrease in the impact depth may be the fact that most of the particle energy causes the generation of lateral cracks, resulting in a decrease in the impact depth for the impact angle of 60°. By analyzing the maximum depth and width of the craters, we find that under the same impact speed, when the impact angle is between 45° and 75°, the material removal rate is the highest and the increase of the removal rate does not cause subsurface damage.

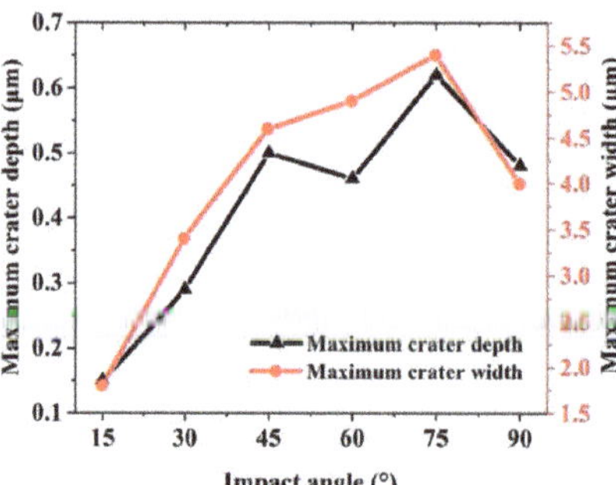

Figure 10. Maximum depth and width of craters under different impact angles.

4.3. Effect of Particle Diameter on Material Removal

Figure 11 displays the surface morphology of optical glass impacted by single-particle alumina with different particle diameters (6, 9, and 12.8 μm) at the impact speed of 50 m/s and the impact angle of 90°. The results show that the particle diameter greatly affects the material removal process. When the particle diameter is 6 μm, the material removal mainly depends on plastic deformation and no brittle crack is generated. However, when the particle diameter is 9 μm, short transverse cracks and radial cracks begin to form, indicating the occurrence of brittle removal, which is even more obvious when the diameter reaches 12.8 μm. Therefore, we can conclude that the increase of the particle diameter affects the material removal mode during the impact process. Furthermore, when the impact particle diameter is 6 μm, the material removal is mainly achieved by plastic mode.

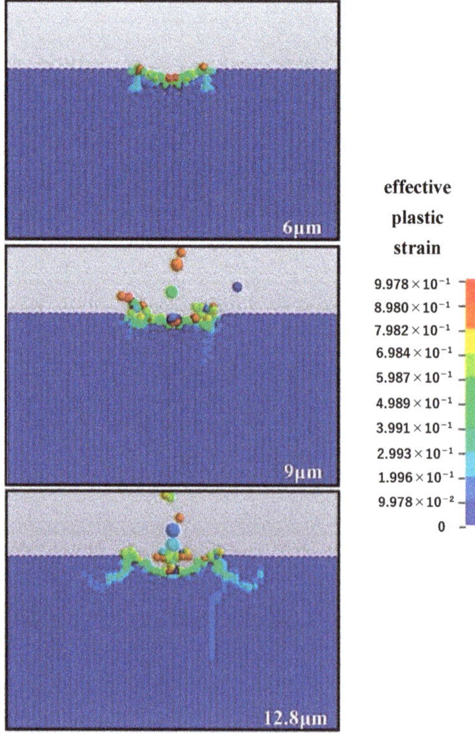

Figure 11. Surface morphology of impacted optical glass under different particle diameters.

Figure 12 quantitatively illustrates the relationship between the maximum depth and maximum width of the craters on the material surface as the particle diameter increases. Under particle diameters of 6, 9, and 12.8 μm, the maximum depths of the craters are 0.35, 0.39, and 0.48 μm respectively, and the maximum widths are 2.9, 3.3, and 4.0 μm, respectively. By analyzing the maximum depth and width of the craters, we found that as the particle diameter increases, the material removal rate increases accordingly, but the increase leads to brittle failure.

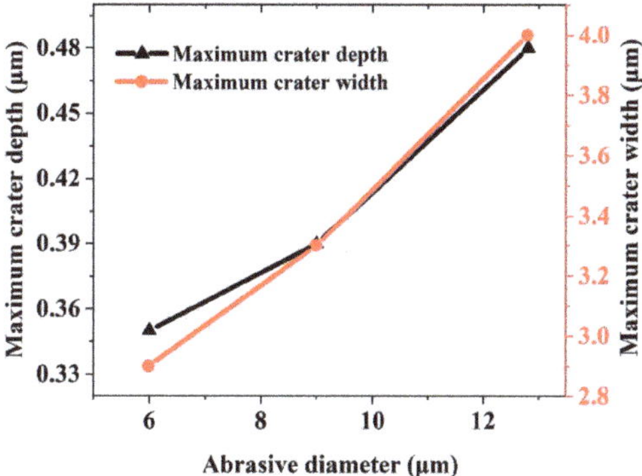

Figure 12. Maximum depth and width of craters under different particle diameters.

4.4. Effect of Impact Parameters on Energy Loss of Particles

In order to understand the erosion mechanism of optical glass from the perspective of energy conversion, the energy loss was calculated by subtracting the rebound kinetic energy from the initial kinetic energy of particles. Figure 13 shows the effects of various impact angles and impact speeds on the energy loss of particles with a diameter of 12.8 μm. As the impact velocity increases, the energy loss also increases. When the impact speed is greater than 40 m/s, the loss begins to increase rapidly, which accounts for the condition that when the impact speed is greater than 40 m/s, brittle cracks occur on the material surface. Figure 13b shows that the energy loss increases when the impact angle increases from 15° to 60°. However, when the impact angle is greater than 60°, a decrease in the energy loss occurs, indicating that when the impact angle is approximately 60°, the material absorbs energy best. The sharp decrease in the energy loss for the impact angle of 75° may be influenced by the reduction in the contact area and the initiation of median cracks. This condition also explains, in terms of energy, that the material removal rate reaches the maximum when the impact angle is between 45° and 75°. Figure 13c shows that as the diameter of the impact particles increases, the energy loss also increases. The relationship between diameter and energy loss may be basically linear, reflecting that the increase in diameter can greatly improve the material removal rate. The results have guiding significance for the research on the polishing mechanism and process optimization.

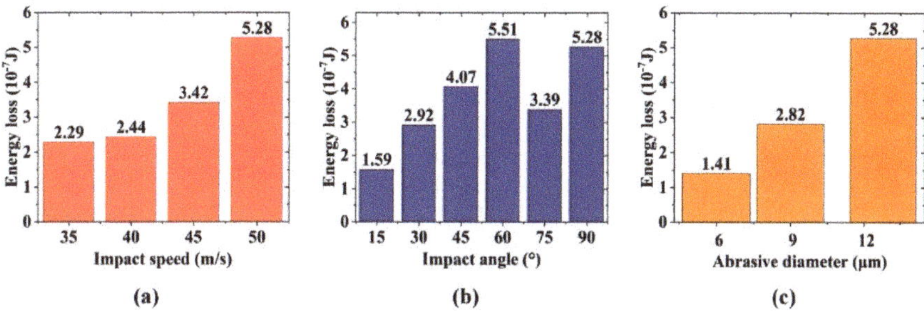

Figure 13. Energy loss changes under different impact parameters: (**a**) impact speed, (**b**) impact angle, and (**c**) particle diameter.

5. Conclusions

In this study, the theoretical model of the single-alumina particle impacting the optical glass was established through SPH. The simulation was conducted under different impact speeds, impact angles, and particle diameters. The correctness of the simulation model was verified by single-particle impact experiments. The conclusions are as follows:

1. The increase of impact speed increases the width and depth of the craters, thereby enhancing the material removal rate. Moreover, when the impact speed is greater than 40 m/s, brittle removal occurs and microcracks are formed around the craters, thereby resulting in material breakage and affecting the surface quality of the material.
2. The dynamic evolution mechanism of crack initiation and propagation in the single-particle impact process of optical glass is clarified. The particles first cause plastic deformation of optical glass. As the impact process continues, transverse cracks are generated around the deformation zone and then gradually expand to the surface of the material. Meanwhile, radial cracks perpendicular to the surface of the material are produced in the subsurface layer and gradually expand to the deeper subsurface layer.
3. With the gradual increase of the impact angle, the transverse cracks that cause a large amount of material removal are first generated. When the impact angle is greater than 75°, radial cracks appear in the subsurface layer of the material. At the same impact speed, when the impact angle is between 45° and 75°, the material removal rate is the highest, and the increase in the material removal rate does not cause damage to the subsurface layer of the material.
4. The increase in particle diameter affects the material removal mode during impact. When the particle diameter is 6 μm, the material is removed by the plastic removal mode. As the particle diameter increases, the material removal rate increases accordingly but leads to brittle failure.
5. The energy loss increases with the increase of impact speed. When the impact speed is greater than 40 m/s, the energy loss begins to increase rapidly. The energy loss also increases when the impact angle increases from 15° to 60°. However, the energy loss begins to decrease when the angle is greater than 60°, which means that the material absorbs energy best at the impact angle of 60°. The increase in particle diameter also increases the kinetic energy loss correspondingly.

Author Contributions: Conceptualization, Z.C.; methodology, Z.C. and S.Y.; software, Z.C. and S.L.; validation, Z.C., S.Y. and S.L.; formal analysis, Z.C. and S.L.; investigation, Z.C. and S.Y.; resources, Z.C.; data curation, Z.C., S.L. and S.Y.; writing—original draft preparation, Z.C.; writing—review and editing, Z.C., S.L., S.Y. and Y.Z.; visualization, Z.C. and Y.Z.; supervision, Z.C.; project administration, S.L.; funding acquisition, Z.C. All authors have read and agreed to the published version of the manuscript.

Funding: This work was supported by the National Natural Science Foundation of China (Grant No. 51905376) and the Laboratory of Science and Technology on Marine Navigation and Control, China State Shipbuilding Corporation (2020010101).

Conflicts of Interest: The authors declare no conflict of interest.

References

1. Gu, W.; Yao, Z.; Liang, X. Material removal of optical glass BK7 during single and double scratch tests. *Wear* **2011**, *270*, 241–246. [CrossRef]
2. Li, Y.; Nan, Z.; Li, H.; Jing, H.; Lei, X.; Chen, X.; Yuan, Z.; Guo, Z.; Jian, W.; Guo, Y. Morphology and distribution of subsurface damage in optical fused silica parts: Bound-abrasive grinding. *Appl. Surf. Sci.* **2011**, *257*, 2066–2073. [CrossRef]
3. Esmaeilzare, A.; Rahimi, A.; Rezaei, S.M. Investigation of subsurface damages and surface roughness in grinding process of Zerodur glass–ceramic. *Appl. Surf. Sci.* **2014**, *313*, 67–75. [CrossRef]
4. André, D.; Laheurte, R.; Darbois, N.; Cahuc, O.; Neauport], J. Subsurface mechanical damage during bound abrasive grinding of fused silica glass. *Appl. Surf. Sci.* **2015**, *353*, 764–773.
5. Ong, N.S.; Venkatesh, V.C. Semi-ductile grinding and polishing of Pyrex glass. *J. Mater. Process. Technol.* **1998**, *83*, 261–266. [CrossRef]

6. Cao, Z.C.; Cheung, C.F. Theoretical modelling and analysis of the material removal characteristics in fluid jet polishing. *Int. J. Mech. Sci.* **2014**, *89*, 158–166. [CrossRef]

7. Haj, M.; Spelt, J.K.; Papini, M. Surface roughness and erosion rate of abrasive jet micro-machined channels: Experiments and analytical model. *Wear* **2013**, *303*, 138–145.

8. Tsai, F.C.; Yan, B.H.; Kuan, C.Y.; Huang, F.Y. A Taguchi and experimental investigation into the optimal processing conditions for the abrasive jet polishing of SKD61 mold steel. *Int. J. Mach. Tools Manuf.* **2008**, *48*, 932–945. [CrossRef]

9. Fan, J.M.; Wang, C.Y.; Wang, J. Modelling the erosion rate in micro abrasive air jet machining of glasses. *Wear* **2009**, *266*, 968–974. [CrossRef]

10. Wang, R.J.; Wang, C.Y.; Wen, W.; Wang, J. Experimental study on a micro-abrasive slurry jet for glass polishing. *Int. J. Adv. Manuf. Technol.* **2017**, *89*, 451–462. [CrossRef]

11. Lawn, B.R.; Swain, M.V. Microfracture beneath point indentations in brittle solids. *J. Mater. Sci.* **1975**, *10*, 113–122. [CrossRef]

12. Li, Q.L.; Wang, J.; Huang, C.Z. Erosion mechanisms of monocrystalline silicon under a microparticle laden air jet. *J. Appl. Phys.* **2008**, *104*, 296. [CrossRef]

13. Sparks, A.J.; Hutchings, I.M. Effects of erodent recycling in solid particle erosion testing. *Wear* **1993**, *162–164*, 139–147. [CrossRef]

14. Bousser, E.; Martinu, L.; Klemberg-Sapieha, J.E. Effect of erodent properties on the solid particle erosion mechanisms of brittle materials. *J. Mater. Sci.* **2013**, *48*, 5543–5558. [CrossRef]

15. Evans, A.G.; Gulden, M.E.; Rosenblatt, M. Impact Damage in Brittle Materials in the Elastic-Plastic Response Regime. *Proc. R. Soc. A Math. Phys. Eng. Sci.* **1978**, *361*, 343–365.

16. Lawn, B.R.; Evans, A.G.; Marshall, D.B. Elastic/Plastic Indentation Damage in Ceramics: The MediadRadial Crack System I. *J. Am. Ceram. Soc.* **1980**, *63*, 574–581. [CrossRef]

17. Zhao, W.A.; Hao, N.; Tian, B.; Hao, C.A.; Ji, Z.A. On the predictive modelling of machined surface topography in abrasive air jet polishing of quartz glass—ScienceDirect. *Int. J. Mech. Sci.* **2019**, *152*, 1–18.

18. Qi, H.; Fan, J.; Wang, J.; Li, H. Impact erosion by high velocity micro-particles on a quartz crystal. *Tribol. Int.* **2015**, *82*, 200–210. [CrossRef]

19. Hadavi, V.; Moreno, C.E.; Papini, M. Numerical and experimental analysis of particle fracture during solid particle erosion, Part II: Effect of incident angle, velocity and abrasive size. *Wear* **2016**, *356–357*, 146–157. [CrossRef]

20. Li, W.Y.; Wang, J.; Zhu, H.; Li, H.; Huang, C. On ultrahigh velocity micro-particle impact on steels—A single impact study. *Wear* **2013**, *305*, 216–227. [CrossRef]

21. Anwar, S.; Axinte, D.A.; Becker, A.A. Finite element modelling of a single-particle impact during abrasive waterjet milling. *Proc. Inst. Mech. Eng. Part J J. Eng. Tribol.* **2011**, *225*, 821–832. [CrossRef]

22. Li, W.; Wang, J.; Zhu, H.; Huang, C. On ultrahigh velocity micro-particle impact on steels—A multiple impact study. *Wear* **2014**, *309*, 52–64. [CrossRef]

23. Behr, R.A.; Kremer, P.A.; Dharani, L.R.; Ji, F.S.; Kaiser, N.D. Dynamic strains in architectural laminated glass subjected to low velocity impacts from small projectiles. *J. Mater. Sci.* **1999**, *34*, 5749–5756. [CrossRef]

24. Aquaro, D.; Fontani, E. Erosion of Ductile and Brittle Materials. *Meccanica* **2001**, *36*, 651–661. [CrossRef]

25. Wang, Y.F.; Yang, Z.G. Finite element model of erosive wear on ductile and brittle materials. *Wear* **2008**, *265*, 871–878. [CrossRef]

26. Benz, W.; Asphaug, E. Simulations of brittle solids using smooth particle hydrodynamics—ScienceDirect. *Comput. Phys. Commun.* **1995**, *87*, 253–265. [CrossRef]

27. Limido, J.; Espinosa, C.; Salaün, M.; Lacome, J.L. SPH method applied to high speed cutting modelling. *Int. J. Mech. Sci.* **2007**, *49*, 898–908. [CrossRef]

28. Huang, Y.; Dai, Z. Large deformation and failure simulations for geo-disasters using smoothed particle hydrodynamics method. *Eng. Geol.* **2014**, *168*, 86–97. [CrossRef]

29. Mingchao, D.U.; Zengliang, L.I.; Dong, X.; Hao, G.; Zhang, Y. Experiment and Simulation Study of Erosion Mechanism in Float Glass due to Rhomboid Particle Impacts. *Int. J. Impact Eng.* **2020**, *139*, 103513.

30. Nishikawa, C.; Mizutani, K.; Zhou, T.F.; Yan, J.W.; Kuriyagawa, T. Investigation of Particle Impact Phenomena in Powder Jet Deposition Process. *Key Eng. Mater.* **2012**, *523–524*, 184–189. [CrossRef]

31. Hao, G.; Dong, X.; Du, M.; Li, Z.; Dou, Z. A comparative study of ductile and brittle materials due to single angular particle impact. *Wear* **2019**, *428–429*, 258–271. [CrossRef]

32. Dong, X.W.; Liu, G.R.; Li, Z.; Zeng, W. A smoothed particle hydrodynamics (SPH) model for simulating surface erosion by impacts of foreign particles. *Tribol. Int.* **2016**, *95*, 267–278. [CrossRef]

33. Dong, X.; Li, Z.; Feng, L.; Sun, Z.; Fan, C. Modeling, simulation, and analysis of the impact(s) of single angular-type particles on ductile surfaces using smoothed particle hydrodynamics. *Powder Technol.* **2017**, *318*, 363–382. [CrossRef]

34. Liu, M.B.; Liu, G.R. Smoothed Particle Hydrodynamics (SPH): An Overview and Recent Developments. *Arch. Comput. Methods Eng.* **2010**, *17*, 25–76. [CrossRef]

35. Monaghan, J.J. Why Particle Methods Work. *Siam J. Sci. Stat. Comput.* **1982**, *3*, 422–433. [CrossRef]

36. Holmquist, T.J.; Johnson, G.R. Characterization and evaluation of silicon carbide for high-velocity impact. *J. Appl. Phys.* **2005**, *97*, 093502. [CrossRef]

37. Guo, X.; Zhai, R.; Shi, Y.; Kang, R.; Guo, D. Study on influence of grinding depth and grain shape on grinding damage of K9 glass by SPH simulation. *Int. J. Adv. Manuf. Technol.* **2020**, *106*, 333–343. [CrossRef]

micromachines

Article

Dynamic Performance of Partially Orifice Porous Aerostatic Thrust Bearing

Muhammad Punhal Sahto [1,2], Wei Wang [1,*], Ali Nawaz Sanjrani [1], Chengxu Hao [1] and Sadiq Ali Shah [3]

[1] School of Mechanical and Electrical Engineering, University of Electronic Science and Technology of China (UESTC), Chengdu 611713, China; punhal76@yahoo.com (M.P.S.); alinawaz.sanjrani@muetkhp.edu.pk (A.N.S.); 13402831920@163.com (C.H.)

[2] Department of Mechanical Engineering, The University of Lahore, Defence Road, Lahore 54000, Pakistan

[3] Department of Mechanical Engineering, Mehran University of Engineering and Technology, Shaheed Zulfiqar Ali Bhutto Campus, Khairpur Mir's 66020, Pakistan; ssssadiqalishah@gmail.com

* Correspondence: wangwhit@163.com

Abstract: The aerostatic thrust bearing's performance under vibration brings certain changes in stiffness and stability, especially in the range of 100 to 10,000 Hz, and it is accompanied by significant increase in fluctuations due to the changes in frequency, and the size of the gas film damping. In this research work, an analysis is carried out to evaluate the impact of throttling characteristics of small size orifice on stiffness and stability optimization of aerostatic thrust bearings. There are two types of thrust bearing orifices such as: partial porous multiple orifice and porous thrust bearings and their effects on variations in damping and dynamic stiffness are evaluated. A simulation based analysis is carried out with the help of the perturbation analysis model of an aerostatic thrust bearing simulation by using FLUENT software (CFD). Therefore, two models of aerostatic thrust bearings—one with the porous and other with partial porous orifice are developed—are simulated to evaluate the effects of perturbation frequencies on the damping and dynamic stiffness. The results reveal a decrease in the amplitude of dynamics capacity with an increase in its frequency, as well as a decrease in the damping of partial porous aerostatic thrust bearings with an increase in the number of orifices. It also reveals an increase in the radius of an orifice with an increment of damping of bearing at the same perturbation frequency and, with an increase in orifice height, a corresponding decrease in the damping characteristics of bearings and in the dynamic stiffness and coefficient of damping of bearing film in the frequency range less than 100 Hz.

Keywords: computational fluid dynamic study; partial porous orifice; dynamic stiffness and damping; fluid flow

Citation: Sahto, M.P.; Wang, W.; Sanjrani, A.N.; Hao, C.; Shah, S.A. Dynamic Performance of Partially Orifice Porous Aerostatic Thrust Bearing. *Micromachines* **2021**, *12*, 989. https://doi.org/10.3390/mi12080989

Academic Editors: Benny C. F. Cheung and Jiang Guo

Received: 1 July 2021
Accepted: 17 August 2021
Published: 20 August 2021

Publisher's Note: MDPI stays neutral with regard to jurisdictional claims in published maps and institutional affiliations.

1. Introduction

Aerostatic bearings have certain characteristics such as minimum friction, rotational accuracy, and ultra precision. Considerable research has been conducted on the performance of aerostatic thrust bearing in a bid to make certain improvements in its existing performance to meet their demands of improved performance in related industries such as Microelectronic, Defense, Semiconductors, Measuring instruments, Aerospace, and Textile. There are two main functions of aerostatic bearings such as reduction of friction and motion errors. Earlier studies have covered main parameters of aerostatic thrust bearings such as stiffness, load-carrying capacity, and static characteristics. However, the geometrical structure of aerostatic thrust is also an important parameter, which may bring certain positive changes in its performance. The bearings with an existence of the air film are capable of running without the friction and achieving greater accuracy of motion [1–3]. The distribution pressure of the film can be improved by bringing certain changes in its structure. Therefore, the revised form of thrust bearing is analyzed, and its performance is compared with a traditional orifice restrictor aerostatic bearing. Several studies were carried out based on improvement of the load capacity, stability, and stiffness, on the basis of proper partitioning of random holes for pressure distribution [4,5]. In this regard, a multi grid coupling dynamics model of 5 DOF was proposed for the spindle of aerostatic. The Thomas and ADI method based on a solution of the Reynold Equation was used along with the restoring force to resolve the problems of the fluctuating air film force with times [6]. Certain numerical and experimental studies have also been conducted since the inception of axial thrust bearings keeping in view its effects on performance on various load capacities [7]. A one-dimensional aerostatic model for circular porous air bearing was developed for onward simulation on the basis of Darcy's flow theory. Computational methods were used with the objectives to optimize both the time and cost parameters [8]. In the proposed study, the equilibrium stability approach was considered, in order to reduce the instability of thrust bearing by the application of damping support. It is pertinent to mention that research was carried out in this regard revealing that the dynamics of thrust bearing improved with the help of flexible damping support on the basis of the comparison of the stiffness of the gas film [9]. In another study, it was concluded that, with the inclusion of pocket and central feed holes into the circular aerostatic thrust bearing, which were supported by rubber O rings, and with an arrangement of the fixed pad, its stability could be enhanced [10]. The Routh–Hurwitz criteria was also developed to estimate the rotor-bearing's stability for optimization of stability [11]. Likewise, Nyquist criteria were applied to investigate the stability of externally pressurized bearings [12]. Certain studies were carried out on high speed bearing operations, which concluded that, with an increase in temperature at high speed, there is a failure of bearing stability and deterioration of the film of fluid [13]. The effects of static characteristics of the porous, single orifice, and multiple orifice types on the performance of aerostatic thrust bearings were also investigated [14]. Dynamic stability studies with spiral grooved thrust bearing were carried out by utilizing two methods. The first method was related with stability of film, and the second method provided a relationship of speed limit of bearing with unconditional stability [15]. Several studies were also conducted to determine the coefficients of the fluid film of a dynamic rotor using a hybrid bearing with various angled injections [16]. The Nyquist's criteria based studies revealed certain limits of stability and its results reflected a relationship between the air gap and supporting structure [17]. Studies were also carried out to analyze the impact of changes in the geometrical shape of the bearing holes on its performance. In such a study, two types of bearings were proposed: one with a hole and the other with an annular groove supply to avoid deflection over the surface of aerostatic porous bearings [18]. Certain investigations were related to the effects of manufacturing errors on the load capacity of bearing and significant exaggeration of static performance due to the changes in the distribution of pressure and thickness of film [19]. Unique procedures and methods were used for the evaluation of effects of pressure depression such as; separation of variables and the boundary-layer equations method. These methods facilitated the evaluation of variable parameters, such as influences of orifice diameter, pressure supply, pressure ratio, and film thickness on the depression of pressure. In such a study, it was concluded that the depression of pressure lessens with reduced film thickness, pressure supply, and enhanced diameter of orifice [20]. Certain studies were based on the optimized algorithm to estimate an equivalent coefficient of damping for the system and simulated transient response of the displacement of the shaft and matched with measured shaft displacement [21]. Other studies were carried out to investigate the impact of the thickness of the material, Young's modulus, and supply of air on the material of porous, and concluded with suggestions of a porous restrictor with a double layer [22]. The Finite Element Method (FEM) and Proportional Division Method (PDM) were proposed in certain research works for computation of an angular stiffness [19]. In another research work, a novel method of modeling for determination of the FSI spindle systems of aerostatic was suggested on the basis of the changes in the structure of air film. A virtual method of weight loading was also projected for the evaluation of the stiffness of the spindle by considering the gravitational eccentricity and deformation of structure, and its reliability was validated through test and trial experiments. The FSI model suggested aerostatic spindle effects of geometrical parameters such as air, film, and structure dimension on performance optimization [23]. The Active Dynamic Mesh Method (ADMM) was also proposed for the investigation of dynamical performance of a double-pad bearing according to the perturbation model. In another work, a model was proposed to investigate the effects of orifice diameter, supply pressure, eccentricity ratio, squeeze number, and the step reply of a double-pad bearing and was followed by its comparison with the Passive Dynamic Mesh Method (PDMM) [24]. The effects of changes in perturbation frequency on the coefficient of damping and dynamic stiffness with EEPG were also investigated [25]. The mathematical calculation method of partially porous aerostatic thrust bearings was carried out by using the Finite Element method [26]. Likewise, CFD Fluent software was used to investigate the aerostatic bearing's working performance in the low vacuum condition [27]. The proposed aerostatic thrust bearing model, with a variable pressure-equalizing groove, revealed a different structure and certain increase in the bearing's stiffness [28]. The use of microgrooves in the double row symmetric radial bearing with the combined external throttling system may decrease the circumferential flow of compressed air and, on the other hand, it will improve the capacity of bearing [29]. Simulation software computational analysis methods, such as the Computational Fluid Dynamics (CFD) method, Finite Element Method (FEM), and Finite Difference Method (FDM), are also used for analysis and calculation of damping and dynamic stiffness on the basis of the solution of partial differential equations. The simulation software provides efficient means for analysis of the effects of certain variables on the performance. However, there are certain demerits of the analytical calculation method, which are related to its accuracy, as the results accuracy is subject to comparing the subsequent results obtained through FDM and FEM, with similar analytical approaches. There are certain complications with the results obtained through these software programs, which are related to the quality of the mesh, convergence criteria, and coordination conditions. This research work provides two types of partial multiple orifice porous aerostatic thrust bearing models with an objective to obtain variations in damping and dynamic stiffness.

2. Methodologies

Simulation software based methods and experimental methods are proposed to evaluate the effects of changes in the variables and its impacts on the performance of aerostatic bearings. Details of such research methods are mentioned in the preceding sections, which are given below.

2.1. CFD Base Dynamic Grid Calculation Method

The CFD method is used in the research work to investigate the aerostatic bearing's characteristics. The equation of Navier–Stokes (N–S) is solved by the CFD method by using the procedure of the 3D transient physical field of axial thrust bearing such as temperature, pressure, and velocity field. Two following types of porous and partial multiple orifice porous thrust bearings are used in this calculation method.

2.2. Fluid Control Equation

In the range of Newtonian fluid, the conservation laws of mass, momentum, and energy can be calculated by the N–S equation. The mass conservation law also called continuity equation is as below [30]:

$$\frac{\partial \rho}{\partial t} + \frac{\partial(\rho u)}{\partial x} + \frac{\partial(\rho v)}{\partial y} + \frac{\partial(\rho w)}{\partial z} = 0 \tag{1}$$

It and can also be written as:

$$\frac{\partial \rho}{\partial t} + div(\rho \vec{u}) = 0 \tag{2}$$

where ρ represents density, t represents time, and the velocity vector components of Cartesian coordinates x, y, and z are u, v, and w, respectively. The microelement is calculated by the sum of all forces and equals the change in momentum of fluid microelements, which is indicated by conservation momentum law and also the second law of Newton. The conservation momentum equation is given below [30], and it can be derived in three directions:

$$
\begin{cases}
\frac{\partial(\rho u)}{\partial t} + \mathrm{div}(\rho \bar{u}u) = -\frac{\partial p}{\partial x} + \frac{\partial \tau_{xx}}{\partial x} + \frac{\partial \tau_{yx}}{\partial y}\frac{\partial \tau_{xx}}{\partial z} + F_x \\
\frac{\partial(\rho u)}{\partial t} + \mathrm{div}(\rho \vec{uv}) = -\frac{\partial p}{\partial y} + \frac{\partial \tau_{xy}}{\partial x} + \frac{\partial \tau_{y}}{\partial y}\frac{\partial \tau_{y}}{\partial z} + F_y \\
\frac{\partial(\rho u)}{\partial t} + \mathrm{div}(\rho \vec{uw}) = -\frac{\partial p}{\partial z} + \frac{\partial \tau_{xz}}{\partial x} + \frac{\partial \tau_{y=}}{\partial y}\frac{\partial \tau_{m}}{\partial z} + F_z
\end{cases}
\tag{3}
$$

where P is pressure of film, τ_{xx}, τ_{xy}, and τ_{xz} are the viscous stress components of fluid, F_x is x-directional force, F_y is y-directional force on body, and F_z is force on the body in the z-direction. In this calculation method, since the physical force is due to gravity, and the z-axis is vertical upward; therefore, $F_x = 0$, $F_y = 0$, and $F_z = -\rho g$. According to energy conservation law, an increased energy rate in the microelement is equal to the net heat flow into microelements plus work done by surface force and physical force on the microelements, which is the thermodynamic first law and is indicated in the following form of equation [30]:

$$
\frac{\partial(\rho T)}{\partial t} + \mathrm{div}(\rho \bar{\imath}T) = \mathrm{div}\left(\frac{k}{c_p}\,\mathrm{grad}\,T\right) + S_T
\tag{4}
$$

where $\mathrm{grad}\,T$ is the temperature gradient, k is heat transfer co-efficient of fluid, c_p is the specific heat capacity, and S_T is viscous dissipation energy [30].

$$
\frac{\partial(\rho \phi)}{\partial t} + \mathrm{div}(\rho \bar{u}\phi) = \mathrm{div}(\Gamma\,\mathrm{grad}\,\phi) + S
\tag{5}
$$

It can become to following form [30]:

$$
\frac{\partial(\rho \phi)}{\partial t} + \frac{\partial(\rho u\phi)}{\partial x} + \frac{\partial(\rho v\phi)}{\partial y} + \frac{\partial(\rho w\phi)}{\partial z}
$$
$$
= \frac{\partial}{\partial x}\left(\Gamma\frac{\partial \phi}{\partial x}\right) + \frac{\partial}{\partial y}\left(\Gamma\frac{\partial \phi}{\partial y}\right) + \frac{\partial}{\partial z}\left(\Gamma\frac{\partial \phi}{\partial z}\right) + S
\tag{6}
$$

where ϕ is the general variable, u, v, and w are velocity components, T is the temperature, Γ is the co-efficient of diffusion, and S is the source term. For the aerostatic thrust bearing with orifices, the pressure distribution in the air gap can be achieved by solving Equation (6). However, as the air passes through the porous material, it is accompanied by a loss in viscosity, inertia, and velocity slip in the boundary layer in between the porous material and air gap. Thus, it is necessary to establish an internal fluid lubrication equation of porous material including viscosity loss, inertia loss, and velocity slip, based on accurate measurement of the characteristic parameters of porous material. The equation of internal fluid lubrication of porous material takes the following final form:

$$
\nabla P_i = -\left(\sum_{j=1}^{3} D_{ij}uv_j + \sum_{j=1}^{3} C_{ij}\frac{1}{2}\rho|v|v_j\right)
\tag{7}
$$

$$
\frac{\partial(\gamma \rho)}{\partial t} + \nabla(\gamma \rho \vec{v}) = 0
\tag{8}
$$

$$
\frac{\partial(\gamma \rho \vec{v})}{\partial t} + \nabla(\gamma \rho \vec{v} \times \vec{v}) = -\gamma \nabla I
\tag{9}
$$

where D_{ij} expresses the matrix co-efficient related to the viscous permeability co-efficient, C_{ij} shows a co-efficient matrix related to the inertia permeability co-efficient, and γ represents the porosity of porous material. The pressure distribution of porous aerostatic thrust bearing can be attained by solving the equation of fluid lubrication and the N–S equation. Likewise, the distribution of pressure in the air gap can also be attained. The methods are used for the computation of the characteristics of flow field of aerostatic thrust bearings such as distribution of pressure in the air gap. Such calculation of FEM, FDM, and FVM are resolved by using the Reynolds equation. Since the equation of Reynolds is the partial differential equation, so that the calculation efficiency of FEM and FDM methods is high, the Three-Region Theory is applied to supplement the Reynolds equation, whereas a drop in pressure is still an issue to express and to be characterized by a slow recovery near the orifice; this is how it provides an imperfect definition of parameters of material property. FLUENT software is utilized in fluid calculation for solution of the transient 3D N–S equation, which is based on FVM, and it is capable of describing characteristic parameters of porous material. FLUENT software solves the pressure distribution characteristics of flow field inside an aerostatic bearing and, on this basis, the other characteristics of the aerostatic thrust bearing such as dynamic stiffness and damping are analyzed. The computer with configurations having Random Access Memory (RAM) of 112 GB, equipped with the graphics of a NVIDIA Quadro K620-GPU(GM107) processor was used to perform this simulation of an aerostatic thrust bearing.

2.3. Computational Models and Grids

The 3D model of the flow calculation domain of the partial orifice aerostatic thrust bearing is used with the same two throttling modes. The fluid domain of orifice throttling is divided into the orifice region and gas film region, as shown in Figure 1. The inlet and outlet pressure are set at the upper surface of an orifice and outer side of film thickness, respectively. The partial porous material is set as porous walls. The wall surface under the gas film region and the partial region is set as wall2, and the region of porous is defined as porous and other surfaces like walls. The fluid calculation domain of partial porous throttling mode is similar to the small orifices, as shown in Figure 1. The throttling area includes the viscous permeability co-efficient, inertial permeability co-efficient, and porosity at the same time, the boundary layers are required to be set between the porous region and the gas film region. This ensures the calculation accuracy of the grids(mesh) of the two throttling areas in the direction of film thickness, and the grids of partial porous in the longitudinal direction are set up to 10 layers and 100 layers, respectively. The modeling, meshing, and geometric characteristics of aerostatic thrust bearing are defined with their differences and similarities which are compared with each other. The design parameters that define the boundary conditions and dimensions of modeling are given in Table 1.

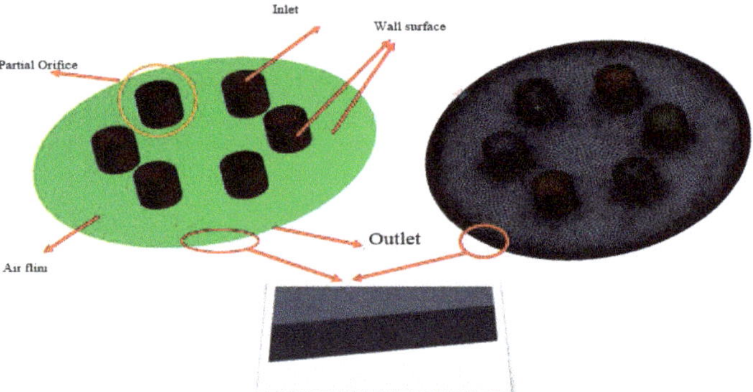

Figure 1. Partial porous orifice aerostatic thrust bearing model.

Table 1. Design parameters and boundary conditions.

Design Variable	Orifice & Porous
Outlet pressure	0 bar
Inlet pressure	5 bar
Operating pressure	101,325 Pa
Temperature	293 K
Dynamic viscosity	1.789×10^{-5} Ns/(m)
Gas density	1.189 kg/m^3
Film thickness	5–15 µm
Bearing outer diameter	80,000 µm
Bearing inner diameter	48,000 µm
Orifices height	2 mm/4 mm/6 mm/8 mm
Orifices radius	0.5 mm/1 mm/1.5 mm/2 mm
Viscous resistance coefficient	1×10^{13} m^{-2}/1×10^{13} m^{-2}/1×10^{13} m^{-2}
Porosity	0.3

The grids are determined through a sweep method by setting the 100 cell numbers for the orifices, and the 1000 numbers are fixed for porous thrust bearing. The hexahedral mesh element type is used, and the size of the mesh is fixed by default. The number of nodes and elements are given in Table 2. The effect of grids on the results of simulation are shown in Table 2 under supply pressure of 0.5 MPa, and the number of elements and nodes are 283,525 and 312,839 respectively. The X, Y, and Z directional lengths are 80 mm, 80 mm, and 8 mm, respectively, with the volume of 2560.4 mm^3. The parameters of numerical convergence in a fluid domain are set as different iteration numbers to attain the results of simulation. However, it was observed that the different iterations are not affecting the results. Therefore, the same iteration numbers are used to save time such as 1×10^{-6} in the residuals of continuity of x, y, and z velocity.

Table 2. Grid division information of aerostatic thrust bearing.

Object Name	Geometry
State	**Fully Defined**
Bounding box	
Length X	80 mm
Length Y	80 mm
Length Z	8 mm
Properties	
Volume	2560.4 mm^3
Scale Factor Value	1
Statistics	
Bodies	9
Active Bodies	9
Nodes	312,839
Elements	283,525
Mesh Metric	None

2.4. Dynamic Grid Method

This method is used for the calculation of static characteristics of aerostatic thrust bearing. The boundary of the fluid domain is fixed, but the grids and nodes are not fixed. The fluid control equation is discretized in the static coordinate system. To analyze the characteristics such as damping and dynamic stiffness of aerostatic bearings, the gas film bearing wall needs to move according to a given perturbation equation because the mesh in the fluid domain changes with time during the calculation of the transient flow field. Three equations of conservation of air film can be stated by Equation (6). The Finite Volume Method (FVM) is used by fluent in discretizing the equation of control. Three basic governing equations in the bearing gas film are expressed by the governing equation in mathematics. The fluid equations are then converted from volume integral to area integral by using a Gauss divergence algorithm. These equations are expressed as:

$$\frac{d}{dt}\int_V \rho\phi dV + \int_{\partial V}\rho\phi(u - u_g)dA$$
$$= \int_{\partial V}\Gamma\nabla\phi dA + \int_V S_\phi dV \tag{10}$$

where S_ϕ is the scalar source term, u_g is the moving velocity, A is the area of control volume boundary, and dv is the closed boundary control volume. The first-order differential is used to get in the term of time differential:

$$\frac{d}{dt}\int_V \rho\phi dV = \frac{(\rho\phi V)^{n+1} - (\rho\phi V)^n}{\Delta t} \tag{11}$$

where n and $n + 1$ show the current and next time steps, respectively, and the relationship satisfies

$$V^{n+1} = V^n + \frac{dV}{dt}\Delta t \tag{12}$$

The derivative of the control volume to time can be expressed as:

$$\frac{dV}{dt} = \int_{\partial V} u_g dV = \sum_j^{n_f} u_{g,j} \cdot A_j \tag{13}$$

where n_f is the number of all boundary faces on the control volume V. By solving Equations (10)–(13), the dynamic motion effect of the boundary on the transient flow field can be attained. Dynamic and smooth mesh methods are used for updating of the meshes. To effectively control mesh quality, the mesh is dynamically updated in time steps by using the Dynamic Layering and Smoothing methods. The dynamic layering method is only applicable to structural grids. The number of grid layers can be increased or decreased on the basis of the change of the height of the grid layer of the moving boundary, due to the large displacement amplitude of the moving boundary. In the smooth methods, mesh number does not change, and mesh boundary is balanced by the spring system, which is appropriate when the moving direction is perpendicular to the boundary, and the amplitude of moving boundary is small. Figure 2 shows the solution process flow chart of dynamic stiffness and damping coefficient of the porous aerostatic bearing.

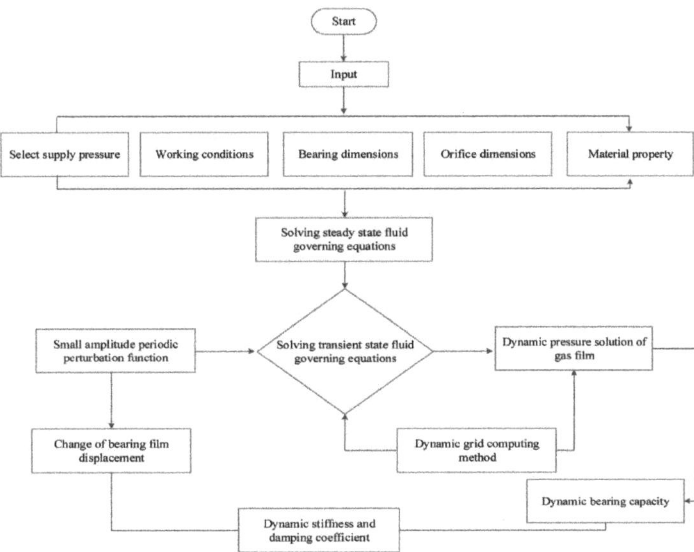

Figure 2. The flow chart determining the dynamic stiffness and damping coefficient.

3. Dynamic Stiffness and Damping Characteristics Analysis

The results are bifurcated in two categories such as those obtained through simulation and experimental methods. Details of parameter analysis are provided in the following sections.

Dynamic Mechanical Properties of Gas Film

In this section, dynamic mechanical properties of gas film are analyzed, as is illustrated in Figure 3. The model of partial porous aerostatic is composed of partial orifices, gas film, and thrust pad, as is shown in Figure 3. In the modeling process, the outside surface of film thickness is defined as outlet pressure, and the upper surface of a partial porous orifice is defined as the pressure inlet. Fluent software is used for estimation of a cloud chart of the pressure distribution, the pressure value at any point, and numerical change curve. The cloud map of the pressure distribution was in the form of a straight line and the precise form of gas flow velocity vector. Finally, the distribution of pressure under the changed structural parameters such as supply of pressure of 0.5 MPa, the atmosphere pressure of 1.03 MPa, and remaining parameters are given in Table 1. The pressure distribution diagram along the radial at the exit of the partial porous orifice is shown in Figure 3, which includes a pressure profile from the center of an aerostatic thrust bearing radially to the periphery of the bearing. The film thickness is 10 μm; the diameter of orifice is 5 mm, and the permeability of porous material is 1×10^{-14} m^2. The simulation results are obtained by Fluent software. Finally, a post-processing tool is used to process the results. The pressure clouds and numerical curves of the partial porous aerostatic bearings are obtained, which are shown in Figure 4. The pressure distribution cloud of partial porous aerostatic bearing is shown in Figure 4a having air supply pressure of 0.5 MPa, gas film thickness 10 μm, permeability of 1×10^{-14} m^2, the height of orifice of $H = 4$ mm, the orifice radius of 1 mm, and the number of orifices $N = 4$. The pressure variation curve of film gap along the x-axis under the same conditions is shown in Figure 4b. Figure 5 shows a velocity vector distribution of gas flow in the gas film formed from the pressure inlet of a partial porous orifice to the film gap. The velocity vector diagram of the whole bearing is shown in Figure 5a. It is clear from Figure 5b that the gas velocity reaches the maximum value at the pressure outlet; then, it gradually decreases and rises rapidly at the position of the partial porous orifice. When it is at the center of the bearing, the gas velocity drops rapidly close to zero because the gas forms a larger vacuum in the center position, making the pressure here reach the maximum value.

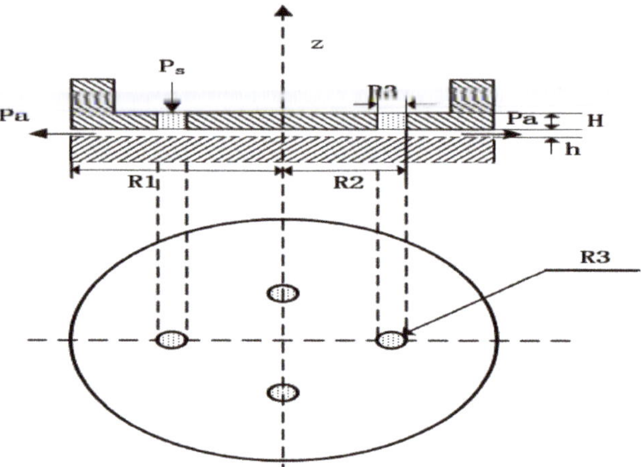

Figure 3. Diagram of partial porous orifice aerostatic thrust bearing.

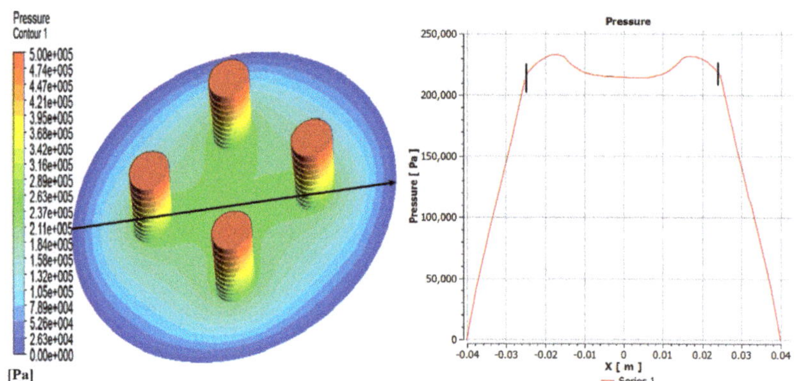

(**a**) Pressure distribution of overall partial porous   (**b**) Pressure distribution of bearing surface

Figure 4. Partial porous orifice pressure distribution.

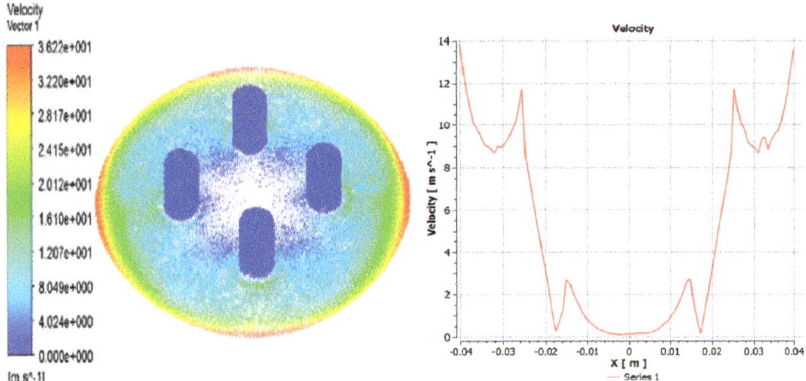

(**a**) Velocity vector diagram of partial porous   (**b**) Velocity distribution of bearing surface

Figure 5. Partial porous orifice velocity distribution.

4. Experimental Setup

The experimental setup was used to evaluate the effects of variation in the parameters on the performance of axial thrust bearing as is shown in Figure 6. The setup includes an air compressor, vibration isolation platform, pressure control valve with pressure gauge, inlet pressure pipe, thrust bearing, sensors, air hose pipe, shell and rotor, simulation module, and a display device. The sole objective of this set up is to obtain the physical values of various parameters of thrust bearing such as displacement, velocity, and acceleration, to get the stability effects.

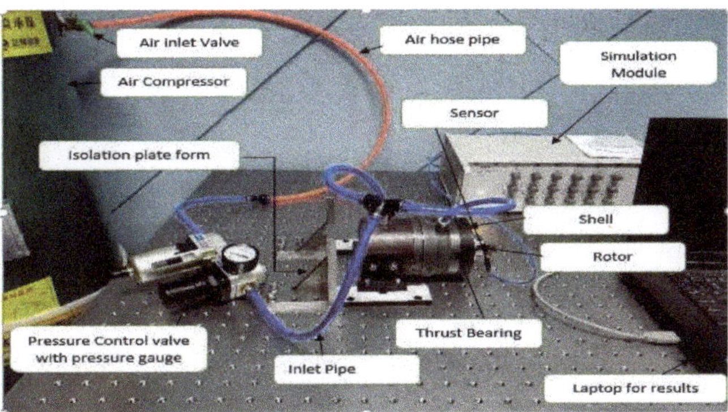

Figure 6. Experimental setups for aerostatic thrust baring.

5. Analysis of Factors Influencing Damping, Static, and Dynamic Stiffness Characteristics

A simulation model is established on the basis of the structural parameters of porous aerostatic bearings. The static pressure distribution of the bearing is obtained by the application of the gas film steady lubrication theory under assumed conditions of small external disturbance. In order to evaluate the effects of an additional external disturbance on the transient flow field in the air film clearance, the level of disturbance is enhanced. Newton's second law is applied to obtain the values of dynamic film pressure distribution, dynamic stiffness, and damping coefficient, and their values on the basis of occurrence of the changes in bearing displacement and bearing capacity. It is assumed in the calculations of dynamic stiffness and damping coefficient of a bearing film that, when the bearing is subject to a harmonic disturbance in the direction of the film, i.e., z-direction, the pressure of the bearing film changes with the changes in the level of external disturbance. If it is also assumed that the film thickness of air is h_0 of the bearing at equilibrium, the resultant change in the amplitude of the simple harmonic disturbance on the bearing surface is Δh and the frequency of the disturbance is ω, then the changes in the air film thickness with respect to time can be expressed by a relationship, shown in the following equation:

$$h = h_0 + \Delta h \tag{14}$$

Since the dynamic film lubrication equation of porous aerostatic bearing for a compressible object was obtained in Section 2, the value of h given in Equation (14) is put into Equation (10), in order to obtain a correlation between $\Delta p / \Delta h$ and ω, which is shown in Equation (15):

$$\begin{aligned}
&2\left\{\frac{\partial}{\partial x}\left[h_0^3 \frac{\partial}{\partial x}\left(\frac{\Delta p}{\Delta h}\right)\right]\right\} + \left\{\frac{\partial}{\partial y}\left[h_0^3 \frac{\partial}{\partial y}\left(\frac{\Delta p}{\Delta h}\right)\right]\right\} \\
&+ 3\left[\frac{\partial}{\partial x}\left(h_0^2 \frac{\partial p_0^2}{\partial x}\right) + \frac{\partial}{\partial y}\left(h_0^2 \frac{\partial p_0^2}{\partial y}\right)\right] = j24\mu\left(h_0 \frac{\Delta p}{\Delta h}\omega + p_0\omega\right)
\end{aligned} \tag{15}$$

Equation (15) shows a nonlinear correlation between $\Delta p / \Delta h$ and perturbed frequency ω. A further solution of Equation (15) leads to the complex number solution as is shown in the equation given below:

$$\frac{\Delta p}{\Delta h} = R(\omega, p_0, h_0) + j\,\text{Im}(\omega, p_0, h_0) \tag{16}$$

where R represents the real part and Im is the imaginary part in the above equation. Since the bearing capacity of a porous aerostatic thrust bearing is the integral sum of the air film pressure in the bearing gap on the entire surface region of the porous, Equation (16) is therefore solved further to obtain the values of the dynamic stiffness, which is given as follows:

$$K^* = \int R(\omega, p_0, h_0)dA + j\int \text{Im}(\omega, p_0, h_0)dA \tag{17}$$

The real part in Equation (17) is the dynamic stiffness K of the bearing, and the imaginary part is the product of the damping factor C and the disturbance frequency. These parts are indicated separately in the following equation:

$$\begin{aligned}
K &= R(\omega, p_0, h_0)dA \\
C &= \int \text{Im}(\omega, p_0, h_0)/\omega dA
\end{aligned} \tag{18}$$

The values of dynamic stiffness and damping coefficients of the gas film is reflected in Equation (18), which shows a relationship to the external disturbance frequency, the gas pressure before the disturbance, and the thickness of the air film. Therefore, the influence of the disturbance amplitude, film thickness, gas supply pressure, and permeability of porous materials on the dynamic stiffness and damping of the bearing are analyzed.

5.1. Influence of the Number of Partial Orifices (N = 4, 5, 6, 7)

The pressure distribution diagram of partial porous aerostatic bearing having air supply pressure of 0.5 MPa, gas film thickness of 10 μm, permeability of 1×10^{-14} m^2, the height of orifice H = 4 mm, the radius of orifice r of 1 mm, and the number of orifices N of 4, 5, 6, 7 is shown in Figure 7. However, Figure 8 shows the changes in pressure distribution of partial porous aerostatic thrust bearing with different orifices' numbers. It reveals that the number of orifices of partial porous aerostatic thrust bearing has an important impact on the pressure of bearing. It shows that, when there are four orifices, the pressure of bearing is 0.23 MPa, which increases to 0.278 MPa due to an increase in the number of orifices to 7, along with increase in the static bearing capacity of the partial porous aerostatic bearing up to a uniform 20.8%, and the uniform distribution of pressure on the surface of thrust plate of a bearing. A relationship between dynamic stiffness and damping curves of partial porous aerostatic thrust bearing at different orifice numbers is analyzed, which is reflected in Figure 9. It reveals that, in the low-frequency range (less than 100 Hz), the dynamic stiffness of the partial porous thrust bearing is constant, but it rises rapidly in the intermediate frequency range of 10^3 to 10^4 Hz, and then slows down gradually in the high-frequency range 10^5 to 10^6 Hz as is reflected in Figure 9a. Likewise, another relationship that exists between dynamic stiffness and perturbation frequencies is also analyzed, which reveals that the dynamic stiffness of bearing remains constant under the same perturbation frequency with an increase in the number of orifices. However, it reveals that the damping of partial porous aerostatic bearing gradually decreases with an increase in external perturbation frequency until its drop to zero, as is shown in Figure 9b. The dynamic damping curves of four kinds of partial porous orifices, in the low-frequency range (less than 100 Hz), indicates a gradual decrease in damping coefficient of partial porous bearing with subsequent increase in the number of orifices. Therefore, according to the comprehensive analysis, it can be concluded that the orifice number has a great influence on the static bearing capacity of the partial porous aerostatic thrust bearing but without influence on the bearing film's dynamic stiffness. It can also be said that, with the increase in the number of orifices, the damping parameters of partial porous bearings become smaller, so the effect of increasing the number of orifices on the dynamic characteristics of bearings is not apparent.

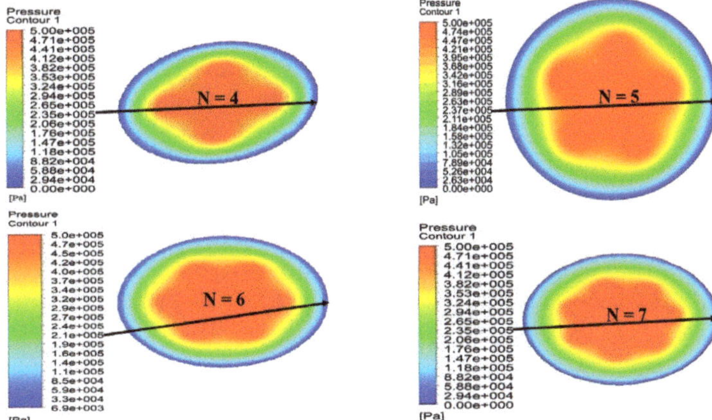

Figure 7. Cloud map of bearing pressure distribution when the number of partial porous orifices is different.

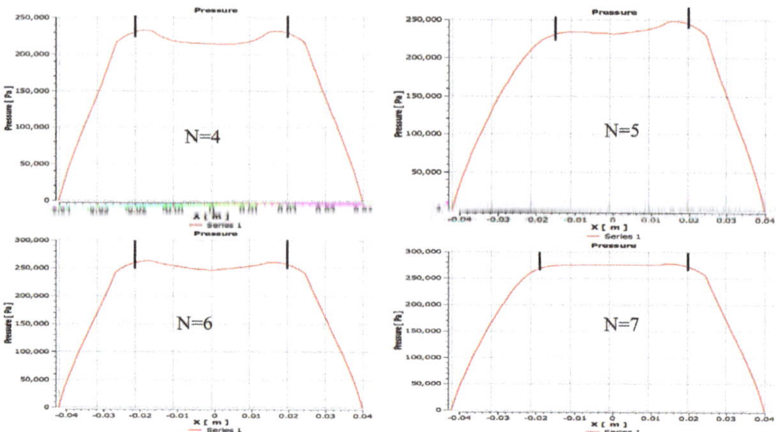

Figure 8. Distribution of bearing pressure with a different number of partial porous orifice.

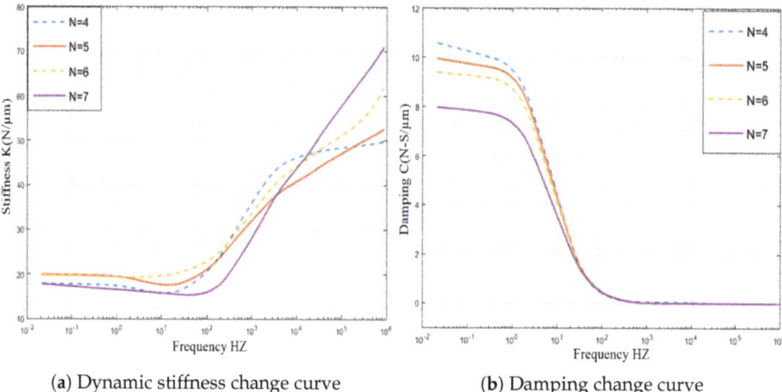

(**a**) Dynamic stiffness change curve   (**b**) Damping change curve

Figure 9. Dynamic stiffness and damping curve of thrust bearing with the different number of partial porous orifices.

5.2. Influence of Radius of Partial Orifice (r = 0.5 mm, 15 mm, 1.5 mm, 2 mm)

At this stage, the effects of changes in the radius of partial porous on pressure distribution are evaluated. These effects are reflected in Figure 10, following the setting of bearings' gas supply pressure of 0.5 MPa, a gas film thickness of 10 μm, a permeability of 1×10^{-14} m^2, the height of orifice H of 4 mm, and number of orifices N of 4 at orifices radii, and r of 0.5 mm, 15 mm, 1.5 mm, 2 mm, respectively. The changes in the bearing pressures with different position are shown in Figure 11. The comparative analysis of the orifice radius of the partial porous aerostatic bearing leads to the proposition of a significant nonlinear increase of pressure distribution of the bearings with an increase of the radius. It also indicates proportional change between the radius of partial porous material and the pressure distribution of the bearings. When the radius r = 0.5 mm, the pressure of the bearing is 0.12 MPa, and when the radius r = 2 mm, the pressure of the bearing increases to 0.35 MPa. It also indicates that, when the radius increases, the distribution of pressure on the partially porous thrust surface is more uniform, which further improves the static bearing capacity and stiffness parameters of an aerostatic bearing. Based on the static carrying capacity of the above-mentioned partial porous orifice aerostatic thrust bearing, the dynamic simulation calculation establishes the influence of the orifice radius on the damping and dynamic stiffness of the film of the partial porous orifice aerostatic thrust bearing, which is depicted in Figure 12. Figure 12a shows that the external perturbation frequency increases, as dynamic stiffness of partial porous orifice bearing firstly decreases slowly, and then rises rapidly in the intermediate frequency range 10^3 to 10^4 Hz. However, it reveals that, in the high-frequency range 10^5 to 10^6 Hz, it gradually slows down, but it still shows a steady rise. Therefore, by comparing the dynamic stiffness curves of four different orifice radii, it can be concluded that the dynamic stiffness of bearing film increases with the increase of orifice radius at the same external perturbation frequency, but, with the rise of external perturbation frequency, the damping of the partial porous bearing film gradually decreases, and finally decreases to zero. It also indicates that, by comparing the dynamic damping parameters of four kinds of orifice radii, it can be concluded that, in the low-frequency range (less than 100 Hz), the bearing damping increases with the increase of orifice radius at the same perturbation frequency as shown in Figure 12b. Therefore, the analysis leads to the conclusion that the orifice radius of the partial porous orifice aerostatic thrust bearing has a great influence on the dynamic stiffness and damping of the bearing film.

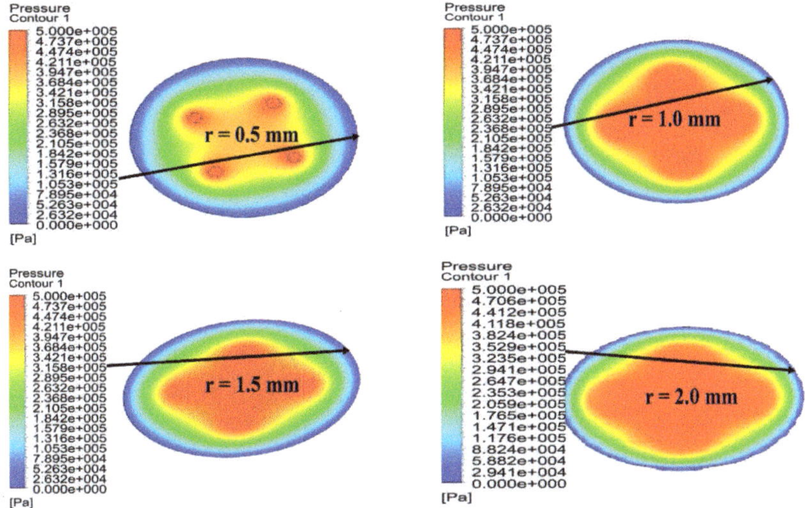

Figure 10. Cloud chart of bearing pressure distribution with a different radius of partial porous orifice.

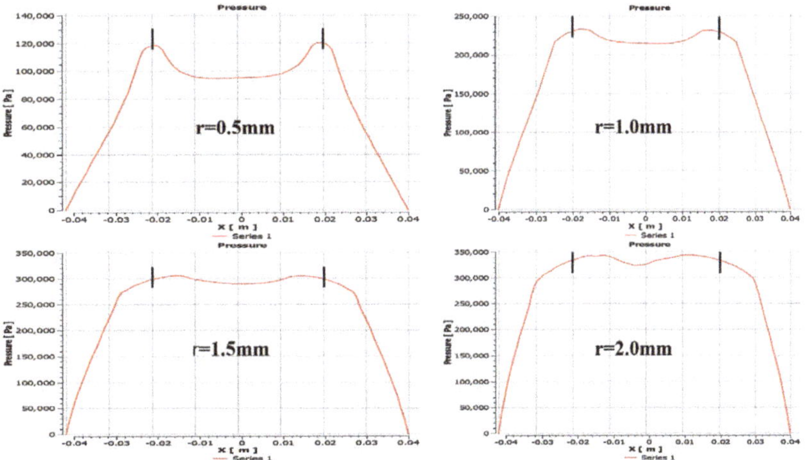

Figure 11. Distribution of bearing pressure with bearing when the radius of the partial porous orifice is different.

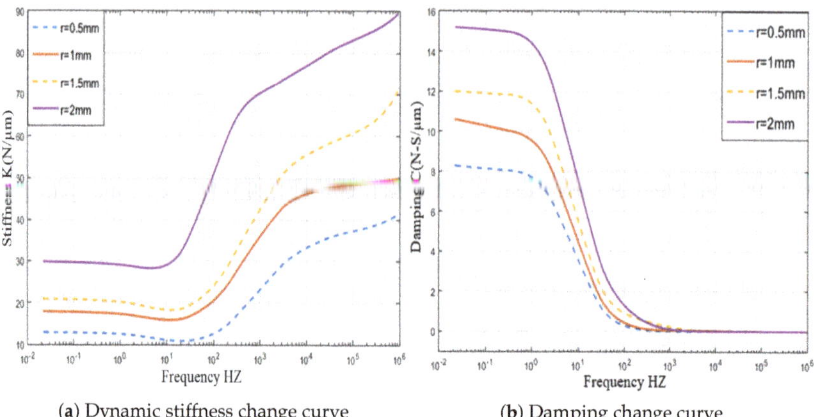

(**a**) Dynamic stiffness change curve (**b**) Damping change curve

Figure 12. Dynamic stiffness and damping curve of thrust bearing with a different radius of partial porous orifices.

5.3. Influence of the Height of the Partial Porous Orifice (H = 2 mm, 4 mm, 6 mm, 8 mm)

In this analysis, a relationship between pressure distribution of porous orifice aerostatic thrust bearing and different orifice heights is carried out. The pressure distribution of partial porous orifice aerostatic thrust bearing under different heights of the orifice is shown in Figure 13, at air supply pressure of 0.5 MPa, the film thickness of 10 μm, permeability of 1×10^{-14} m^2, the orifice radius r of 1 mm, and the number of partial porous orifices of 6 and the height of the orifices H of 2 mm, 4 mm, 6 mm, and 8 mm, respectively. The changes are reflected in Figure 14, which depicts the pressure distribution of a partial porous aerostatic bearing under different orifice heights. This analysis leads to the conclusion that the height of the orifice of the partial porous aerostatic bearing has a great impact on the pressure distribution of the bearing. It indicates that, when the height of the orifice is $H = 2$ mm, the pressure of the bearing is 0.42 MPa, and, when the height of the orifice is 8 mm, the 0.265 MPa is the pressure of the bearing, and the bearing static carrying capacity drops by 58% with an increase of orifice height, which indicates that, with increasing orifice height, the pressure of the partial porous aerostatic bearing decreases, and, with the increase of the height of the orifice, there is a downward trend. The dynamic stiffness and damping of partial porous orifice aerostatic thrust bearing at different orifice heights are calculated using dynamic simulation, as is shown in Figure 15. According to these calculations, as the external perturbation frequency increases, the dynamic stiffness of a partial porous orifice bearing in the low frequency (less than 100 Hz) and medium frequency (10^3–10^4 Hz) remains constant initially and then it rises gradually in the high-frequency range (10^5–10^6 Hz), as illustrated in Figure 15a—and, likewise, the dynamic stiffness of the bearing under different orifice heights, as compared with the partial porous aerostatic bearing analyzed. According to these calculations with increasing height of an orifice, the dynamic stiffness of the bearing under the same disturbance frequency presents a significant decline, in which the height of the orifice is 8 mm, as is shown in Figure 15b. Therefore, it can be concluded that the damping of the partial porous aerostatic thrust bearing decreases gradually with the increase of the external perturbation frequency, and the final decline tends to zero. Therefore, it can be concluded that, when orifice heights increase, bearing damping characteristics in the low-frequency range (less than 100 Hz) decrease significantly. From the comprehensive analysis of Figure 15, it can be observed that the height of porous orifice has an obvious influence on the dynamic stiffness and damping of the partial porous aerostatic thrust bearing, as the dynamic stiffness and damping coefficient of bearing film decrease with the increase of orifice height.

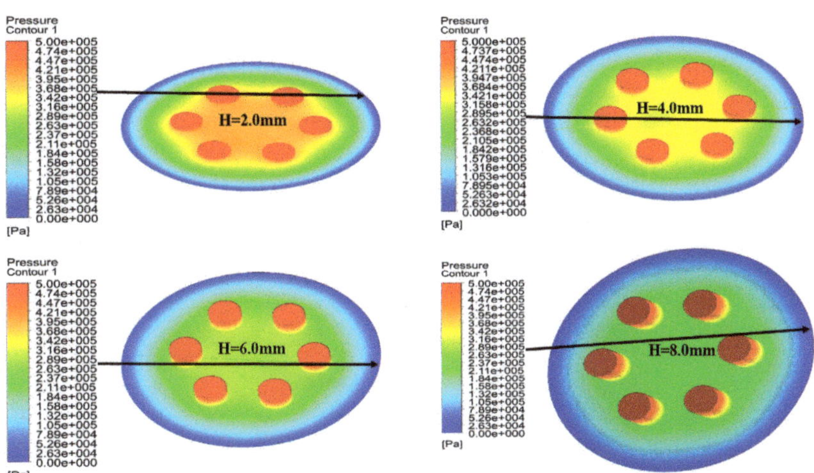

Figure 13. Cloud distribution pressure of the bearing with different height of the partial porous orifice.

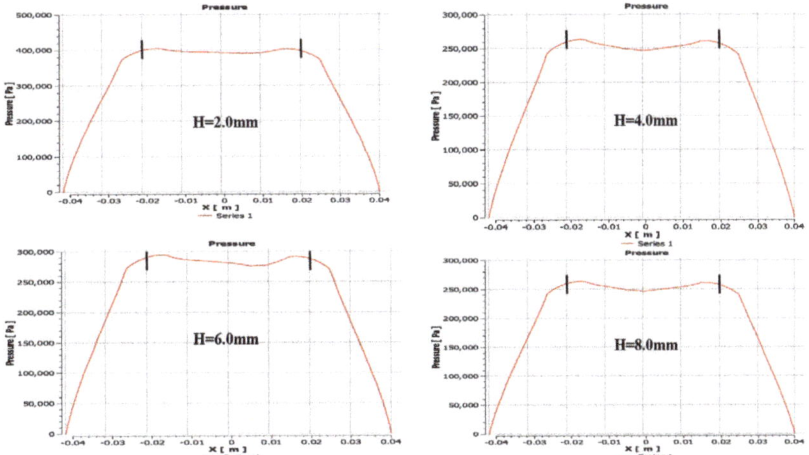

Figure 14. Bearing pressure distribution with bearing when the height of the partial porous orifice is different.

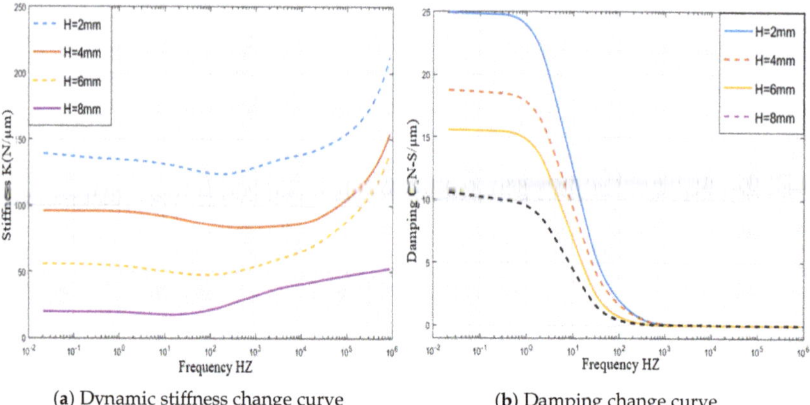

(**a**) Dynamic stiffness change curve   (**b**) Damping change curve

Figure 15. Dynamic stiffness and damping curve of thrust bearing with different height of partial porous orifice.

6. Validation of Numerical Results through Experiments

Equation (16) verifies the results of dynamic stiffness numerically based on real and imaginary parts of the equation, whereas, in Equation (17), the two parts are further elaborated to determine the dynamic stiffness, damping factor, and the disturbance frequency. Equation (18) correlates the relationship with the above-mentioned equations by changing the variables such as disturbance amplitude, film thickness, gas supply pressure, and permeability of porous materials as analyzed for the calculation of damping and dynamic stiffness of the aerostatic bearing. Further changes in the time interval and its effects on acceleration and inlet pressure are analyzed at this stage of the research process. Initially, it was set for 10 periods with a frequency started from 3 s along with a constant increase in a time interval of 0.002 s. The interval changes, the acceleration changes, and the inlet pressure changes are shown in Figure 16. Since the reflected acceleration is constant, the vibration is stable and is controlled at the side of the shell. The rotor has little variation from the central limit (CL), and vibration displacement is around 2 μm while the thickness of the gas film is about 20 μm.

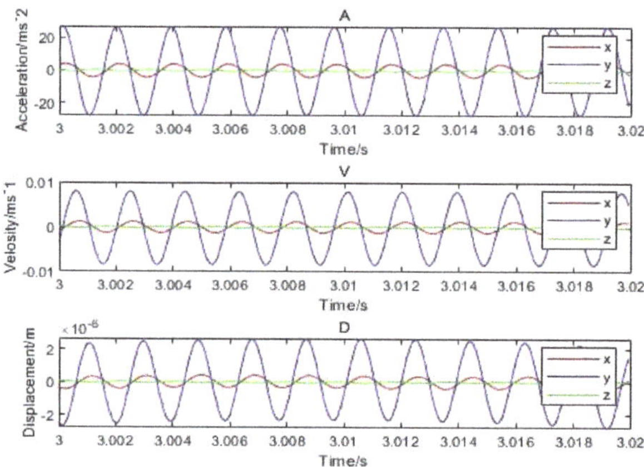

Figure 16. Dynamic acceleration, velocity, and displacement of rotor thrust bearing with (630 Hz) perturbation frequencies (h = 20 μm and Ps = 0.62 MPa).

It can be inferred that the direction of x and y has the highest vibration and that the displacement amplitude is about 10 percent of the gas film. The acceleration, velocity, and displacement of bearing shell with the inlet pressure of 0.62 MPa, the amplitudes of 0.9 m/s^2 and 0.5 m/s^2 in the direction of x, and the amplitude in the z-direction, respectively, are illustrated in Figure 17. The thickness of the gas film was set as 20 µm, and the vibration displacement as 0.5 µm. It was observed that the amplitude of the shell was 20 times less than the amplitude of the rotor due to decrease in the effect of vibration on the shell. The changes in the amplitude of vibrations due to changes in inlet pressure are shown in Figure 18, which indicates different vibration conditions of the rotor in different directions. Accordingly, the pneumatic vibration occurred at 0.6 MPa of pressure, and the vibration started at maximum amplitude and at the inlet value of pressure 0.68 MPa, at 3.5 s, and it ended at 10 s. The only change in vibration of bearing was observed at the inlet pressure within the time span of 10 s. It was also observed that the amplitude changed with the value of inlet pressure, while the frequency was constant at 630 Hz. Certain changes in the amplitude of vibrations with inlet pressure were observed, which are illustrated in Figure 19, which indicates that, with the increase in inlet pressure, there is some increase in the amplitude of vibration. The vibration state of the shell was observed in three directions. It was noticed that a pneumatic type of vibration occurred at the pressure of 0.6 MPa, whereas the highest amplitude was observed at the 0.68 MPa pressure inlet value as the vibration started at 3.5 s, and it ended after a time span of 10 s. It was also observed that, despite increase in the inlet pressure value, there was no change in the value of frequency, which was 630 Hz. The amplitude of the shell was found to be lower than that of the rotor at the same frequency. In order to evaluate the effects on acceleration due to changes in the position of bearing, the bearing was set up at the vertical position, and the results of the rotor were obtained, which are illustrated in Figure 20a. It was observed that the acceleration was maximum in the direction of y and minimum in the direction of x at the inlet supply pressure of 0.62 MPa, while displacement of 0.25 µm and the thickness of the film of 20 µm, respectively. It was also observed that the vibration of the rotor in a horizontal position was greater than the rotor put in a vertical position, as is shown in Figure 20b, as is visible in the appearance of increase in the acceleration amplitude of the rotor with a subsequent rise in inlet pressure. Therefore, the high amplitude was recorded at the inlet pressure from 0.62 MPa to 0.68 MPa along with the beginning of vibration at 3.5 s and its finishing at 10 s. The effects of changes in inlet pressures on acceleration amplitude were also analyzed at a later stage. It was observed that, at 624.2905 Hz frequency, amplitude changes with the increase of inlet pressure, as is reflected in Figure 21a, which indicates that the acceleration amplitude increases by 0.48 m/s^2 with the rise in inlet pressure. It was also observed that the displacement amplitude was maximum at 7.06 s, but it decreased when time increased to 7.068 s. The relationship between frequency of the shell and acceleration amplitude were further analyzed, which are illustrated in Figure 21b, and it shows that the acceleration amplitude was 0.089219 m/s^2 at frequency of 624.2905 Hz. The vibration of the shell started from 3.5 s and ended at 10 s. It was also noticed that the vibration of the isolation platform was in xy directions, and the acceleration amplitude increased to 0.55 m/s^2 with an increase in pressure at the time of 7.052 s and reduced after 7.07 s. The effects of time span changes on the displacement amplitude were also analyzed. It was observed that the displacement amplitude was maximum at 7.05 s and minimum at 7.07 s, which is shown in Figure 22a. The vibration of the platform begins from 3.5 s and ends at 10 s as depicted in Figure 22b.

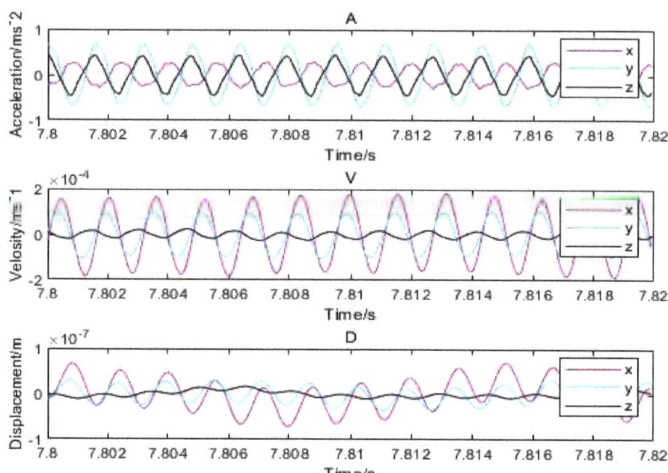

Figure 17. Dynamic acceleration, velocity, and displacement of shell bearing with (630 Hz) perturbation frequencies: ($h = 20$ μm and $Ps = 0.62$ MPa).

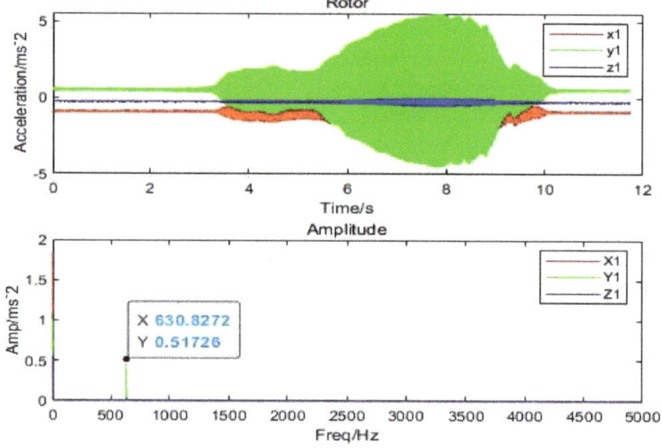

Figure 18. Amplitude of rotor of aerostatic thrust bearing with (630.8272 Hz) perturbation frequencies: ($h = 20$ μm and $Ps = 0.62$ MPa).

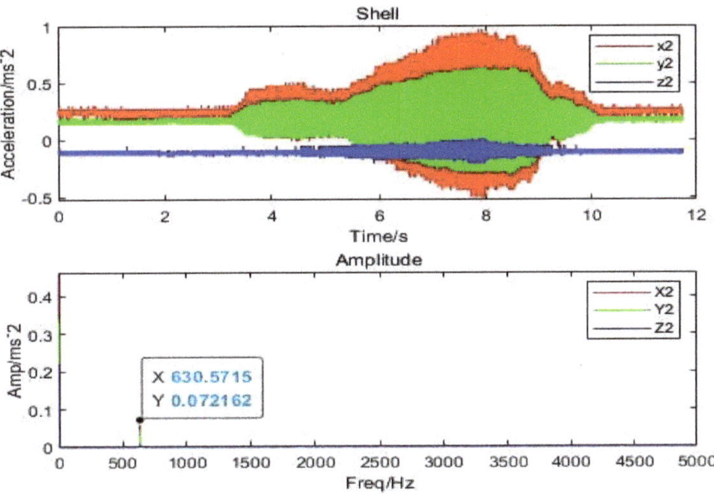

Figure 19. Amplitude response of shell of aerostatic thrust bearing with (630 Hz) perturbation frequencies: ($h = 20$ μm and $Ps = 0.62$ MPa).

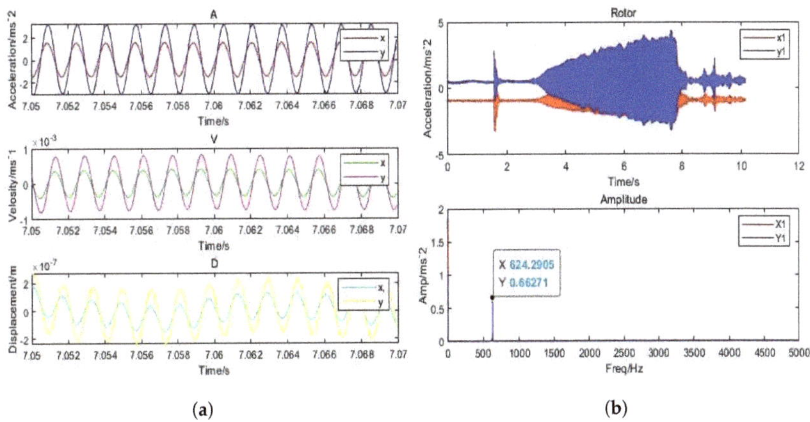

(a) (b)

Figure 20. (**a**) Dynamic acceleration response of rotor of aerostatic thrust bearing with (630 Hz) perturbation frequencies; (**b**) amplitude response of rotor of aerostatic thrust bearing with (630 Hz) perturbation frequencies. ($h = 20$ μm and $Ps = 0.62$ MPa).

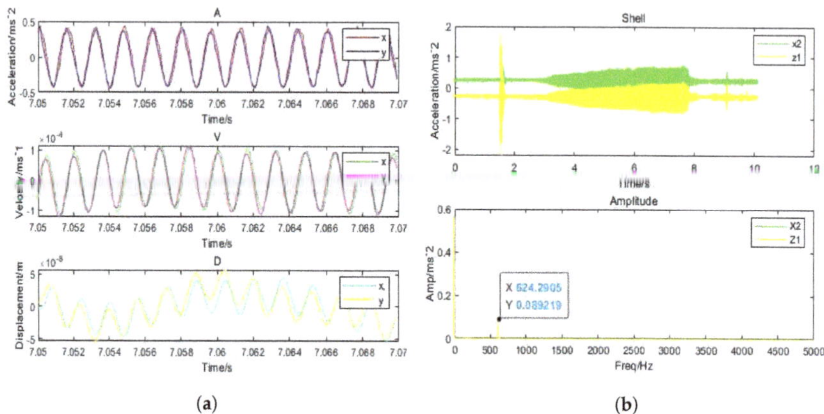

Figure 21. (**a**) Dynamic acceleration of shell with (630 Hz) perturbation frequencies; (**b**) amplitude of shell aerostatic thrust bearing with (630 Hz) perturbation frequencies: ($h = 20$ μm and $Ps = 0.62$ MPa).

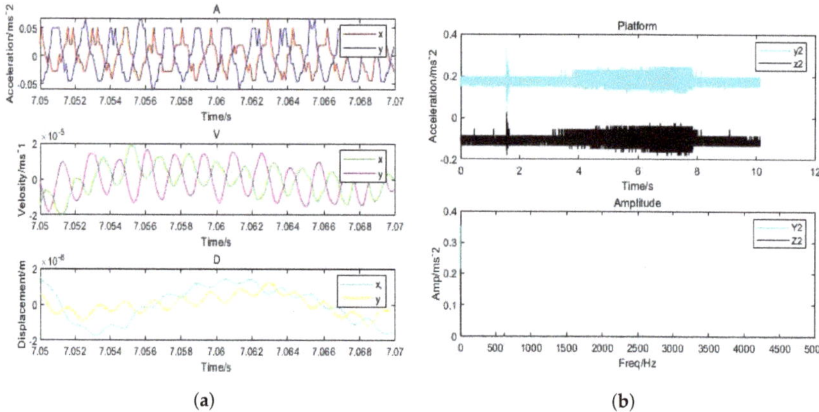

Figure 22. (**a**) dynamic acceleration of shell with (630 Hz) perturbation frequencies; (**b**) amplitude of shell aerostatic thrust bearing with (630 Hz) perturbation frequencies: ($h = 20$ μm and $Ps = 0.62$ MPa).

7. Conclusions

The simulated and experimental results reveal optimization of the dynamic stiffness and damping characteristics of the aerostatic porous thrust and partial porous orifice type bearings on account of changes in design and operating parameters such as orifice diameter, orifice height, material permeability coefficient, and orifice distribution. The experimental results of the dynamic stiffness and damping characteristics of the proposed parameters of a porous and partial porous aerostatic thrust bearing are compared with the simulated results to verify the accuracy of the simulation results on dynamic stiffness and damping characteristics. The dynamic stiffness and damping parameters reveal a nonlinear correlation between dynamic stiffness and damping characteristics at less than a 100 Hz frequency level. However, in the range of 100 Hz to 10,000 Hz, the amplitude of dynamic carrying capacity fluctuations shows a linear relationship. It indicates that, with a change in frequency, there is a change in the size of the gas film damping along with the variations in the dynamic bearing capacity of the gas film surface of the porous aerostatic thrust bearing in 10 cycles. Therefore, it can be concluded that, in the range of 100 Hz to 10,000 Hz with an increase in frequency, there is a significant decrease in amplitude of dynamic bearing capacity fluctuation, which indicates the influence of throttling on dynamic stiffness and damping characteristics of aerostatic thrust bearings film due to changes in the diameter of partial porous orifice aerostatic thrust bearing. It may also be concluded that a larger radius of partial porous aerostatic thrust bearing produces greater dynamic stiffness and damping of the partial porous aerostatic thrust bearing. It can also be concluded that a nonlinear relationship exists between orifice heights and damping characteristics of bearings at a frequency less than 100 Hz.

Author Contributions: Conceptualization, M.P.S.; methodology, M.P.S.; original draft writing, M.P.S.; software, M.P.S.; supervision, W.W.; experimental data, W.W.; review and editing, W.W.; writing review, A.N.S.; review experimental results, C.H.; critical review, commentary, and revision of the research paper, S.A.S. All authors have read and agreed to the published version of the manuscript.

Funding: This work is fully supported by NSAF U1830110 and Sichuan Province Science and Technology project 2020JDRC0010.

Conflicts of Interest: The authors declare no conflict of interest.

Nomenclature

A	area of control volume boundary
S_T	viscous dissipation energy
C_{ij}	inertia permeability co-efficient
S	source term
c_p	specific heat capacity
S_ϕ	scalar source term
D_{ij}	viscous permeability co-efficient
$\mathrm{grad}\,T$	temperature gradient
F_x	force in x-direction on body
T	temperature
F_y	force in y-direction on body
t	time
F_z	z-directional force on body
u	velocity vector in x-direction
H_z	frequency
v	velocity vector in y-direction
H	height of orifice
dV	closed boundary control volume
k	heat transfer co-efficient of fluid
w	velocity vector in z-direction
n_f	number of all boundary faces
x, y, z	Cartesian coordinates
N	orifice number
Γ	co-efficient of diffusion
P_s	supply pressure
γ	porosity of porous material
P_a	atmosphere pressure
ρ	gas density
r	radius of orifice
τ	viscous stress of fluid

References

1. Chen, D.; Huo, C.; Cui, X.; Pan, R.; Fan, J.; An, C. Investigation the gas film in micro scale induced error on the performance of the aerostatic spindle in ultra-precision machining. *Mech. Syst. Signal Process.* **2018**, *105*, 488–501. [CrossRef]
2. Zhang, S.; To, S.; Wang, H. A theoretical and experimental investigation into five-DOF dynamic characteristics of an aerostatic bearing spindle in ultra-precision diamond turning. *Int. J. Mach. Tools Manuf.* **2013**, *71*, 1–10. [CrossRef]
3. Gao, Q.; Chen, W.; Lu, L.; Huo, D.; Cheng, K. Aerostatic bearings design and analysis with the application to precision engineering: State-of-the-art and future perspectives. *Tribol. Int.* **2019**, *135*, 1–17. [CrossRef]
4. Schenk, C.; Buschmann, S.; Risse, S.; Eberhardt, R.; Tünnermann, A. Comparison between flat aerostatic gas-bearing pads with orifice and porous feedings at high-vacuum conditions. *Precis. Eng.* **2008**, *32*, 319–328. [CrossRef]

5. Cui, H.; Wang, Y.; Yue, X.; Huang, M.; Wang, W. Effects of manufacturing errors on the static characteristics of aerostatic journal bearings with porous restrictor. *Tribol. Int.* **2017**, *115*, 246–260. [CrossRef]
6. Chen, G.; Chen, Y. Multi-Field Coupling Dynamics Modeling of Aerostatic Spindle. *Micromachines* **2021**, *12*, 251. [CrossRef]
7. KWAN, Y.P.; Corbett, J. Porous aerostatic bearings: An updated review. *Wear* **1998**, *222*, 69–73. [CrossRef]
8. Plante, J.S.; Vogan, J.; El-Aguizy, T.; Slocum, A.H. A design model for circular porous air bearings using the 1D generalized flow method. *Precis. Eng.* **2005**, *29*, 336–346. [CrossRef]
9. Boffey, D. A study of the stability of an externally-pressurized gas-lubricated thrust bearing with a flexible damped support. *J. Lubr. Technol.* **1978**, *100*, 364–368. [CrossRef]
10. Boffey, D.; Desai, D. An experimental investigation into the rubber-stabilization of an externally-pressurized air-lubricated thrust bearing. *J. Lubr. Technol.* **1980**, *102*, 65–70. [CrossRef]
11. Jia, C.; Pang, H.; Ma, W.; Qiu, M. Analysis of dynamic characteristics and stability prediction of gas bearings. *Ind. Lubr. Tribol.* **2017**, *69*, 123–130. [CrossRef]
12. Blondeel, E.; Snoeys, R.; Devrieze, L. Dynamic stability of externally pressurized gas bearings. *J. Lubr. Technol.* **1980**, *102*, 511–519. [CrossRef]
13. Suryawanshi, S.R.; Pattiwar, J.T. A Technical Review on Design & Thermal Behavior of Non-circular Hydrodynamic Journal Bearing using CFD Technique. *Spvryan'S Int. J. Eng. Sci. Technol. (SEST)* **2015**, *2*, 1–5.
14. Sahto, M.P.; Wang, W.; Imran, M.; He, L.; Li, H.; Weiwei, G. Modelling and Simulation of Aerostatic Thrust Bearings. *IEEE Access* **2020**, *8*, 121299–121310. [CrossRef]
15. Constantinescu, V.; Galetuse, S. On the dynamic stability of the spiral-grooved gas-lubricated thrust bearing. *J. Tribol.* **1987**, *109*, 183188. [CrossRef]
16. Jolly, P.; Hassini, M.A.; Arghir, M.; Bonneau, O. Identification of stiffness and damping coefficients of hydrostatic bearing with angled injection. *Proc. Inst. Mech. Eng. Part J J. Eng. Tribol.* **2013**, *227*, 905–911. [CrossRef]
17. Plessers, P.; Snoeys, R. Dynamic stability of mechanical structures containing externally pressurized gas-lubricated thrust bearings. *J. Tribol.* **1988**, *110*, 271–278. [CrossRef]
18. Yoshimoto, S.; Kohno, K. Static and dynamic characteristics of aerostatic circular porous thrust bearings (effect of the shape of the air supply area). *J. Tribol.* **2001**, *123*, 501–508. [CrossRef]
19. Cui, H.; Wang, Y.; Yue, X.; Huang, M.; Wang, W.; Jiang, Z. Numerical analysis and experimental investigation into the effects of manufacturing errors on the running accuracy of the aerostatic porous spindle. *Tribol. Int.* **2018**, *118*, 20–36. [CrossRef]
20. Zhang, J.; Zou, D.; Ta, N.; Rao, Z. Numerical research of pressure depression in aerostatic thrust bearing with inherent orifice. *Tribol. Int.* **2018**, *123*, 385–396. [CrossRef]
21. Peixoto, T.F.; Daniel, G.B.; Cavalca, K.L. Experimental Estimation of Equivalent Damping Coefficient of Thrust Bearings. In Proceedings of the International Symposium on Dynamic Problems of Mechanics, São Sebastião, Brazil, 5–10 March 2017; pp. 17–29.
22. Wang, W.; Cheng, X.; Zhang, M.; Gong, W.; Cui, H. Effect of the deformation of porous materials on the performance of aerostatic bearings by fluid-solid interaction method. *Tribol. Int.* **2020**, *150*, 106391. [CrossRef]
23. Gao, Q.; Lu, L.; Chen, W.; Chen, G.; Wang, G. A novel modeling method to investigate the performance of aerostatic spindle considering the fluid-structure interaction. *Tribol. Int.* **2017**, *115*, 461–469. [CrossRef]
24. Zhuang, H.; Ding, J.; Chen, P.; Chang, Y.; Zeng, X.; Yang, H.; Liu, X.; Wei, W. Numerical Study on Static and Dynamic Performances of a Double-Pad Annular Inherently Compensated Aerostatic Thrust Bearing. *J. Tribol.* **2019**, *141*, 051701. [CrossRef]
25. Zhao, X.L.; Dong, H.; Fang, Z.; Chen, D.D.; Zhang, J.A. Study on dynamic characteristics of aerostatic bearing with elastic equalizing pressure groove. *Shock Vib.* **2018**, *2018*, 6142386. [CrossRef]
26. Fourka, M.; Bonis, M. Comparison between externally pressurized gas thrust bearings with different orifice and porous feeding systems. *Wear* **1997**, *210*, 311–317. [CrossRef]
27. Hongxia, Z.; Yuntang, L.; Yingxiao, L.; Dandan, L. CFD Investigation on the Performance of Aerostatic Thrust Bearing with Exhaust Slots Used in Low-vacuum Condition. In Proceedings of the 1st International Conference on Mechanical Engineering and Material Science, Shanghai, China, 28–30 December 2012.
28. Dong, H.; Zhao, X.L.; Zhang, J.a. Static characteristic analysis and experimental research of aerostatic thrust bearing with annular elastic uniform pressure plate. *Adv. Mech. Eng.* **2015**, *7*, 1687814015575454. [CrossRef]
29. Kodnyanko, V.; Shatokhin, S.; Kurzakov, A.; Pikalov, Y.; Pikalov, I.; Grigorieva, O.; Strok, L.; Brungardt, M. Numerical Modeling on the Compliance and Load Capacity of a Two-Row Aerostatic Journal Bearing with Longitudinal Microgrooves in the Inter-Row Zone. *Appl. Sci.* **2021**, *11*, 5714. [CrossRef]
30. Cui, H.; Wang, Y.; Yang, H.; Zhou, L.; Li, H.; Wang, W.; Zhao, C. Numerical analysis and experimental research on the angular stiffness of aerostatic bearings. *Tribol. Int.* **2018**, *120*, 166–178. [CrossRef]

micromachines

Article

Research on Multi-Physics Coupling Simulation for the Pulse Electrochemical Machining of Holes with Tube Electrodes

Zhaolong Li [1,2], **Bingren Cao** [2] **and Ye Dai** [1,2,*]

1   Key Laboratory of Advanced Manufacturing Intelligent Technology of Ministry of Education, Harbin University of Science and Technology, Harbin 150080, China; lizhaolong@hrbust.edu.cn
2   School of Mechanical and Power Engineering, Harbin University of Science and Technology, Harbin 150080, China; 1920110144@stu.hrbust.edu.cn
*   Correspondence: daiye312@163.com; Tel.: +86-0451-8639-0588

Abstract: Electrical parameters of the power supply are significant factors affecting the accuracy and stability of the electrochemical machining (ECM). However, the electric field, flow velocity and temperature in the machining area are difficult to measure directly under the influence of the power supply. Therefore, taking the film cooling hole as an example, the multi-physics coupling simulation analysis of the ECM is performed on the basis of Faraday's law and fluid heat transfer mathematical model. The machining characteristics of the direct current and pulse ECM are compared through simulation. The results show that the pulse ECM improves the distribution of temperature and current density in the machining area. The period has little effect on the temperature, current density and side removal rate. The side removal rate increases with the increase of the duty ratio and lateral gap. Increasing of the duty ratio and decreasing of the lateral gap will increase the temperature and current density. Increasing the inlet pressure accelerates the frequency of renewal of heat and electrolysis products, which can reduce the single side gap. The experience of the ECM holes verifies the results of the simulation. The accuracy and stability of the ECM of holes are enhanced by optimizing the duty ratio, lateral gap and inlet pressure.

Keywords: electrochemical machining; accuracy; stability; period; duty ratio; lateral gap; inlet pressure

Citation: Li, Z.; Cao, B.; Dai, Y. Research on Multi-Physics Coupling Simulation for the Pulse Electrochemical Machining of Holes with Tube Electrodes. *Micromachines* **2021**, *12*, 950. https://doi.org/10.3390/mi12080950

Academic Editors: Benny C. F. Cheung and Jiang Guo

Received: 19 July 2021
Accepted: 10 August 2021
Published: 11 August 2021

Publisher's Note: MDPI stays neutral with regard to jurisdictional claims in published maps and institutional affiliations.

1. Introduction

With the rapid development of accuracy and miniaturization in the aerospace, automotive and medical industries, small holes of special shape and groove structures have appeared on many mechanical parts. For example, in order to meet the characteristics of lightweight and high temperature resistance of turbine blades, high temperature resistant cemented carbide [1–3] is used as the main material of blades and the film cooling holes [4,5] are machined to improve the properties of thermal dissipation. However, traditional machining is difficult to machine microstructures in high temperature resistant cemented carbides. Electrical discharge machining [6,7] and laser machining [8,9] will form the hot recast layer and micro-cracks on the metal surface. These methods will affect the accuracy and stability of holes machining.

The ECM is based on the principle of electrochemical dissolution of the anode in the electrolyte [10–13]. The surface quality of the machined small holes is good, and there is no stress concentration and surface hardening layer. The diameter of film cooling holes is usually between 1–3 mm. The tiny structure prevents the placement of sensors in the machining area to collect temperature, flow velocity and current density distribution. The heater leased by the electrochemical reaction and electric current raises the temperature of the machining area [14,15]. The flowing electrolyte acts as a conductive medium to provide an ECM environment. The renewal of electrolysis products and heat is related to the flow field distribution of the electrolyte [16]. In order to improve the accuracy and stability of ECM of small holes, experts and scholars at home and abroad have studied the distribution characteristics of the electric, flow and temperature field in the machining area under the influence of the power supply. Chai et al. [17] studied the distribution of the volume fraction of hydrogen and the temperature in the machining area of the cooling hole with an external direct current power supply. Based on the finite element electric field model of the direct current power supply, Li et al. [18] obtained the mathematical relationship between the hole forming law and the machining time under different mask diameters. Wang et al. [19] improved the flow field and machining positioning by synchronizing the pulse power supply and low frequency vibration, which reduced the stray corrosion on the sidewall of the diamond hole. T. Sathish [20] used the parameter optimization to study the effect of the duty ratio on the roundness, taper and surface roughness of the titanium alloy holes through experiments. Manpreet [21] obtained the optimal direct current voltage suitable for ECM of silicon wafers by taking overcut and taper as the output quality characteristics. Ma et al. [22] used the pulse electrochemical method to optimize the surface roughness of non-circular holes after EDM, which was decreased from 4.227 μm to 0.229 μm.

The research structure of this article is as follows: Firstly, the mathematical model of multi-physics coupling simulation of ECM small holes is described in detail in Section 2. The machining properties of the direct current and pulse ECM are compared, and the distribution of the temperature, flow and electric field with pulse ECM are studied in Section 3 according to the established geometric model of the film cooling hole. In Section 4, the changes of the temperature, current density and side removal rate under the influence of the period, duty ratio, lateral gap and pressure are introduced in detail. Finally, the experiment of ECM of small holes in nickel-based alloys was performed to verify the simulation results and optimize the duty ratio and inlet pressure. The optimized parameters can improve the machining accuracy and stability of film cooling holes.

2. Multi-Physics Coupling Mathematical Model and Theoretical Analysis

ECM is a complex process using electrochemical principle, which involves the interaction between the electric field, flow field, heat conduction, chemical reaction and structure. Figure 1 intuitively and clearly describes the multi-physics coupling relationship in the ECM process. In this article, the pulse power supply is used for ECM of small holes. Conductivity is used as a "bridge" to connect the electric field and the flow field, so as to achieve the characteristics of the three-field coupling of heat, electricity and flow. The temperature, current density and side removal rate in the machining area under the influence of various physical fields are analyzed. In order to simplify the above model, this article makes the following assumptions:

(1) Polarization is not considered during the ECM process, and the surface of the anode material is always assumed to be in an activated state.
(2) The bubbles generated by the cathode and anode are ignored.
(3) The conductivity of the electrolyte depends only on the change of the temperature.
(4) The relative position between the electrode and the workpiece remains unchanged.

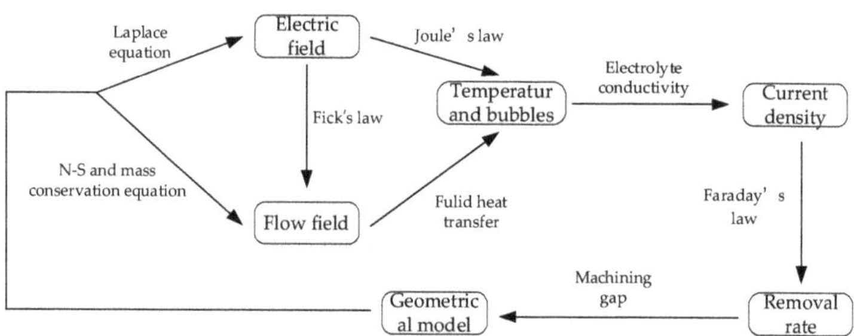

Figure 1. Schematic diagram of multi-physics coupling relationship.

The electrode is not only the metal connecting the positive and negative electrodes of the power supply, but also the electrode system formed by the interface of the positive and negative electrodes and their adjacent electrolyte in the tubular electrode ECM. The details are shown in Figure 2.

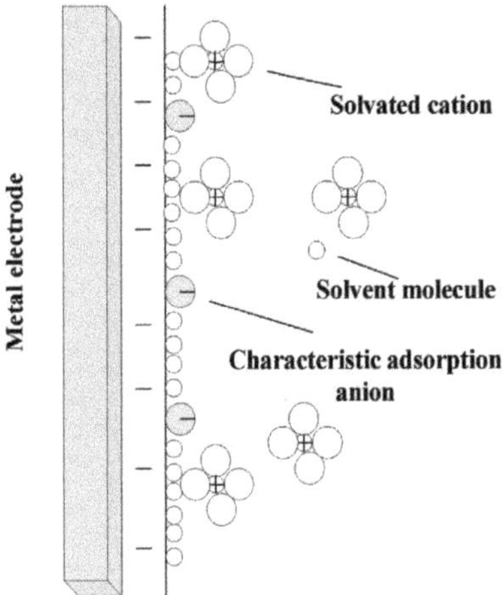

Figure 2. Schematic diagram of electrode system.

When the workpiece is electrolyzed, the workpiece is made of high-temperature resistant nickel-based alloy containing carbon, silicon and other non-metallic substances. Thus, many metal and nonmetal ions are oxidized at the anode to form ions. The specific chemical equation is as follows:

$$M \rightarrow M^{n+} + ne \tag{1}$$

These ions enter the electrolyte, which are subjected to strong electric field and strong flow field inside the machining gap. They are continuously diffused towards the end face of the tubular electrode and oxidized on the cathode surface. After adsorption and deposition, it forms a nonmetallic nonconductive film layer to increase the frequency of short circuit.

2.1. Electric Field Model

When the ECM is in a stable state, the machining gap is maintained in a stable range. The equation of lateral gap:

$$\Delta h = \frac{\eta \omega \sigma (U - \delta E)}{V_i} = \frac{\sigma (U - \delta E)}{i} \tag{2}$$

$$V_i = \eta \omega i \tag{3}$$

Assuming the homogeneity of electrolyte, the potential distribution $\varphi(x,y)$ in the machining gap can be expressed by the Laplace equation. The current density at any point in the electric field satisfies Ohm's law.

$$\nabla^2 \varphi = \frac{\partial^2 \varphi}{\partial x^2} + \frac{\partial^2 \varphi}{\partial y^2} = 0 \tag{4}$$

$$i = -\sigma \cdot \nabla \varphi \tag{5}$$

where U(V) is the processing voltage, δE(V) is the voltage drop between the cathode and anode, V_i(m/s) is the material removal rate, σ(S/m) is the conductivity of the electrolyte, η

is the current efficiency, $\omega(\text{mm}^3/(\text{A}\cdot\text{min}))$ is the volume electrochemical equivalent, $\varphi(V)$ is the electric potential and $i(\text{A}/\text{m}^2)$ is the local current density.

2.2. Flow Field Model

Assuming that the size of the bubble is small, the effect of the bubble on the electrolyte flow can be ignored. For incompressible viscous fluids, the fluid flow in the turbulent state is restricted by the Navier-Stokes equation.

$$\rho\left(\frac{\partial u}{\partial t} + u\cdot\nabla u\right) = -\nabla p + \mu\Delta u \tag{6}$$

$$\rho\nabla\cdot u = 0 \tag{7}$$

where $\rho(\text{kg}/\text{m}^3)$ is the electrolyte density, $u(\text{m}/\text{s})$ is the flow velocity, $p(\text{Pa})$ is the pressure, $\mu\text{g}/(\text{m}\cdot\text{s}))$ is the dynamic viscosity of the electrolyte.

Based on the change of the flow field with time, the "k-ε" turbulence model in the Reynolds average (RANS) equation is used to solve the turbulent energy "k" and the turbulent dissipation rate "ε" in the electrolyte flow process.

$$\rho\frac{\partial k}{\partial t} + \rho\cdot u\cdot\nabla k = \nabla\cdot\left[\left(\mu + \frac{\mu T}{\sigma_k}\right)\nabla k\right] + P_k - \rho\varepsilon \tag{8}$$

$$\rho\frac{\partial\varepsilon}{\partial t} + \rho\cdot u\cdot\nabla\varepsilon = \nabla\cdot\left[\left(\mu + \frac{\mu T}{\sigma_\varepsilon}\right)\nabla\varepsilon\right] + C_{\varepsilon 1}\frac{\varepsilon}{k}P_k - C_{\varepsilon 2}\rho\frac{\varepsilon^2}{k} \tag{9}$$

where P_k is the generating term of turbulent energy, σ_k and σ_ε are Prandtl numbers corresponding to "k" and "ε" with values of 1.0 and 1.3, $C_{\varepsilon 1}$ and $C_{\varepsilon 2}$ are model constants with values of 1.44 and 1.92.

2.3. Temperature Field Model

The temperature distribution of the electrolyte is calculated by solving the internal energy, which satisfies the convection-diffusion equation.

$$\rho C_P\frac{\partial T}{\partial t} + \rho C_P\mu\cdot\nabla T = \nabla\cdot(k_0\cdot\nabla T) + Q_i \tag{10}$$

$$Q_i = i\cdot\nabla\varphi \tag{11}$$

where $C_p(\text{J}/(\text{kg}\cdot\text{K})$ is the heat capacity of electrolyte and k_0 $(\text{W}/(\text{m}\cdot\text{K})$ is the heat conductivity coefficient. $Q_i(\text{W}/\text{m}^3)$ is the Joule heat generated by the current.

The flow field and electric field affect the temperature field distribution, and the conductivity of the electrolyte is affected by the temperature. The relationship is:

$$\sigma_T = \sigma(1 + \gamma(T - T_0)) \tag{12}$$

where γ is the temperature correlation coefficient, $T_0(\text{K})$ is the initial temperature.

3. Simulation Analysis of the Temperature, Flow and Electric Field

3.1. The Establishment of Geometric Model

Figure 3 shows the geometric model of the multi-physics coupling simulation of the film cooling hole. The solution area is composed of the geometry of the electrode, contour of the workpiece and flow channel of the electrolyte. The boundaries $\Gamma 2$ and $\Gamma 3$ are the inner wall and front-end surface of the cathode. The boundary $\Gamma 4$ is the outer wall of the cathode coated with an insulating layer. The boundaries $\Gamma 7$, $\Gamma 8$ and $\Gamma 9$ are anodes. The boundaries $\Gamma 1$ and $\Gamma 6$ are the inlet and outlet of the electrolyte. The boundary $\Gamma 5$ is the wall.

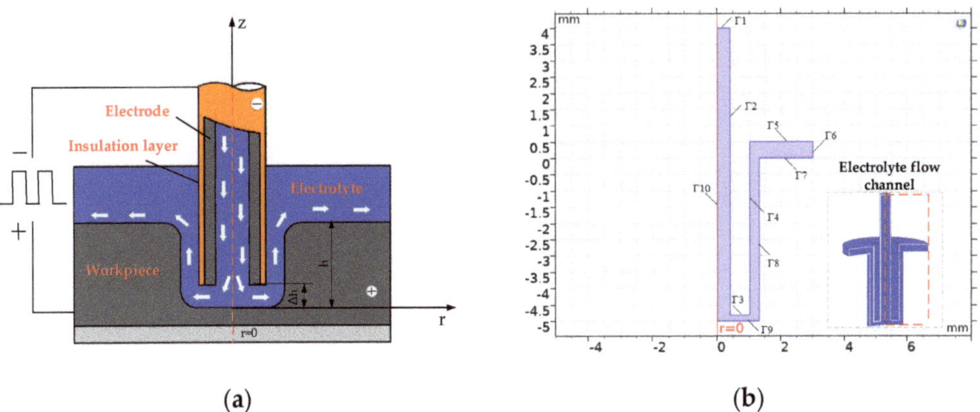

Figure 3. Geometric model of the film cooling hole machining; (**a**) schematic diagram of the machining zone; (**b**) electrolyte flow channel model and boundary definitions.

3.2. The Effect of Power Supply on the Temperature and Current Density

When the electrode and workpiece are connected to an external power supply, Joule heat is generated in the machining area. Excessive local temperature will cause electrolyte evaporation. The increase of the temperature affects the conductivity of the electrolyte and causes uneven dissolution of the workpiece surface. The pulse power supply (the pulse on is 24 V, the pulse off is 0.001 V) has a period of 1000 μs and a duty ratio of 0.5. $U(t)$ is the pulse voltage. U is the pulse voltage amplitude. T is the pulse period.

$$U(t) = \begin{cases} flc1hs\left(\sin\left(2*pi*\frac{t}{T}\right), \frac{T}{2}\right) * U & t \in \left(0, \frac{nT}{2}\right) \\ 0.001 & t \in \left(\frac{nT}{2}, nT\right) \end{cases} \tag{13}$$

The detailed boundary conditions of the simulation model are shown in Table 1. Figure 4a shows the temperature distribution of the machining area under the direct current and pulse ECM. The abrupt change of temperature occurs at the junction of the inner-outer walls and the end face of the electrode. In the steady state, the highest temperature of the direct current machining area is 323 K, while the pulse machining area is 303 K. Figure 4b shows that the current density distribution on the surface of the workpiece is similar to the temperature distribution. In the machining area with a radius of 1 mm, the current density range is reduced from 1.95×10^6 A/m^2 with the direct current to 0.91×10^6 A/m^2 with the pulse. Therefore, the pulse ECM improves the temperature distribution by interval discharge and reduces the current density difference on the workpiece surface.

Table 1. Parameters of the simulation model.

Parameters	Values
Density of electrolyte (ρ)	1200 (kg/m^3)
Dynamic viscosity of electrolyte (μ)	0.001 (Pa/s)
Heat capacity of electrolyte (C_p)	4200 (J/(kg·K)
Initial conductivity (σ)	12 (S/m)
Heat conductivity coefficient (k_0)	0.65 (W/(m·K))

Table 1. *Cont.*

Parameters	Values
Pulse period (T)	800, 1000, 1200, 1400 (μs)
Duty ratio (1)	0.2, 0.4, 0.5, 0.7
Lateral gap (Δh)	0.12, 0.15, 0.18, 0.21 (mm)
Inlet pressure (p_0)	0.2 (MPa)
Temperature correlation coefficient (γ)	0.16
Volume electrochemical equivalent (ω)	2 mm^3 /(A·min))

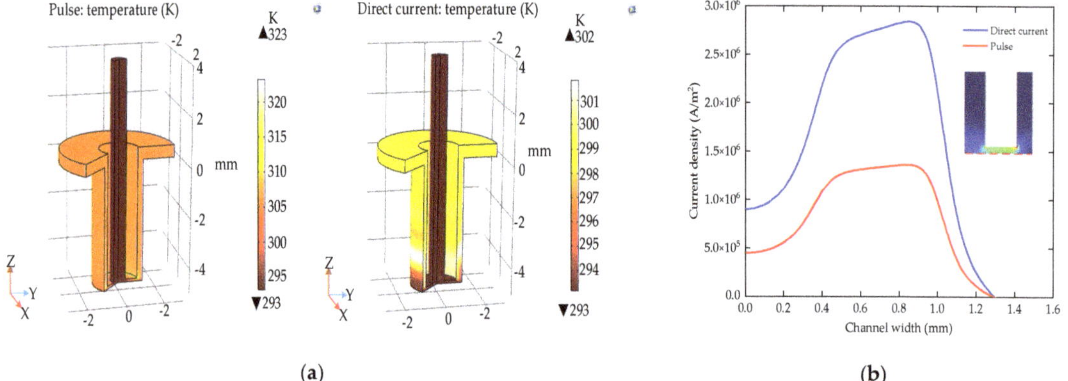

Figure 4. The temperature and current density with different power supply; (**a**) the temperature of the machining area; (**b**) the current density on the anode surface.

To sum up: Compared with the direct current ECM, the pulse ECM has better performance in the ECM of small holes. On the one hand, the pulse ECM avoids the effect of high temperature on the machining accuracy; on the other hand, it is beneficial to reduce the short circuit caused by uneven dissolution.

4. Physical Field Distribution with Pulsed Power Supply

4.1. Temperature Field

In order to study the effect of pulse ECM on the temperature, four reference points are selected on the electrode end face at equal intervals (the area with large temperature changes). The process parameter is a voltage of 24 V (the period is 1000 μs, the duty ratio is 0.5, the lateral gap is 0.12 mm, the pressure is 0.2 MPa and the depth is 5 mm).

Figure 5a shows the temperature change rules of reference points over time. The temperature reaches the maximum in the first period. After about 5 periods, it drops to a stable state. Figure 5b shows that the temperature in the pulse ECM machining area increases to 318 K within 1/4 period, and the temperature decreases to 303 K within 3/4 period. During the pulse off, the discharge between the electrodes and the electrochemical reaction are suspended which causes no current to flow in the circuit. The cumulative effect of Joule heat generated by the current and the heat of electrochemical reaction is reduced. Therefore, the pulse ECM can reduce the influence of temperature on electrical conductivity and current density, which is beneficial to improve the stability of ECM.

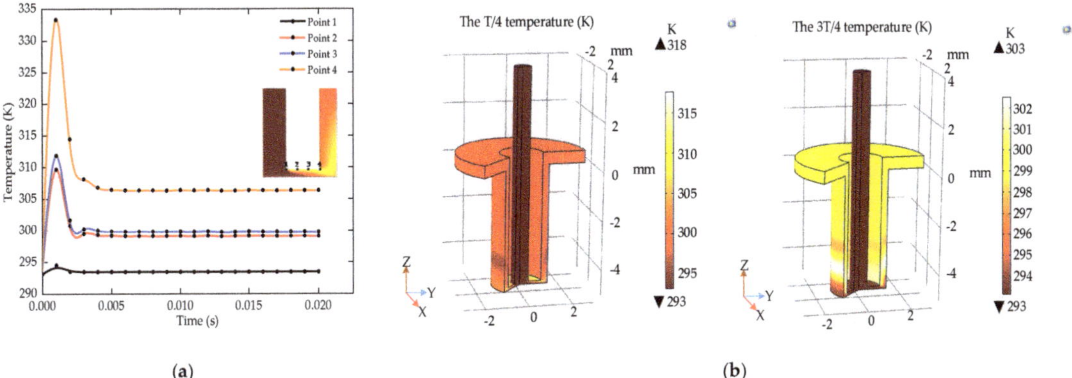

Figure 5. The temperature distribution of the pulse ECM; (**a**) the temperature change rules of reference points; (**b**) the temperature during a pulse period.

4.2. Flow Field

In order to study the influence of the pulse ECM on the flow velocity of electrolyte, four reference points are selected equidistantly on the workpiece surface directly opposite the electrode end surface, where the flow velocity varies greatly. The process parameter is a voltage of 24 V (the period is 1000 μs, the duty ratio is 0.5, the lateral gap is 0.12 mm, the pressure is 0.2 MPa and the depth is 5 mm).

The velocity of reference points increases rapidly in the first period and gradually increases to a stable state after about four periods, as shown in Figure 6a,b shows that the electrolyte flow velocity varies less in a period of the pulse ECM. However, during the pulse off, the discharge between the electrodes is stopped to allow sufficient time for the electrolyte to take away the Joule heat generated by the current and the electrolysis product of electrochemical reaction. Therefore, the pulse ECM can speed up the renewal of the electrolyte and improve the machining accuracy.

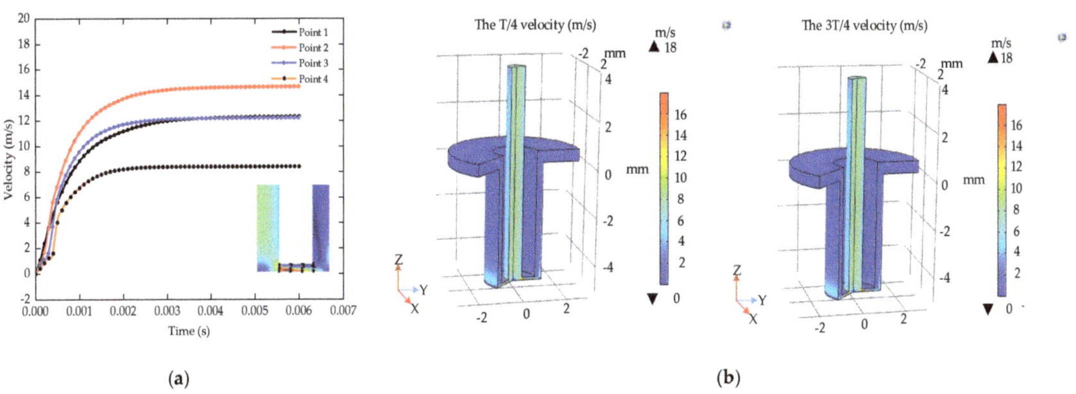

Figure 6. The velocity distribution of the pulse ECM; (**a**) the velocity change rules of reference points; (**b**) the velocity during a pulse period.

4.3. Electric Field

According to the simulation analysis of the temperature and flow field with the pulse ECM, the stable state is about 5 periods later. The process parameter is a voltage of 24 V (the period is 1000 μs, the duty ratio is 0.5, the lateral gap is 0.12 mm, the pressure is 0.2 MPa and the depth is 5 mm).

Figure 7 shows the current density on the workpiece surface within 1/4 and 3/4 period during a stable machining period. In the machining area with a radius of 1 mm, the current density range at the T/4 is 1.7×10^6 A/m^2. The current density range of the 3T/4 is 1500 A/m^2. The range of current density in a period is 0.91×10^6 A/m^2. The current density on the workpiece surface is larger in the center and smaller on both sides. It is easy to form a "bulge" structure at the bottom during the ECM. The pulse ECM reduces the current density range, which is beneficial to the uniform dissolution of the workpiece. Therefore, the pulse ECM can reduce the short circuit caused by the "bulge" structure and improve the stability of the machining.

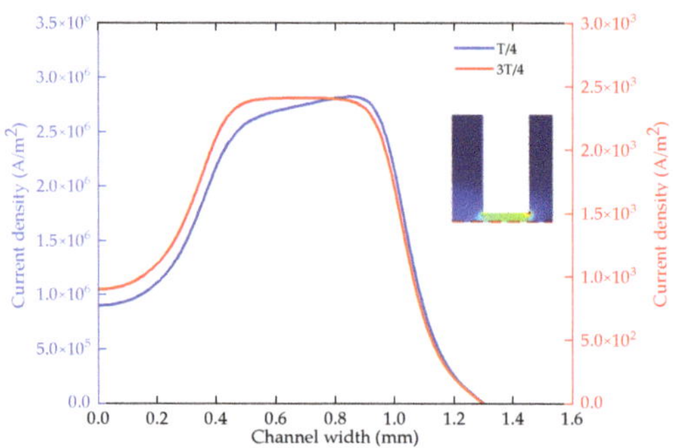

Figure 7. The current density of the pulse ECM.

5. Influence of Process Parameters

5.1. The Effect of Pulse Period

The pulse period is 800–1400 μs (the voltage is 24 V, the duty ratio is 0.5, the lateral gap is 0.12 mm, the pressure is 0.2 MPa, the depth is 5 mm). In the steady state, the temperature distribution of the electrode end surface is basically the same with different pulse periods, as shown in Figure 8a. The temperature has two steps at 0.4 mm and 1 mm. The highest temperature is 306 K. With different pulse periods, the average current density of a period on the workpiece surface is basically the same, as shown in Figure 8b. The maximum current density is about 1.36×10^6 A/m^2.

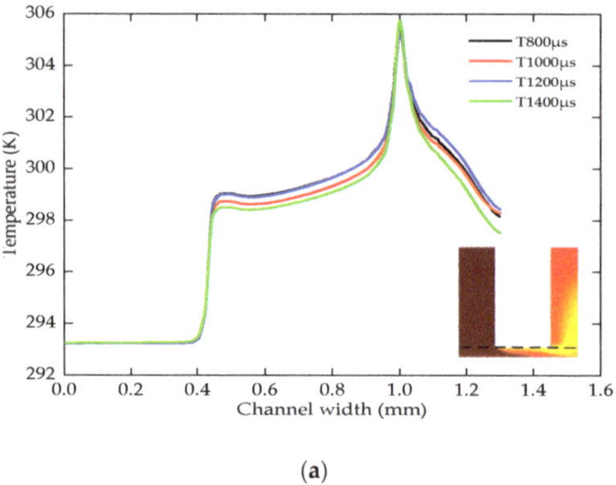

(**a**)

Figure 8. *Cont.*

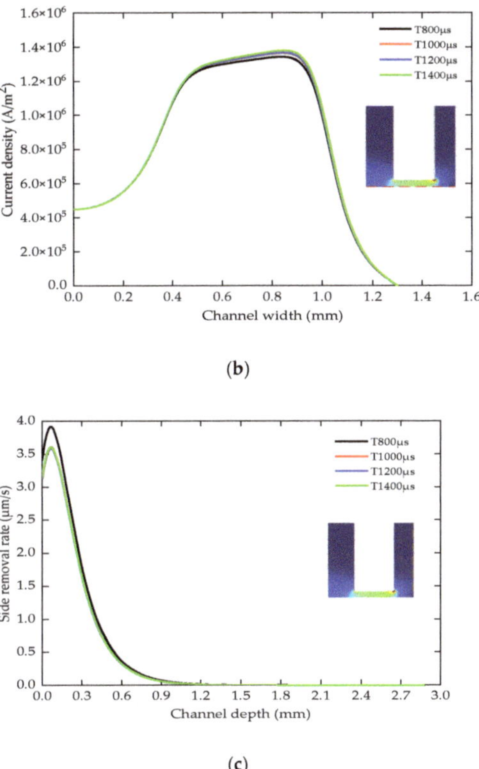

(b)

(c)

Figure 8. The temperature, current density and side removal rate of various pulse periods; (**a**) the change of the temperature; (**b**) the change of the current density; (**c**) the change of the side removal rate.

Figure 8c shows the side removal rate of the work piece. The maximum side removal rate for a period of 800 μs is approximately 3.9 μm/s. The side removal rate in the other cycles are basically the same, and the maximum side removal rate is 3.5 μm/s. At the same duty ratio, the proportion of discharge time within the whole pulse ECM process remains unchanged. While more heat and electrolytic products are generated during the pulse on, there is also enough time to update the Joule heat and product within the pulse off. Therefore, the temperature, current density and side removal rate fluctuate in a small range. The period change has little effect on machining accuracy and stability in the pulse ECM.

5.2. The Effect of Duty Ratio

The duty ratio is 0.2–0.7 (the voltage is 24 V, the period is 1000 μs, the lateral gap is 0.12 mm, the pressure is 0.2 MPa, the depth is 5 mm). In the steady state, the temperature distribution of the electrode end surface is shown in Figure 9a. As the duty ratio increases, the two temperature steps also become larger. The maximum temperature increased from 297 K to 313 K. The rise of the temperature will increase the conductivity of the electrolyte, which in turn affects the distribution of the current density in the entire machining area. Figure 9b shows the current density on the workpiece surface. Due to the uneven distribution of temperature and different duty ratios, the maximum current density increases from 0.4×10^5 A/m^2 to 2.06×10^6 A/m^2 in the machining area with a radius of 1 mm. Excessive current density range will cause uneven dissolution of the surface of the work piece.

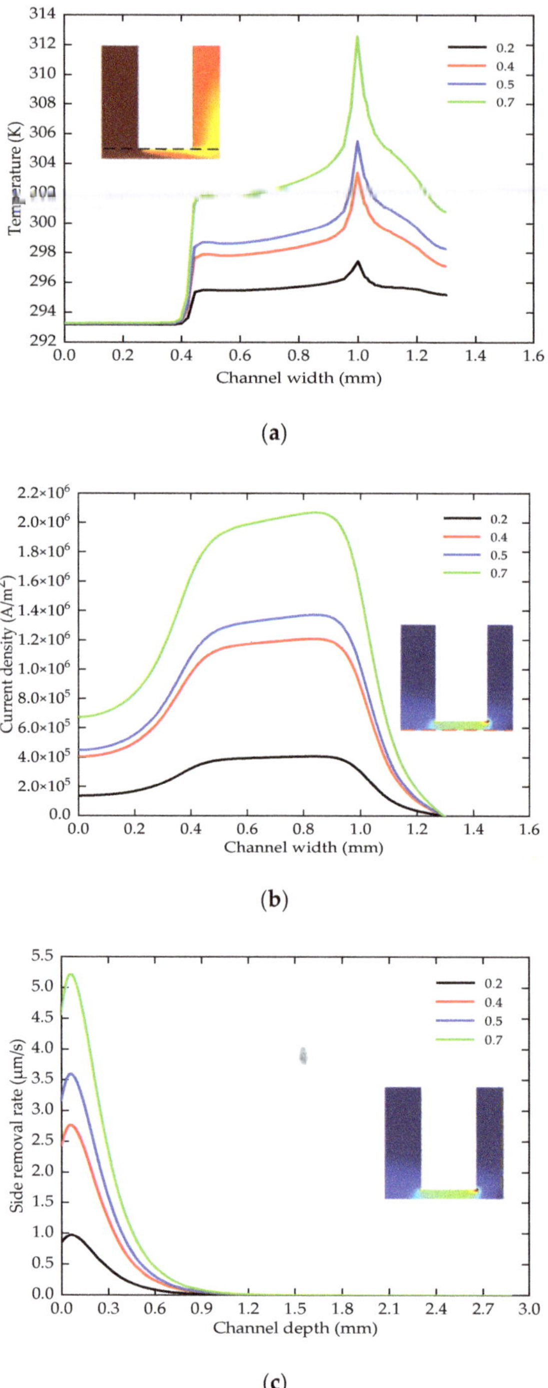

Figure 9. The temperature, current density and side removal rate of various duty ratios; (**a**) the change of the temperature; (**b**) the change of the current density; (**c**) the change of the side removal rate.

It can be seen from Figure 9c that the side removal rate and dissolution area change greatly. When the duty ratio is 0.7, the maximum side removal rate is 5.1 µm/s and the dissolution area is 1.5 mm. However, when the duty ratio is 0.2, the maximum side removal rate is 1 µm/s, and the dissolution area is 0.9 mm. If the duty ratio is too small, the machining efficiency will decrease. The duty ratio is too large, and the increase of temperature and side wall dissolution area will affect the machining accuracy. Therefore, the medium duty ratio is conducive to improving the machining accuracy and stability.

5.3. The Effect of Lateral Gap

The lateral gap is 0.12–0.21 mm (the voltage is 24 V, the period is 1000 µs, the duty ratio is 0.5, the pressure is 0.2 MPa, the depth is 5 mm). The increase of the lateral gap reduces the temperature step at a radius of 1 mm. The maximum temperature of the electrode end surface decreases from 307 K to 299 K in the steady state, as shown in Figure 10a. As the lateral gap increases, Figure 10b shows that the current density range decreases from 0.92×10^6 A/m^2 to 0.43×10^6 A/m^2 in the machining area with a radius of 1 mm.

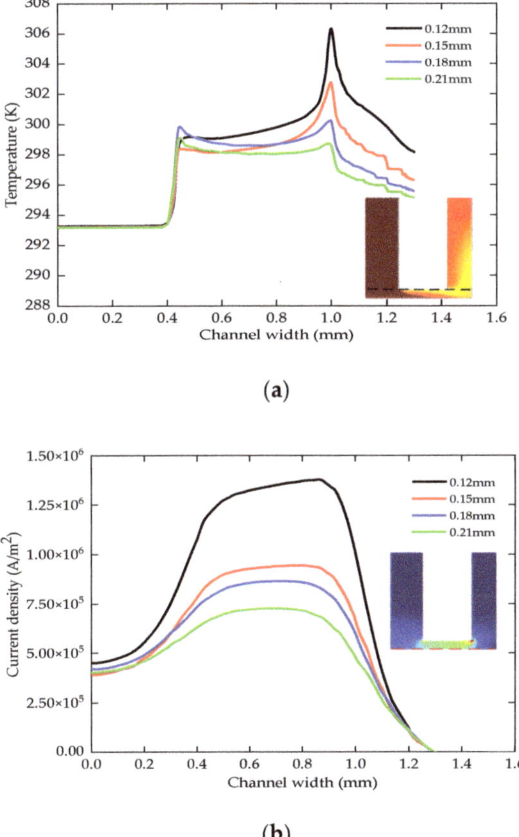

(**a**)

(**b**)

Figure 10. *Cont.*

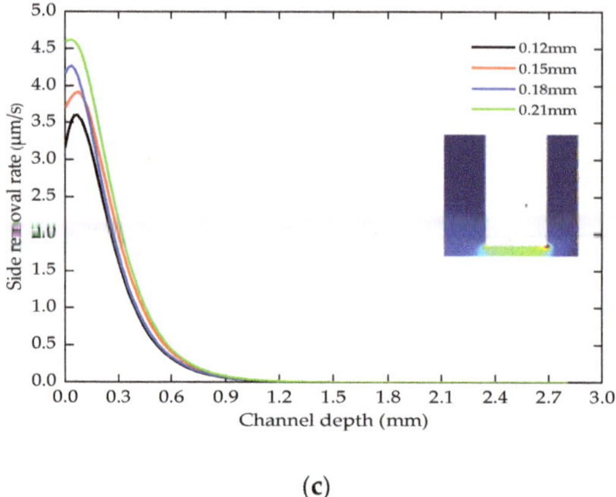

(**c**)

Figure 10. The temperature, current density and side removal rate of various lateral gaps; (**a**) the change of the temperature; (**b**) the change of the current density; (**c**) the change of the side removal rate.

It can be seen from Figure 10c that the side removal rate increases uniformly according to the increase of the lateral gap, and the dissolution area does not change significantly. Therefore, the larger lateral gap accelerates the electrolyte and product updates per unit time. It is conducive to improving the temperature and current density distribution in the machining area and can ensure the stability of machining.

5.4. The Effect of Inlet Pressure

The machining depth is 5–20 mm (the voltage is 24 V, the period is 1000 μs, the duty ratio is 0.5, the pressure is 0.2 MPa, the lateral gap is 0.18 mm). Three cut-off lines were selected along the flow direction of the electrolyte to obtain the distribution of the average flow velocity of the electrolyte at various machining depths, as shown in Figure 11.

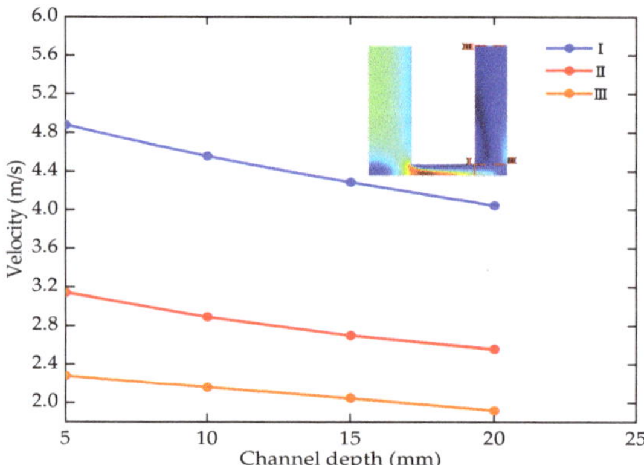

Figure 11. Electrolyte flow velocity with different machining depths.

The average velocity decreases with the increase of electrolyte flow distance and the machining depth. The electrochemical dissolution of materials generates a large amount of heat and electrolysis products in the machining area. The decrease of flow velocity will reduce the heat and product update rate. In serious cases, the direct contact between the electrode and the workpiece results in a short circuit which affects the stability of the machining. Therefore, it is essential to improve the flow field distribution of machining gap.

The inlet pressure is 0.2–0.32 MPa (the voltage is 24 V, the period is 1000 µs, the duty ratio is 0.5, the machining depth is 5–20 mm, the lateral gap is 0.18 mm). The flow velocity changes in the machining area are shown in Figure 12.

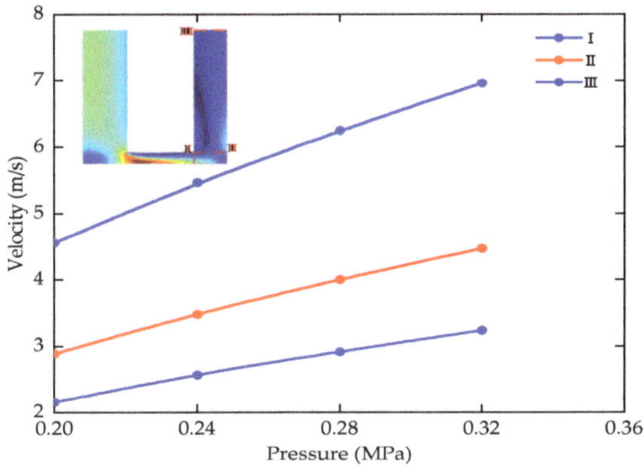

Figure 12. Electrolyte flow velocity with the different inlet pressure.

The average flow velocity of the electrolyte at the cut-off line increases steadily when the inlet pressure is appropriately increased according to different machining depths. If the inlet pressure is too small, the electrolyte flow velocity is too low to take away the electrolytic products in time. The inlet pressure is too large, and the rapid increase of flow velocity is easy to cause vortex. Therefore, appropriately increasing the inlet pressure can improve the flow field distribution and the stability of the ECM of small holes.

6. Experimental Results and Discussion

6.1. Experimental Equipment

The multi-physics coupling simulation results and the optimization of machining parameters are verified by experiments of the ECM holes. The schematic of the machining system is shown in Figure 13, including electrolyte update system, pulse power supply and CNC system. The workpiece is driven by a stepping motor to move along the X–Y plane. The tool electrode connected to the spindle feeds uniformly along the motor-driven linear guide in the Z direction. The electrolyte is pumped to the machining area through the pressure in the flow channel. The negative and positive electrodes of the pulse power supply are, respectively, connected to the tool electrode and the workpiece. All experiments were performed using this systemin this study.

6.2. Experimental Materials

The tool electrode is a titanium tube coated with PTFE, and the thickness of the insulating layer is about 40 µm. The tool electrode has the inside diameter of 1.8 mm and outside diameter of 2.1 mm. The electrolyte is a 13% nitric acid solution. The voltage of the pulse power supply is 24 V. The workpiece material is a high-temperature resistant GH4169 nickel-based alloy, as shown in Table 2.

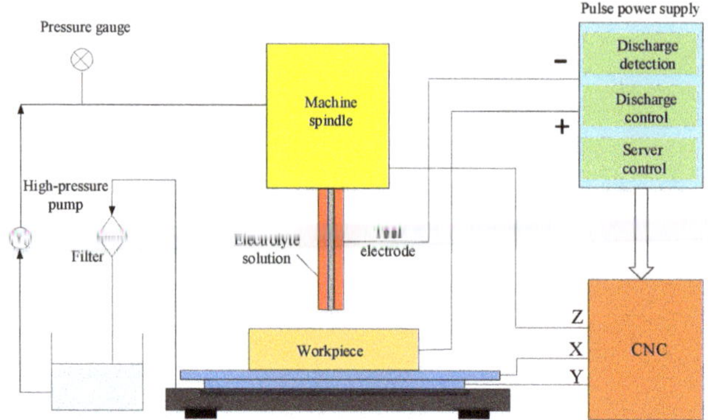

Figure 13. Schematic diagram of ECM system.

Table 2. Composition table of high temperature nickel-based alloy GH4169.

Composition	N_i	C_r	Mo	Cu	Ti	Al	Nb	C
Percentage	55	21	3.3	0.3	1.15	0.7	5.5	0.06

The sampling morphology of the entrance of the ECM small holes is shown in Figure 14a. The axial section of the machined workpiece is shown in Figure 14b, which describes the change of aperture morphology.

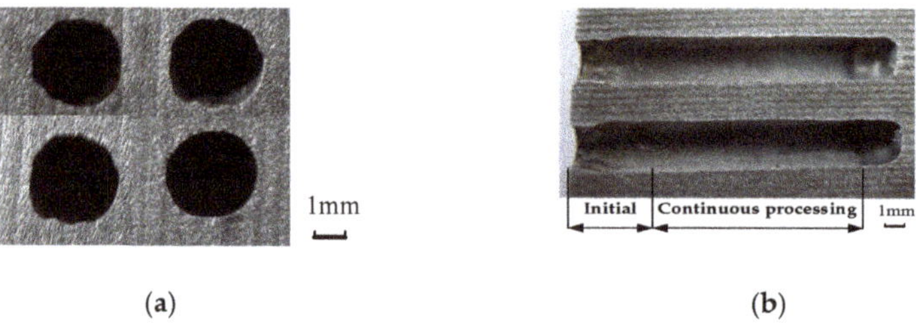

(a)   (b)

Figure 14. The morphology of the machined holes; (**a**) profile of small holes inlet end; (**b**) profile of keyhole section.

By analyzing Figure 14, it can be found that the entrance end is approximately circular. Since the electrolyte is sprayed to the workpiece surface by the tubular electrode in the initial machining stage and the scattering area is wide, the gap electric field is stray. It results in the difference between the section morphology of the inlet end and that of the stable machining section.

In order to study the relationship between the morphology of the inlet end and the machining gap and the electric field, a comparative experiment was carried out, as shown in Table 3. A tool electrode feed rate of 0 m/s is a special case. The initial machining gap is set. The tool electrode remains stationary throughout the ECM process from the beginning to the end of the reaction.

Table 3. Validation experiments.

No	Ton (µs)	Toff (µs)	V (mm/min)	gs (mm)
1	1850	500	0	3.0
2	1850	500	1.02	2.5
3	1850	500	1.14	2.4
4	1850	500	1.26	2.3
5	2000	500	0	3.1
6	2000	500	1.02	2.6
7	2000	500	1.14	2.5
8	2000	500	1.26	2.4

6.3. Analysis of Processing Quality

Experiments compare the dimensional accuracy and machining efficiency of the ECM holes at various duty ratios and inlet pressure. The single side gap and linear removal rate are used as evaluation indicators. For the accuracy of the comparison test, other variables are defined as constants.

The single side gap is the difference between the radius of the machining hole and the radius of the tool electrode.

$$D = \frac{1}{n}\sum_{i=1}^{n}(r_h - r_0) \tag{14}$$

where r_h(mm) is the radius of the machining hole, r_0(mm) is the radius of the tool electrode, n is the number of measurements of the diameter of the hole along the depth direction.

The linear removal rate is the depth of the electrochemically machined hole per unit time.

$$L = \frac{m}{\rho * t * \overline{A}} \tag{15}$$

where m(kg) is the mass of removal, ρ(kg/m^3) is the density of the workpiece, t(s) is the machining time and $\overline{A}$(m^2) is the average value of the cross-section of the small hole.

The specific machining current collected in the test in Table 3 is shown in Figure 15a,b. The change law of acquisition current with time is the same with different process parameters.

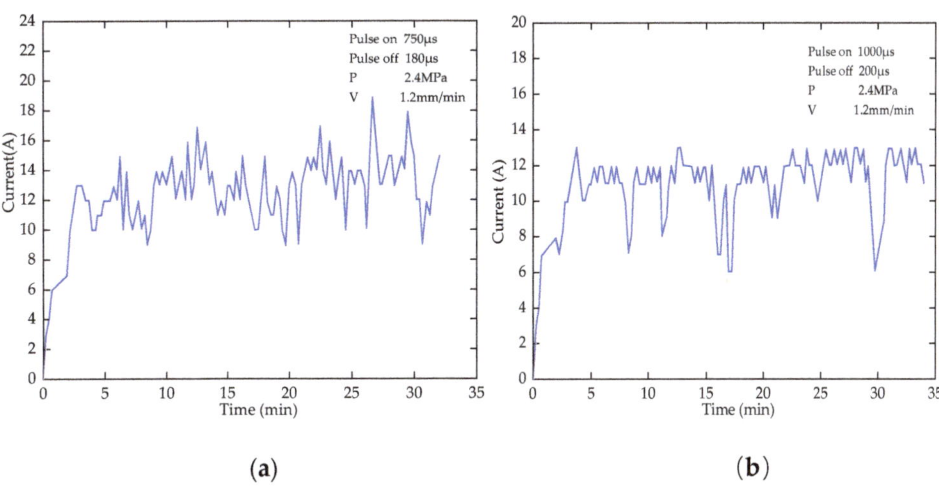

Figure 15. Current acquisition diagram; (**a**) sampling current with pulse on 750 µs and pulse off 180 µs; (**b**) sampling current with pulse on 1000 µs and pulse off 200 µs.

Analysis shows that the inlet current fluctuates. The current tends to be constant as the machining proceeds. According to Sander's equation, the details are shown as follows:

$$\frac{i\tau^{\frac{1}{2}}}{C_o^{*}} = \frac{nFAD_O^{\frac{1}{2}}\pi^{\frac{1}{2}}}{2} \tag{16}$$

where:

I Current (A)
τ—Transition time of potential (S)
F—Faraday constant (C)
A—Electrolysis Area (cm^2)
Do—Diffusion coefficient of electrolytic reaction material (cm^2/s)
C_o^{*}—Surface concentration of the reactive substance at the electrode (mol/L).

When the machining enters a stable state, the processing current I is inversely proportional to τ and directly proportional to C_o^{*}, D_o and A. The parameters are relatively stable and the machining current tends to be constant. According to the analysis of Figure 15, the stationary stage of machining conforms to the description of current in the De Sang equation. In the machining inlet stage, the instability of C_o^{*}, D_o and A will result in the unstable current and the uneven morphology of the inlet section.

6.4. Experimental Results and Discussion

Figure 16 shows the single side gap and linear removal rate of the hole with duty ratios of 0.64–0.82. Process parameters: The inlet pressure is 2.7 MPa, the pulse period is 1000 µs, the depth is 40 mm, the lateral gap is 0.2 mm. As the duty ratio increases, the single side gap increases from 0.08 mm to 0.24 mm, and the linear removal rate increases from 1.2 mm/min to 1.63 mm/min. In the simulation analysis, the current density of the workpiece surface and the side removal rate also increase with the increase of the duty ratio. Therefore, the simulation results and experiments are mutually verified, and choosing a medium duty ratio is beneficial to improve the machining accuracy.

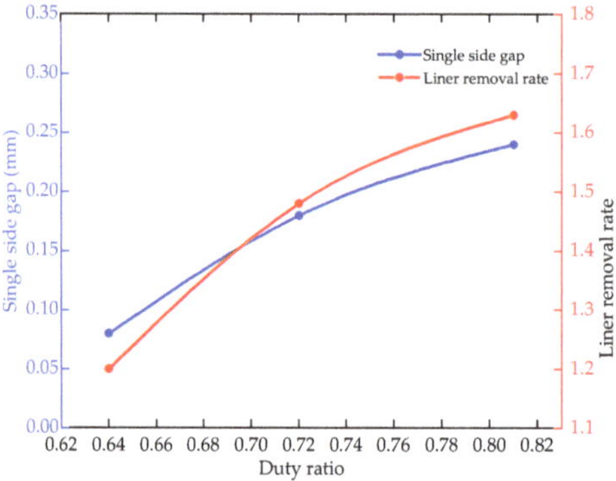

Figure 16. The single side gap and linear removal rate at various duty ratios.

Figure 17 shows the single side gap and linear removal rate of the hole with the inlet pressure of 1.8–2.7 MPa. Process parameters: the pulse period is 1000 µs, the duty ratio is 0.64, the depth is 40 mm and the lateral gap is 0.2 mm. The initial inlet pressure is 1.8 Mpa. For every 10 mm increase in machining depth, the inlet pressure increases by 0.3 MPa. The single side gap is reduced from 0.19 mm to 0.08 mm. The linear removal rate is reduced from 1.51 mm/min to 1.2 mm/min. In the simulation analysis, the electrolyte flow velocity increases with the increase of pressure. The faster flow velocity accelerates the removal of Joule heat and products, which reduces the uneven dissolution of the sidewall caused by the high temperature. Therefore, increasing the inlet pressure is beneficial to improve the forming accuracy of holes for larger machining depths.

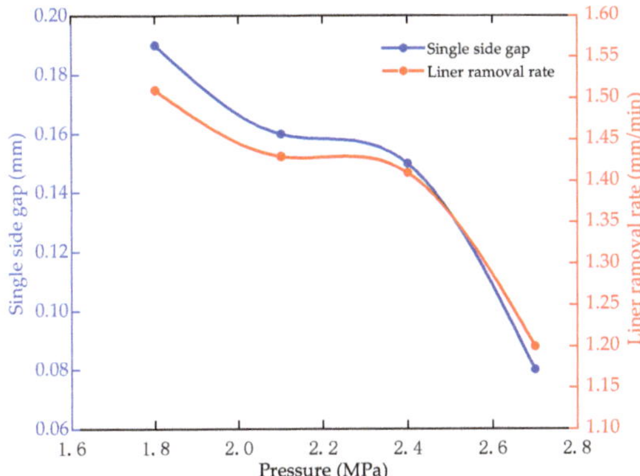

Figure 17. Single side gap and linear removal rates with different duty ratios.

7. Conclusions

In this article, a study of multi-physics coupling simulation of film cooling holes is established. The stability and machining accuracy of the direct current and pulse ECM are compared. Through the research and experimental verification of the electrical parameters (period, duty ratio) and process parameters (lateral gap, pressure) of the pulse ECM, the following conclusions are drawn:

(1) The current density is more evenly distributed and the temperature stage in the machining area is reduced with the pulse ECM.

(2) The temperature, current density and side removal rate gradually increase with the increase of the duty cycle but are not affected by the period. The use of medium duty ratio helps to improve the machining accuracy of holes.

(3) The larger lateral gap can reduce the range of current density and accelerate the renewal of Joule heat and electrolytic products.

(4) Increasing the inlet pressure is beneficial to improve the flow field distribution of machining gap and reduce the single side gap, which improve the accuracy and stability of ECM holes.

(5) The reasons for the differences in the morphology of the entrance section of the machined small holes were analyzed. The electrolyte injection range and initial machining gap of the initial stage affect the electric field distribution in the machining area.

Author Contributions: Conceptualization, Z.L., Y.D. and B.C.; methodology, Z.L., Y.D. and B.C.; validation, Z.L., Y.D. and B.C.; formal analysis, Z.L., Y.D. and B.C.; investigation, Z.L., Y.D. and B.C.; resources, Z.L., Y.D. and B.C.; data curation, Z.L. and B.C.; writing—original draft preparation, Z.L. and B.C.; writing—review and editing, Z.L. and B.C.; visualization, Z.L. and B.C.; supervision, Z.L. and B.C.; project administration, Z.L. and Y.D.; funding acquisition, Z.L. and Y.D. All authors have read and agreed to the published version of the manuscript.

Funding: This work was supported by the National Natural Science Foundation of China, grant number 52075134.

Data Availability Statement: Data is contained within the article.

Acknowledgments: The authors would like to thank Key Laboratory of Advanced Manufacturing Intelligent Technology of Ministry of Education for helpful discussions on topics related to this work. The authors are grateful to B.C. for his help with the preparation of Figures in this paper.

Conflicts of Interest: The authors declare no conflict of interest.

Abbreviations

ECM   Electrochemical Machining
CNC   Computer Numerical Control
PTFE   Polytetrafluoroethylene

References

1. Zhang, C.; Xu, Z.; Hang, Y.; Xing, J. Effect of solution conductivity on tool electrode wear in electrochemical discharge drilling of nickel-based alloy. *Int. J. Adv. Manuf. Technol.* **2019**, *103*, 743–756. [CrossRef]
2. Wang, G.Q.; Zhang, Y.; Li, H.S.; Tang, J. Ultrasound-assisted through-mask electrochemical machining of hole arrays in ODS superalloy. *Materials* **2020**, *13*, 5780. [CrossRef] [PubMed]
3. Niu, S.; Qu, N.; Yue, X.; Li, H. Effect of tool-sidewall outlet hole design on machining performance in electrochemical mill-grinding of Inconel 718. *J. Manuf. Process.* **2019**, *41*, 10–22. [CrossRef]
4. Cheng, C.-P.; Wu, K.-L.; Mai, C.-C.; Yang, C.-K.; Hsu, Y.-S.; Yan, B.-H. Study of gas film quality in electrochemical discharge machining. *Int. J. Mach. Tools Manuf.* **2010**, *50*, 689–697. [CrossRef]
5. Zhang, Y.; Xu, Z.Y.; Zhu, Y.; Zhu, D. Machining of a film-cooling hole in a single-crystal super alloy by high-speed electro-chemical discharge drilling. *Chin. J. Aeronaut.* **2016**, *29*, 560–570. [CrossRef]
6. Chung, D.K.; Lee, K.H.; Jeong, J.; Chu, C.N. Machining characteristics on electrochemical finish combined with micro EDM using deionized water. *Int. J. Precis. Eng. Manuf.* **2014**, *15*, 1785–1791. [CrossRef]
7. Tao, X.; Liu, Z.; Qiu, M.; Tian, Z.; Shen, L. Research on an EDM-based unitized drilling process of TC4 alloy. *Int. J. Adv. Manuf. Technol.* **2018**, *97*, 867–875. [CrossRef]
8. Mishra, S.; Yadava, V. Modelling of Hole Taper and Heat Affected Zone Due to Laser Beam Percussion Drilling. *Mach. Sci. Technol.* **2013**, *17*, 270–291. [CrossRef]
9. Sun, A.X.; Chang, Y.B.; Liu, H.J. Numerical simulation of laser drilling and electrochemical machining of metal micro-hole. *Optik* **2019**, *181*, 92–98. [CrossRef]
10. Coteata, M.; Schulze, H.-P.; Slătineanu, L. Drilling of Difficult-to-Cut Steel by Electrochemical Discharge Machining. *Mater. Manuf. Process.* **2011**, *26*, 1466–1472. [CrossRef]
11. Das, A.K.; Saha, P. Machining of circular micro holes by electrochemical micro-machining process. *Adv. Manuf.* **2013**, *1*, 314–319. [CrossRef]
12. Yao, J.; Chen, Z.T.; Nie, Y.J.; Li, Q. Investigation on the electrochemical machining by using metal reinforced double insulating layer cathode. *Int. J. Adv. Manuf. Technol.* **2016**, *89*, 2031–2040. [CrossRef]
13. Zou, H.; Yue, X.; Luo, H.; Liu, B.; Zhang, S. Electrochemical micromachining of micro hole using micro drill with non-conductive mask on the machined surface. *J. Manuf. Process.* **2020**, *59*, 366–377. [CrossRef]
14. Liu, G.D.; Tong, H.; Li, Y.; Zhong, H.; Tan, Q.F. Multiphysics research on electrochemical machining of micro holes with internal features. *Int. J. Adv. Manuf. Technol.* **2020**, *110*, 1527–1542. [CrossRef]
15. Tang, L.; Feng, X.; Zhao, G.G.; Li, Q.L.; Zhao, J.S.; Ren, L. Cathode cross tank and return hole optimization design and experiment verification of electrochemical machining closed integral impeller outside flow channels. *Int. J. Adv. Manuf. Technol.* **2018**, *97*, 2921–2931. [CrossRef]
16. Zhao, J.S.; Wang, F.; Liu, Z.; Zhang, X.L.; Gan, W.M.; Tian, Z.J. Flow field design and process stability in electrochemical machining of diamond holes. *Chin. J. Aeronaut.* **2016**, *29*, 1830–1839.
17. Chai, M.X.; Li, Z.Y.; Yan, H.J.; Huang, Z.X. Flow field characteristics analysis of interelectrode gap in electrochemical machining of film cooling holes. *Int. J. Adv. Manuf. Technol.* **2020**, *112*, 525–536. [CrossRef]
18. Li, H.S.; Zhang, C.; Wang, G.Q.; Qu, N.S. Study of the hole-formation process with different mask diameters via through-mask ECM. *Int. J. Electrochem. Sci.* **2018**, *13*, 3006–3022. [CrossRef]
19. Wang, F.; Yao, J.; Kang, M. Electrochemical machining of a rhombus hole with synchronization of pulse current and low-frequency oscillations. *J. Manuf. Process.* **2020**, *57*, 91–104. [CrossRef]
20. Sathish, T. Experimental investigation of machined hole and optimization of machining parameters using electrochemical machining. *J. Mater. Res. Technol.* **2019**, *8*, 4354–4363. [CrossRef]
21. Singh, M.; Singh, S.; Kumar, S. Experimental Investigation for Generation of Micro-Holes on Silicon Wafer Using Electro-chemical Discharge Machining Process. *Silicon* **2019**, *12*, 1683–1689. [CrossRef]
22. Ma, N.; Yang, X.L.; Gao, M.Q.; Song, J.L.; Liu, G.L.; Xu, W.J. A study of electro discharge machining-pulse electrochemical machining combined machining for holes with high surface quality on super alloy. *Adv. Mech. Eng.* **2015**, *7*, 1–11. [CrossRef]

micromachines

Article

Interaction Mechanism of Thermal and Mechanical Field in KDP Fly-Cutting Process

Chenhui An [1,*], Ke Feng [2], Wei Wang [2], Qiao Xu [1], Xiangyang Lei [1], Jianfeng Zhang [1], Xuelian Yao [1] and Haibo Li [1]

[1] Laser Fusion Research Center, CAEP, Mianyang 621900, China; xuqiao@vip.sina.com (Q.X.); leixiangyang2@163.com (X.L.); xxxxxx5726@vip.sina.com (J.Z.); xlyao93@163.com (X.Y.); haiya126@sina.com (H.L.)
[2] School of Mechanical and Electrical Engineering, University of Electronic Science and Technology, Chengdu 611731, China; duying0108@163.com (K.F.); wangwhit@163.com (W.W.)
* Correspondence: hplaser@126.com

Abstract: As an important nonlinear optical material, potassium dihydrogen phosphate (KDP) crystal is used in high-power laser beams as the core element of inertial confinement fusion. It is the most general method of single point diamond fly-cutting (SPDF) to produce high precision and crack-free KDP surfaces. Nevertheless, the cutting mechanism of such material remains unclear, and therefore needs further analysis. Firstly, the stress field, cutting force and cutting temperature under different working conditions are calculated by a KDP crystal cutting simulation model. Then, the rules and the cause of change and interaction mechanisms of force and temperature are analyzed by comparing the measurement experiments with simulations. Furthermore, the causes of chip formation and micro-cracks on the machined surface are analyzed based on thermo-mechanical coupling and chip morphology. The conclusion can be deduced: Although the temperature has not reached the phase transition temperature during the finishing process, under high cutting speeds and large unformed chip thickness, such as semi-finishing and roughing, the temperature can reach up to 180 °C or higher, and KDP crystals are very likely to phase transition—chip morphology also verifies this phenomenon.

Keywords: potassium dihydrogen phosphate (KDP) crystal; single point diamond fly-cutting; interaction mechanism; chip morphology; phase transition temperature

Citation: An, C.; Feng, K.; Wang, W.; Xu, Q.; Lei, X.; Zhang, J.; Yao, X.; Li, H. Interaction Mechanism of Thermal and Mechanical Field in KDP Fly-Cutting Process. *Micromachines* **2021**, *12*, 855. https://doi.org/10.3390/mi12080855

Academic Editors: Benny C. F. Cheung and Jiang Guo

Received: 22 June 2021
Accepted: 16 July 2021
Published: 21 July 2021

Publisher's Note: MDPI stays neutral with regard to jurisdictional claims in published maps and institutional affiliations.

1. Introduction

Potassium dihydrogen phosphate (KH_2PO_4, KDP) is a unique artificial crystal that can grow to a diameter greater than 500 mm, and it has excellent nonlinear optical properties [1]; therefore, it is the only nonlinear optical element material that can be used in high-energy laser beams [2]. However, due to its disadvantageous properties such as deliquescence, fragmenting and strong anisotropy, it is less possible to process the KDP crystal by grinding, polishing and other traditional methods [3,4]. Hence, single point diamond fly-cutting is the most feasible process method for KDP crystal, and because of the high requirement of the crystal elements such as nanoscale roughness and micron level surface accuracy, the ultra-precision fly-cutting machine tools are adapted and the fine finishing process is analyzed [5]. A single point diamond fly-cutting machine with a large diameter cutting head, which can effectively avoid the mentioned disadvantages and reduce the effect of the material anisotropy influence, is the most common machining method for large caliber KDP crystal components [6].

In the KDP crystal fly-cutting process, the cutting speed could reach up to 10 m/s, which is higher than that in ordinary turning processes. Under such a high cutting speed, on one hand, the crystal material near the cutting zone will incur a large plastic deformation, and on the other hand, it causes severe friction between cutting tools and workpieces. The plastic deformation and severe friction would produce heat, which would increase the material's temperature near the cutting zone [7,8], resulting in some surface defects such as phase transition, recrystallization and micro-cracks, especially for the KDP crystal of which the phase transition temperature is only 182.3 °C [9].

Lo [10] established a finite element model to represent the cutting process by using diamond tools with different rake angles, taking the influence of the edge radius into account. Zong [11,12] established a finite element model to simulate the entire diamond turning process, and found an optimal selection of tool geometries for the KDP cutting process, i.e., −25° rake angle and 8° clearance angle [13]. A new constitutive model was used by Wang [14] to simulate the cutting process of KDP crystal, which combines the anisotropic elastic and pressure-dependent plastic model and is performed to investigate the influence of cutting parameters on the brittle ductile transition depth and cutting force. In [15], the authors took research on the thermal field of DKDP crystal cutting and explored the influence of process parameters on the cutting temperature. However, there are few studies on the thermo-mechanical coupling phenomenon in the KDP crystal cutting process.

At the same time, there are many scholars involved in the research on material properties. Zhang [16] tested the mechanical parameters of KDP crystals and found that brittle failure is easier to appear in [100] crystal orientation than [001] crystal orientation, and the compressive strength is much higher than tensile strength. In 2015, Ding [9] studied the high temperature thermal behavior and thermal dehydration reaction of KDP crystal, and the main conclusion is that the KDP crystal is still tetragonal and has no phase transition of the monoclinic phase at 183 °C, and then the dehydration reaction begins to decompose at 207 °C. Hou [17] used XRD (X-ray diffraction) to analyze the residual stress of the subsurface of KDP crystal. It is found that the residual stress in the vertical [112] plane is higher than other planes at different cutting depths, and its value is about 30 MPa, which is about 7 times that of the parallel [112] plane residual stress. In 2018, Huang [18] studied the KDP crystal cracking caused by temperature inhomogeneity, and found that when the crystal has an internal temperature difference of 4 °C in a small spatial scale of about tens of millimeters, the maximum residual heat stress would be very close to or even exceed the tensile strength of 6.67 MPa [19], which made the crystals easier to crack.

There are many kinds of methods that can be used to measure the online temperature in the machining process, and the most common method is using embedded thermocouple probes [16]. Thus, this kind of probe must be embedded in the measured position, and they need a high coefficient of heat conduction of the measured objects, both of which cannot be fulfilled in the KDP cutting process [20,21]. In the KDP crystal fly-cutting process, the linear velocity of the diamond tool is above 10 m/s, and measuring the temperature distribution in the KDP cutting process is extremely difficult due to the small shear area and low heat production [21]. Therefore, a kind of high-speed infrared camera is necessary in this situation which can meet both requirements of remote measurement and high temporal resolution. The ImageIR® 5300 has a sampling frequency as high as 12 kHz, and its maximum resolution is 0.02 °C, which is the ideal equipment for this experiment.

In short, although lots of research works are published about the cutting force and the temperature change in the KDP crystal fly-cutting process, few studies have combined the temperature and stress analysis together, although both of which would have effects on the material removal mechanism. This paper is based on the anisotropic finite element model of KDP crystal to study the interaction of the thermal and mechanical field in KDP crystal cutting and its influence on chip morphology. It is the main method by combining simulations on the stress field, temperature field and cutting force with experimental verification results. Studying the interaction of the thermal and mechanical field is helpful to understand the material removal mechanism, verify the material changes in the cutting process, and reveal the phase transformation and surface defects that may be caused by the heating produced in the fly-cutting process.

2. Simulation

2.1. Simulation Modeling

As is shown in Figure 1a, in the KDP fly-cutting process, the workpiece is placed horizontally and fed at a constant velocity, while the spindle with the diamond cutting tool rotates at a given speed. Along the main cutting direction, the section of the removed workpiece material produced by the arc edge tool is circular, which can be regarded as the superposition of many sections with different cutting depths, so the three-dimensional problem can be transformed into a series of two-dimensional plane problems with certain thicknesses. In Figure 1b, the area enclosed by the three points of $O'A'B'$ is an undeformed chip.

$$t_c = R - \sqrt{R^2 + f^2 - 2f\sqrt{2Ra_p - a_p^2}} \tag{1}$$

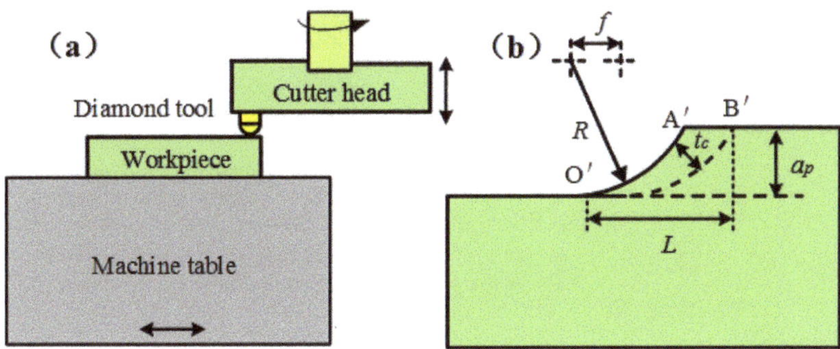

Figure 1. (**a**) Fly-cutting schematic diagram; (**b**) Undeformed chip thickness.

The max value of the unformed chip thickness t_c can be calculated according to Equation (1), where R is the nose radius and the value is 5 mm, f is feed and the value is 10 μm/rev according to the actual situation, and a_p is the cutting depth.

Figure 2 shows the finite element simulation model of KDP crystal. The KDP crystal workpiece is rectangular with a size of 100 μm × 30 μm. The 3600 four-node plane strain units (CPE4) are used to complete the mesh. The bottom of the KDP crystal is completely fixed in a horizontal manner. Compared with the soft KDP crystal, the diamond tool is far harder and is often regarded as a rigid body and moves to the left of the workpiece at a certain speed. The parameters of the diamond tool are as follows: the edge radius is 120 nm, the tool clearance angle is 18°, and the tool rake angle is −45°. The 300 four-node plane strain units are used to mesh the diamond tool.

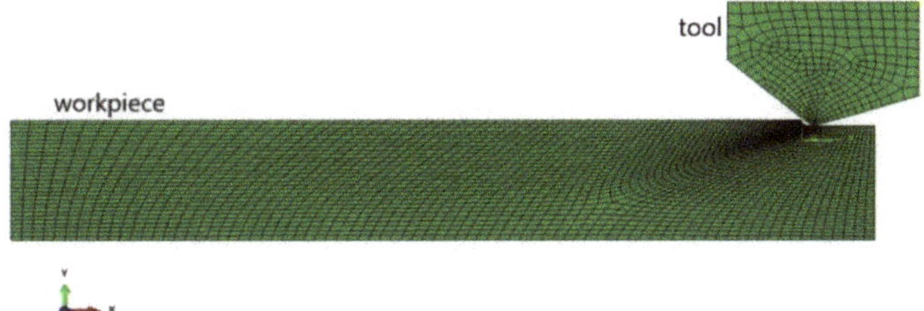

Figure 2. KDP crystal cutting process finite element simulation model.

2.2. Material Properties

2.2.1. Diamond

The KDP crystal fly-cutting process mainly involves two different materials. The processed material is KDP crystal, which is characterized by low hardness and high brittleness, and the tool material is natural diamond, which is the hardest material in the world. The hardness of diamond is more than 10 times that of KDP crystal. The specific material properties of natural diamond are shown in Table 1.

Table 1. Natural diamond material properties.

Physical Quantity	Value
Elastic modulus (Gpa))	1114
Poisson ratio	0.07
Linear expansion coefficient ($\mu m/(m \cdot {}^\circ C)$)	1.18
Specific heat capacity ($J/(Kg\,{}^\circ C)$)	507.9
Thermal conductivity ($W/(m\,{}^\circ C)$)	2000
Density (kg/m^3)	3520

2.2.2. KDP Crystal

Since the KDP crystal is an anisotropic material, the structure of the KDP crystal is a tetragonal system, and the matrix of the elastic phase is composed of six mutually independent parameters to characterize the stress–strain relationship of the KDP crystal. The stress–strain relationship inside the KDP crystal material can be expressed by the stiffness matrix of the material, as shown in Equation (2):

$$
\begin{bmatrix} \sigma_x \\ \sigma_y \\ \sigma_z \\ \tau_{xy} \\ 0 \\ 0 \end{bmatrix} =
\begin{bmatrix} C_{11} & C_{12} & C_{12} & & & \\ C_{12} & C_{11} & C_{13} & & & \\ C_{13} & C_{13} & C_{33} & & & \\ & & & C_{44} & & \\ & & & & C_{66} & \\ & & & & & C_{66} \end{bmatrix}
\begin{bmatrix} \varepsilon_x \\ \varepsilon_y \\ 0 \\ \gamma_{xy} \\ 0 \\ 0 \end{bmatrix}
\tag{2}
$$

There, ε_x, ε_y, and γ_{xy} are the strain components, and $C_{11}\sim C_{66}$ are the elastic constants of KDP crystal.

Table 2 shows the KDP crystal stiffness coefficient values, which can express the properties inside the KDP crystal.

Table 2. Stiffness constant value [22].

C_{11}	C_{12}	C_{13}	C_{33}	C_{44}	C_{66}
71.6 GPa	−6.3 GPa	14.9 GPa	56.4 GPa	12.5 GPa	6.2 GPa

When the KDP material is in the plastic stage, the final model of the material constitutive equation can be derived from Equation (3):

$$
\begin{cases}
\sigma = 471.227\varepsilon^{0.1721}\,(\sigma \geq 181.725\,MPa) \\
\sigma_{ture} = \sigma(1+\varepsilon) \\
\varepsilon_p = \ln(1+\varepsilon) - \sigma_{ture}/E
\end{cases}
\tag{3}
$$

where σ_{ture} is the true rheological stress and ε_p is plastic deformation [10].

Temperature is accompanied by the cutting process, so temperature analysis is an integral part of the finite element simulation of KDP crystal fly-cutting. The heat transfer in the shear zone would change the plastic deformation; therefore, the surface micro structure may be changed. The diamond tool is a rigid body relative to the KDP crystal and only heat transfer is calculated, and the KDP crystal workpiece is a deformable body, and the thermal and mechanical effects are analyzed in the simulation. In the heat transfer analysis of the KDP crystal, the heat conduction and convection calculations are performed. Relative thermal parameters of diamond tools and Type II KDP crystals, for example, the specific heat capacity C, thermal conductivity α, thermal expansion coefficient λ and thermal emissivity k, are listed in Table 3.

Table 3. Thermal parameters of diamond tools and type II KDP crystals.

Parameters	Diamond	KDP Crystal		
$\alpha[\mu m/(m°C)]$	2000	α_x	α_y	α_z
		16.1	16.1	29.0
$\lambda(10^{-6}/°C)$	1.18	λ_x	λ_y	λ_z
		2.0	2.0	3.0
$C[J/(kg°C)]$	502	$-120 + 4.56T - 7.38(10^{-3})T^2 + 6.59(10^{-6})T^3 - 3.05(10^{-9})T^4 + 5.72(10^{-13})T^5$ $(273\,K < T < 1500K)$		
k	0.03	0.36		

Figure 3 shows the temperature in different directions—the cutting depth is 2 μm (the max value of the unformed chip thickness is 297 nm) and the line velocity of the diamond tool is 10 m/s. (a) The cutting direction is [100] crystal orientation, and the temperature is 90 °C. (b) The cutting direction is [001] crystal orientation, and the temperature is 96 °C, indicating that the anisotropy has a certain influence on the cutting temperature of the KDP crystal, and it is better to cut along [100] crystal orientation as a lower hit is being produced in this direction.

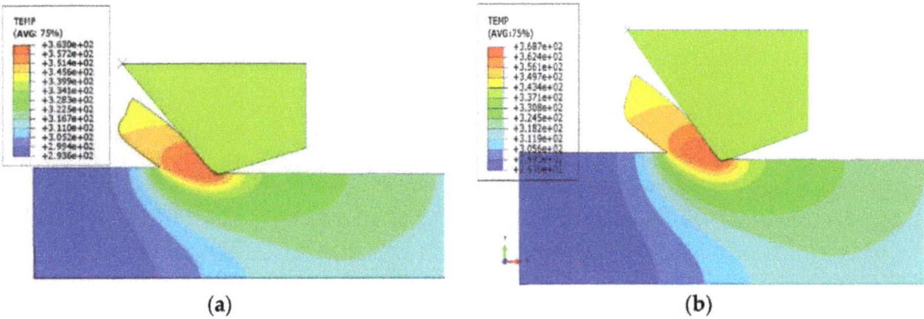

|(a)|(b)|

Figure 3. Temperature distribution along different crystal orientations: (**a**) [100] crystal orientation, (**b**) [001] crystal orientation.

3. Experiment

3.1. Cutting Experiment

As shown in Figure 4, in order to measure the force and temperature in the process of KDP crystal, a vertical ultra-precision turning machine tool was used—the spindle rotation error is 0.1 μm, the feed repetition accuracy is 0.1 μm, and the spindle speed can reach 8000 r/min.

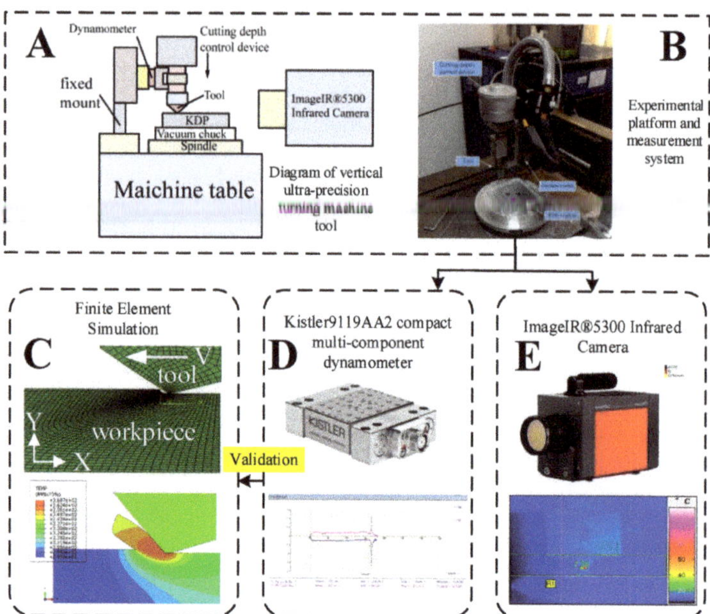

Figure 4. Experimental platform and measurement system: (**A**) Diagram of vertical ultra-precision turning machine tool; (**B**) Experimental platform and measurement system; (**C**) Finite Element Simulation; (**D**) Kistler9119AA2 compact multi-component dynamometer; (**E**) ImageIR®5300 Infrared Camera.

The tool selected for the ultra-precision cutting machine tool was a diamond arc-edge tool. The parameters of the tool are consistent with those in the simulation model, as shown in Table 4.

Table 4. Diamond tool parameters.

Rake Angle	Flank Angle	Nose Radius	Edge Radius
−45°	18°	5 mm	120 nm

The unformed chip thickness and cutting speed both need to be analyzed in order to obtain the effect of single process parameters on KDP crystal fly-cutting, and orthogonal experiments of cutting speed and unformed chip thickness are used to complete the influence of single parameters on the cutting force and cutting temperature. The experimental parameters were designed as shown in Table 5.

Table 5. Experimental design parameters.

Parameters	Value
cutting speed	3 m/s, 5 m/s, 8 m/s, 10 m/s, 15 m/s
feed	10 μm/rev
the max value of the unformed chip thickness (cutting depth)	297 nm (2 μm), 429 nm (5 μm), 622 nm(10 μm)

In the KDP crystal fly-cutting process, the process parameters are small, such as the μm range cutting depth, and the cutting forces are always less than 0.1 N; therefore, a high accuracy measurement device for the cutting force is required. Figure 4D shows the Kistler9119AA2 compact multi-component dynamometer used in this experiment. The dynamometer is compact in design, has high sensitivity and natural frequency, and has small temperature error. It is very suitable for measuring the change in cutting force in ultra-precision machining processes.

In the KDP crystal fly-cutting process, the diamond tool rotates with the spindle at a high speed, the line velocity of the cutter can exceed 10 m/s, and the contact period between the tool and KDP crystal is very short, so it is necessary to use a temperature measurement system with a high frequency response. The IR®5300 infrared camera used in this experiment is shown in Figure 4E. Its sampling frequency is as high as 12 kHz, its maximum resolution is 0.02 °C, and it can measure the weak temperature changes in high-speed cutting areas.

3.2. KDP Chip Morphology Observation Experiment

Studying the chip morphology formed by high-speed cutting is helpful to explain the chip formation mechanism. The chip forming process is the plastic deformation of the workpiece material sliding along the shear surface after being pushed by the tool rake face. Chip shape can explain the deformation mechanism and process in the cutting process, and indirectly explain the quality of the workpiece surface. Based on the observation of chip morphology, the influence of process parameters and temperature on chip formation can be analyzed by combining the simulation and experimental results.

In this experiment, the NeoScope Jcm-5000 desktop scanning electron microscope was used to observe the KDP chip morphology. The surface morphology of the chips after finishing, semi-finishing and roughing was observed, respectively, and the process parameters are consistent with those in the simulation and measurement experiments.

4. Results

4.1. Results Comparison of Cutting Force

Figure 5 shows the cutting force measurement chart at the cutting speed of 3 m/s and the unformed chip thickness of 429 nm. F_x is the horizontal cutting force and F_z is the vertical cutting force. The data between the two dotted lines in the figure are the cutting force measurement results from the cutting tool to the workpiece in the cutting process. From the figure, we can calculate the average values of the horizontal and vertical cutting forces, which are 0.237 N and 0.205 N, respectively.

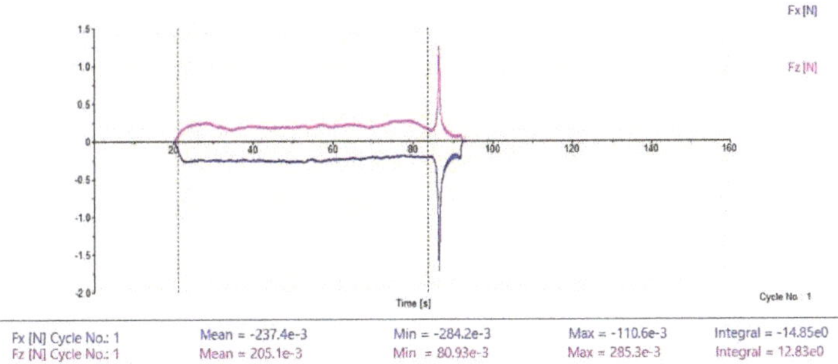

Figure 5. Cutting force measurement chart.

Figure 6a shows the comparison of cutting forces at different cutting speeds under the max unformed chip thickness of 429 nm. In the figure, it can be found that: (1) cutting speed ranged from 3 m/s to 10 m/s, both F_z and F_x decreased by 20–30% with the increase in cutting speed, meanwhile the simulation results and experimental results have the same trend and the numerical values are close, which shows that the simulation is in line with the actual situation. (2) When the cutting speed increased from 10 m/s to 15 m/s, the variation trends of the measured and simulated cutting force were different. The simulated cutting force F-S_{im} decreased with the increase in the cutting speed, but the experimental cutting force F-E_{xp} increased. The main reason is that the chip generation rate is higher than the chip discharge rate, and the cutting force is affected by the chip accumulation in the actual process; in addition, with the increase in machine speed, the stability of the machine tool will become worse, which affects the measurement of cutting force.

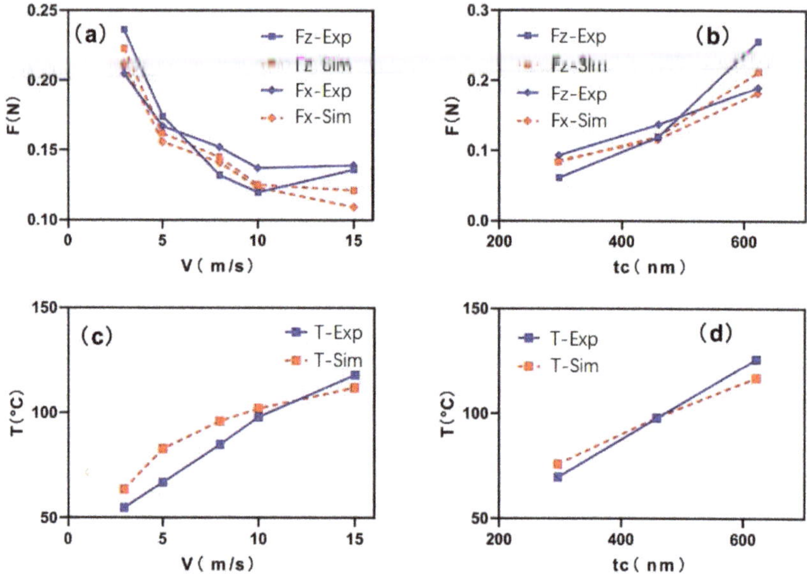

Figure 6. Variation of cutting force and temperature under different cutting conditions. (**a**) Comparison of cutting forces at different cutting speeds under the same unformed chip thickness of 429 nm. (**b**) The comparison of cutting forces at different unformed chip thicknesses under the cutting speed of 10 m/s. (**c**) Comparison of cutting temperatures at different cutting speeds and unformed chip thickness of 429 nm. (**d**) Comparison of cutting temperature at different cutting and depth cutting speed of 10 m/s.

Figure 6b shows the comparison of cutting forces at different unformed chip thicknesses under the cutting speed of 10 m/s. In the figure, F_z and F_x increased with the increase in depth. When the cutting speed remains constant, the measured and simulated values of cutting force at different cutting depths are basically the same trend, and the numerical values are close, which shows that the finite element simulation model is in line with the actual situation and therefore can be convinced.

4.2. Results Comparison of Cutting Temperature

Figure 7a,b show the cutting temperature images of KDP crystal taken by the ImageIR®300 infrared thermal camera. When the cutting speeds are 5 m/s and 10 m/s, respectively, the unformed chip thickness remains 429 nm. The maximum values of the measured temperatures are 70 °C and 98 °C, respectively.

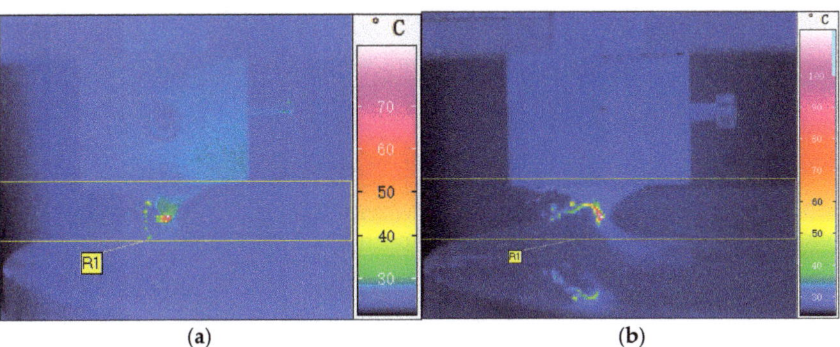

Figure 7. Cutting temperature images: (**a**) Cutting speed 5 m/s, (**b**) Cutting speed 10 m/s.

Figure 6c shows the comparison of cutting temperature results at different cutting speeds under the unformed chip thickness of 429 nm. As shown in the figure, firstly, as cutting speed increased, the cutting temperature increased by 260%. However, when the cutting speed increased to a certain extent, the growth rate of cutting temperature slowed down and the influence of cutting speed weakened gradually. Secondly, the maximum difference between the simulation temperature and the experimental results is less than 10 °C, which shows that the accuracy of the finite element simulation model can be accepted. Thirdly, when the unformed chip thickness was 429 nm and the cutting speed was 15 m/s, the simulation temperature of the chip reached 112 °C and the actual temperature reached 118 °C. Under such a high temperature and high speed deformation, the internal structure of the KDP crystal may change with the observation in Section 4.5, so high-speed cutting such as 15 m/s should be avoided as much as possible in the actual process.

Figure 6d shows the comparison of cutting temperature results at different unformed chip thicknesses under the cutting speed of 10 m/s. As shown in the figure, with the increase in unformed chip thickness, the cutting temperature increased by 74–90%, and the trend of chip temperature change is basically the same. Then, the simulation temperature of chips is close to the experimental results, and the maximum difference is less than 10 °C, which also validates the accuracy of the simulation model.

4.3. Influence of Cutting Speed

Figure 8a–e show the stress distribution at an unformed chip thickness of 429 nm (the cutting depth is 5 µm) and cutting speeds of 3 m/s, 5 m/s, 8 m/s, 10 m/s and 15 m/s. The compressive stress distribution is mainly found in the chip, and the maximum compressive stress occurs near the boundary between the workpiece and the chip along the shear direction, and the values are 309 MPa, 254 MPa, 243 MPa, 240 MPa, and 232 MPa, respectively. The maximum cutting stress at 15 m/s decreases by nearly 24% compared with that at 3 m/s.

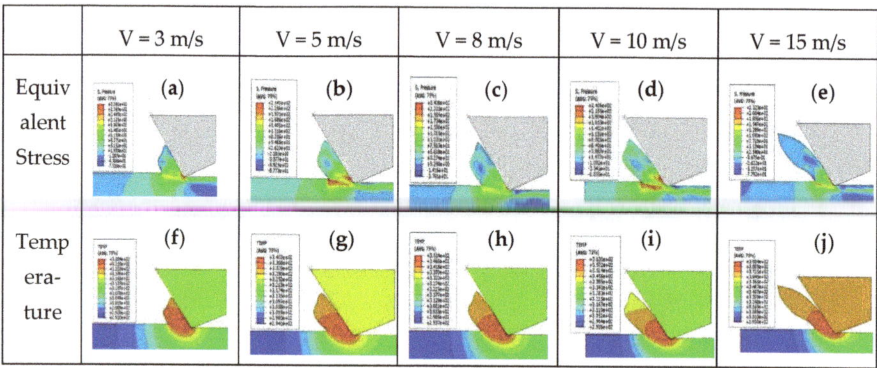

Figure 8. Stress distribution and temperature distribution of KDP crystal at different cutting speeds with unformed chip thickness of 429 nm: (**a**) Equivalent Stress of V = 3 m/s; (**b**) Equivalent Stress of V = 5 m/s; (**c**) Equivalent Stress of V = 8 m/s; (**d**) Equivalent Stress of V = 10 m/s; (**e**) Equivalent Stress of V = 15 m/s; (**f**) Temperature of V = 3 m/s; (**g**) Temperature of V = 5 m/s; (**h**) Temperature of V = 8 m/s; (**i**) Temperature of V = 10 m/s; (**j**) Temperature of V = 15 m/s.

Figure 8f,j show the temperature distribution of KDP crystals with an unformed chip thickness of 429 nm and different cutting speeds. The cutting speeds are 3 m/s, 5 m/s, 8 m/s, 10 m/s, and 15 m/s, respectively. The maximum temperatures of the chips in the figure are 55 °C, 67 °C, 85 °C, 98 °C, and 112 °C, respectively.

In Figure 8, when the unformed chip thickness was 429 nm, as the cutting speed increased: (1) The cutting force at 15 m/s decreased by nearly 25% compared with that at 3 m/s, and the maximum cutting stress decreased about 20% from 3 m/s to 15 m/s. (2) Cutting temperature increased from 58 °C to 112 °C, which increased by 93%. (3) In the process, the cutting force decreased with the higher cutting speed, and the most likely explanation can be deduced: Most chips contacted with the rake face, and with cutting speed increasing, the temperature gradually increased and the material softened, then the workpiece material was easier to remove and the friction between chip and tool decreased. So as cutting continued, the material softening level increased and the cutting force tended to be stable gradually, which was related to cutting temperature, so that the growth rate of temperature decreased and the cutting temperature became stable gradually. (4) After KDP crystal processing, there is tensile stress on the subsurface, which is about 3 MPa~10 MPa. (5) There is also a residual temperature of 30 °C~50 °C on the machined surfaces.

4.4. Influence of Unformed Chip Thickness

Figure 9a–c show a stress distribution with unformed chip thicknesses of 297, 429 and 622 μm, respectively, when cutting speed is 10 m/s. Similar to the situation shown in Figure 5, the compressive stress is distributed in the chips on the tool rake face and in the material near the shear zone. The maximum compressive stress on the shear plane is 235 MPa, 240 MPa and 280 MPa, respectively, and the maximum compressive stress at an unformed chip thickness of 10 μm increases by 20% compared with that at the unformed chip thickness of 297 nm.

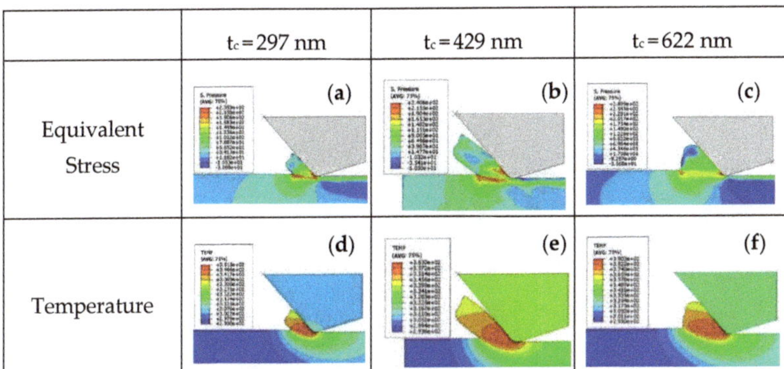

Figure 9. Stress distribution and temperature distribution at different unformed chip thicknesses at cutting speed of 10 m/s: (**a**) Equivalent Stress of tc = 297 nm; (**b**) Equivalent Stress of tc = 429 nm; (**c**) Equivalent Stress of tc = 622 nm; (**d**) Temperature of tc = 297 nm; (**e**) Temperature of tc = 429 nm; (**f**) Temperature of tc = 622 nm.

Figure 9d–f show the KDP crystal temperature distribution with unformed chip thicknesses of 297 nm, 429 nm and 622 nm, respectively, when cutting speed is 10 m/s. The maximum temperatures of the chips are 76 °C, 98 °C, and 117 °C.

In Figure 9, (1) when the cutting speed is 10 m/s, as the unformed chip thickness increases, the cutting force at 10 μm is 140% higher than that at 297 nm; however, the maximum compressive stress increases by only 20%. (2) The maximum chip temperature increases by 28% when the unformed chip thickness is 429 nm compared with 2 μm, while the maximum chip temperature increases by 19% when the unformed chip thickness is 622 nm compared with 429 nm. As the depth of the cut increases, the amount of chips increases, and the amount of chips rubbed against the tool will remain in a relatively stable range to a certain extent, and the frictional heat generation will be stable, so the temperature growth rate will decrease.

4.5. Chip Morphology

Figure 10a shows the KDP chip produced by finishing. The max value of the unformed chip thickness is about 297 nm and the cutting speed is 10 m/s. It shows the cutting temperature is about 70 °C. It is not difficult to find that the chip is continuous and strip-like, and the surface is smooth and complete, so that low temperature has little influence on the KDP crystal structure, and such chip morphology is ideal (Figure 10b,c).

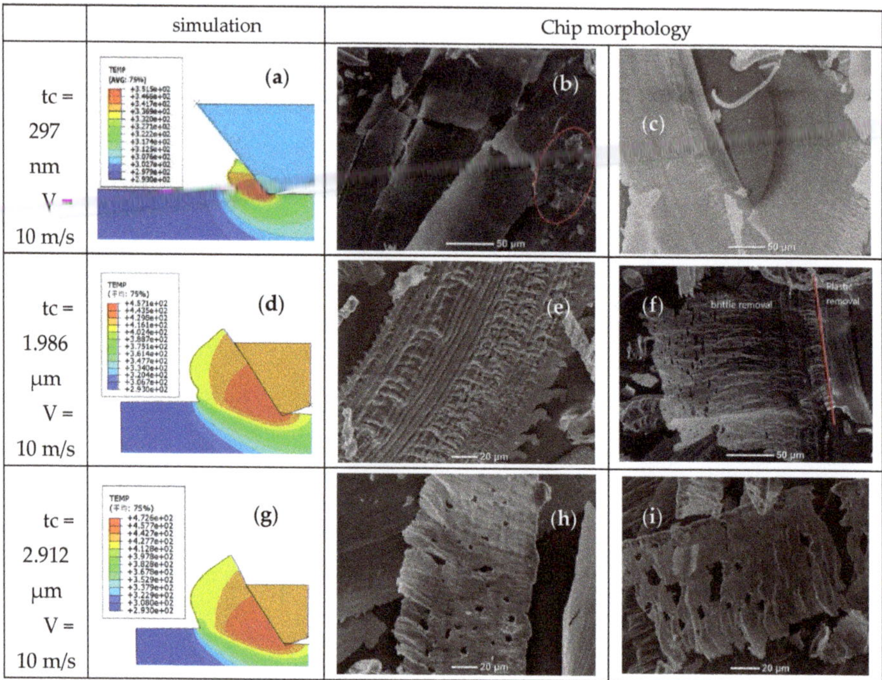

Figure 10. Chip morphology of KDP crystal at different processing stages: (**a**) Temperature of tc = 297 nm, V = 10 m/s; (**b,c**) Finishing chip morphology; (**d**) Temperature of tc = 1.986 μm, V = 10 m/s; (**e,f**) semi-finished chip morphology; (**g**) Temperature of tc = 2.912 μm, V = 10 m/s; (**h,i**) Roughing chip morphology.

The max value of the unformed chip thickness is about 1.986 μm. From the simulation results in Figure 10d, the cutting temperature is 184 °C. In Figure 10e, we can see that the chip is continuous, but the surface is wrinkled, which means the material is softened and can be more easily deformed, and the chips are produced quickly and accumulated obviously, so the softening chips become wrinkled in the discharging process. In Figure 10f, there is an obvious brittle plastic transformation, and the right side of the chip has a completely smooth appearance because the removal mode of this chip is plastic material removal, while on the left side of the chip, the chip exhibits obvious wrinkles and pore-like morphology because the removal mode is brittle material removal. The main reason for this phenomenon is that the actual undeformed cutting thickness can be increased along the arc-edge tool, which is illustrated in Figure 1b. In one process, the thickness of the two sides of the chip is different. The material on the side whose chip thickness is less than the depth of brittle–plastic transition is removed plastically, and the quality of the chip is better, while the material on the side whose chip thickness is greater than the depth of brittle–plastic transition is mainly removed brittly; therefore, the chip morphology is incomplete and the quality is poor.

The max value of the unformed chip thickness is about 2.912 μm. From the simulation results in Figure 10g, the cutting temperature is about 199 °C. In Figure 10h,i, it can be seen that the chip shows not only a small segment with obvious fractures, but the surface has obvious wrinkles with serious damage, and at the same time, the burrs on the edge of the chip are melted at a high temperature. In the case of high speed and large unformed chip thickness such as rough processing, the chip temperature can be higher than the phase transition temperature, which makes the material easier to deform and phase transition, so the chip is easier to perforate and melt.

5. Discussion

In this paper, the force and heat changes under different working conditions were analyzed by finite element simulation and experimental research, the chips' morphology under actual working conditions was observed, and then the mechanism of chip formation was analyzed. Next, the structure of KDP crystal can be analyzed and discussed through the existing results and simulations.

First of all, in the simulation and experimental results, under the cutting parameters we used in the KDP finishing, the temperature can reach only 140 °C, which is less than the dehydration temperature of KDP crystal (about 182.3 °C). However, in the KDP semi-finishing and roughing, the unformed chip thickness is 5~9 times higher than that in finishing, so the temperature can reach 184 °C~199 °C, and KDP crystals are more likely to dehydrate or deform.

Secondly, in the KDP crystal cutting, taking the cutting speed of 15 m/s and the max value of the unformed chip thickness of 429 nm as an example, the chip temperature measurement value reaches 112 °C, and the temperature difference within the crystal easily exceeds 4 °C. In Figure 8e, the max value of the tensile stress is 7.792 MPa, which exceeds the tensile strength limit of 6.67 MPa. In summary, this shows that the value of tensile stress is likely to be bigger than the tensile strength of KDP crystal under the combined action of the stress produced by crystal removal and the thermal stress of temperature, resulting in micro-cracks on the machined surface and sub-surface.

Thirdly, the phenomenon that chip morphology changes obviously in Figure 10e,f,h,i is explained as follows: in the cutting process, the cutting force and temperature act together on the KDP crystal, causing material properties such as hardness and specific heat capacity to change. Then, these changes in material properties will affect the cutting force and cutting temperature to achieve a stable state. In this state, the cutting stress is greater than the compressive strength of the material, and high-temperature chips break. At the same time, the local temperature of the chip has exceeded the phase transition temperature of the KDP crystal, and the material will easily generate a dehydration reaction, and even change the structure and morphology. After the chip is separated from the workpiece, the temperature gradually decreases. The residual stress makes the internal stress of the chip and the processed surface extremely unbalanced, likely causing secondary damage. Thus, microcracks may appear on the surface of the workpiece, and the chips will have perforations and edge melting.

6. Conclusions

In this paper, the cutting process of KDP crystal is studied by combining finite element simulations, measurement experiments and morphological observation experiments. The main conclusions of this paper are as follows:

1. In this paper, the KDP crystal fly-cutting simulation model is established, the stress field, cutting force and cutting temperature are calculated, and the comparison experiment is carried out to verify the validity of the model. The maximum difference between the simulation temperature and the experimental results is less than 10 °C, which shows that the accuracy of the finite element simulation model can be accepted.
2. The chip morphology of KDP crystal has been observed. Based on the finite element simulation results of the stress field and temperature field, the causes of chip morphology formation and surface cracks of the workpiece have been analyzed.
3. When the max value of the unformed chip thickness is 429 nm, and cutting speed is 15 m/s, the measured chip temperature reaches approximately 110 °C, and at the time there is also a residual temperature of 30 °C~50 °C on the machined surface and a tensile stress of 7.792 MPa, possibly resulting in micro-cracks on the machined surface and sub-surface.
4. In the case of semi-finishing and roughing, the chip temperature can reach 184 °C or higher, and it is quite possible for the KDP crystal to undergo a phase change reaction in the fly-cutting process, which makes the material easier to deform, and the chip

is easier to perforate and melt, while higher residual temperatures can more easily damage the machined surface.

Author Contributions: Conceptualization, C.A.; validation, J.Z.; formal analysis, K.F.; investigation, C.A.; writing—original draft preparation, X.Y.; writing—review and editing, H.L.; visualization, X.Y.; supervision, X.L.; project administration, Q.X.; funding acquisition, W.W. All authors have read and agreed to the published version of the manuscript.

Funding: Supported by NSAF U1830110, and Sichuan Province Science & Technology Project 2020JDRC0010.

Conflicts of Interest: The authors declare no conflict of interest.

References

1. Guo, D.; Jiang, X.; Huang, J.; Wang, F.; Liu, H.; Zu, X. Effect of UV Laser Conditioning on the Structure of KDP Crystal. *Adv. Condens. Matter Phys.* **2014**, *2014*, 238–244. [CrossRef]
2. Chen, M.; Li, M.; Jiang, W.; Xu, Q. Influence of period and amplitude of microwaviness on KH[sub 2]PO[sub 4] crystal's laser damage threshold. *J. Appl. Phys.* **2010**, *108*, 43109. [CrossRef]
3. Katagiri, M.; Namba, Y. Optical Surface Generation of KDP Inorganic Nonlinear Optical Crystals by Ultraprecision Surface Grinding. *J. Jpn. Soc. Precis. Eng.* **1999**, *65*, 888–892. [CrossRef]
4. Menapace, J.A.; Ehrmann, P.R.; Bickel, R.C. Magnetorheological finishing (MRF) of potassium dihydrogen phosphate (KDP) crystals: Nonaqueous fluids development, optical finish, and laser damage performance at 1064 nm and 532 nm. *Proc. SPIE* **2009**, *7504*, 750414. [CrossRef]
5. Chen, D.; Chen, J.; Wang, B. A hybrid method for crackless and high-efficiency ultraprecision chamfering of KDP crystal. *Int. J. Adv. Manuf. Technol.* **2016**, *87*, 293–302. [CrossRef]
6. An, C.; Zhang, Y.; Xu, Q.; Zhang, F.; Zhang, J.; Zhang, L.; Wang, J. Modeling of dynamic characteristic of the aerostatic bearing spindle in an ultra-precision fly cutting machine. *Int. J. Mach. Tools Manuf.* **2010**, *50*, 374–385. [CrossRef]
7. Paturi, U.M.R.; Narala, S.K.R. Constitutive flow stress formulation, model validation and FE cutting simulation for AA7075-T6 aluminum alloy. *Mater. Sci. Eng. A* **2014**, *605*, 176–185. [CrossRef]
8. Zhou, M.; Zou, L. Tool wear mechanism of diamond cutting of ferrous metals in frictional wear experiments. *Opt. Precis. Eng.* **2013**, *21*, 1786–1794. [CrossRef]
9. Ding, J.; Zhao, Y.; Wang, S.; Gu, Y.; Cui, H.; Liu, H.; Xu, G. Investigation on high temperature behavior and thermal dehydration kinetics of KDP Crystal. *J. Synth. Cryst.* **2015**, *44*, 235–241. [CrossRef]
10. Lo, S.-P. An analysis of cutting under different rake angles using the finite element method. *J. Mater. Process. Technol.* **2000**, *105*, 143–151. [CrossRef]
11. Zong, W.J.; Li, D.; Cheng, K.; Sun, T.; Liang, Y.C. Finite element optimization of diamond tool geometry and cutting-process parameters based on surface residual stresses. *Int. J. Adv. Manuf. Technol.* **2007**, *32*, 666–674. [CrossRef]
12. Zong, W.; Sun, T.; Li, D.; Cheng, K.; Liang, Y. FEM optimization of tool geometry based on the machined near surface's residual stresses generate. *J. Mater. Process. Technol.* **2006**, *180*, 271–278. [CrossRef]
13. Zong, W.J.; Li, Z.Q.; Zhang, L.; Liang, Y.C.; Sun, T.; An, C.H.; Zhang, J.F.; Zhou, L.; Wang, J. Finite element simulation of diamond tool geometries affecting the 3D surface topography in fly cutting of KDP crystals. *Int. J. Adv. Manuf. Technol.* **2013**, *68*, 1927–1936. [CrossRef]
14. Wang, S.; An, C.; Zhang, F.; Wang, J.; Lei, X.; Zhang, J. Simulation research on the anisotropic cutting mechanism of KDP crystal using a new constitutive model. *Mach. Sci. Technol.* **2017**, *21*, 202–222. [CrossRef]
15. An, C.; Feng, K.; Wang, W.; Xu, Q.; Lei, X.; Zhang, J.; Wang, S. Study on thermal field in fly-cutting process of DKDP crystal. *Int. J. Adv. Manuf. Technol.* **2019**, *103*, 3013–3024. [CrossRef]
16. Zhang, Q.Y.; Liu, D.J.; Wang, S.L.; Zhang, N.; Mu, X.M.; Sun, Y. Mechanical Parameters Test and Analysis for KDP Crystal . *J. Synth. Cryst.* **2009**, *38*, 1313–1319.
17. Hou, N.; Zhang, Y.; Zhang, L.; Zhang, F. Assessing microstructure changes in potassium dihydrogen phosphate crystals induced by mechanical stresses. *Scr. Mater.* **2016**, *113*, 48–50. [CrossRef]
18. Huang, P.; Wang, S.; Wang, D.; Liu, H.; Liu, G.; Xu, L.; Huang, P. Study on the cracking of a KDP seed crystal caused by temperature nonuniformity. *CrystEngComm* **2018**, *20*, 3171–3178. [CrossRef]
19. Barrow, G.A. A review of experimental and theoretical techniques for assessing cutting temperatures. *Ann. CIRP Ann. Manuf. Technol.* **1973**, *22*, 203–211.
20. Stephenson, D.; Ali, A. Tool Temperatures in Interrupted Metal Cutting. *J. Eng. Ind.* **1992**, *114*, 127–136. [CrossRef]
21. Chen, W.-C.; Tsao, C.-C.; Liang, P.-W. Determination of temperature distributions on the rake face of cutting tools using a remote method. *Int. Commun. Heat Mass Transf.* **1997**, *24*, 161–170. [CrossRef]
22. Fang, T.; Lambropoulos, J. Microhardness and Indentation Fracture of Potassium Dihydrogen Phosphate (KDP). *J. Am. Ceram. Soc.* **2002**, *85*, 174–178. [CrossRef]

 *micromachines*

MDPI

Article

High-Precision Machining Method of Weak-Stiffness Mirror Based on Fast Tool Servo Error Compensation Strategy

Zelong Li [1,2,3], Yifan Dai [1,2,3], Chaoliang Guan [1,2,3,*], Jiahao Yong [4], Zizhou Sun [1,2,3] and Chunyang Du [1,2,3]

[1] College of Intelligence Science and Technology, National University of Defense Technology, 109 Deya Road, Changsha 410073, China; d20153105@163.com (Z.L.); dyf@nudt.edu.cn (Y.D.); 18302996932@163.com (Z.S.); 15573125954@sina.cn (C.D.)

[2] Hunan Key Laboratory of Ultra-Precision Machining Technology, Changsha 410073, China

[3] Laboratory of Science and Technology on Integrated Logistics Support, National University of Defense Technology, 109 Deya Road, Changsha 410073, China

[4] 61175 Troops, The Chinese People's Liberation Army, Naijing 210049, China; jiahaoyong1996@163.com

* Correspondence: chlguan@nudt.edu.cn; Tel.: +86-138-0841-3943

Abstract: Weak-stiffness mirrors are widely used in various fields such as aerospace and optoelectronic information. However, it is difficult to achieve micron-level precision machining because weak-stiffness mirrors are hard to clamp and are prone to deformation. The machining errors of these mirrors are randomly distributed and non-rotationally symmetric, which is difficult to overcome by common machining methods. Based on the fast tool servo system, this paper proposes a high-precision machining method for weak-stiffness mirrors. Firstly, the clamping error and cutting error compensation strategy is obtained by analyzing the changing process of the mirror surface morphology. Then, by combining real-time monitoring and theoretical simulation, the elastic deformation of the weak-stiffness mirror is accurately extracted to achieve the compensation of the clamping error, and the compensation of the cutting error is achieved by iterative machining. Finally, a weak-stiffness mirror with a thickness of 2.5 mm was machined twice, and the experimental process produced a clamping error with a peak to valley (PV) value of 5.2 μm and a cutting error with a PV value of 1.6 μm. The final machined surface after compensation had a PV value of 0.7 μm. The experimental results showed that the compensation strategy proposed in this paper overcomes the clamping error of the weak-stiffness mirror and significantly reduces cutting errors during the machining process, achieving the high precision machining of a weak-stiffness mirror.

Keywords: weak-stiffness mirror; fast tool servo; clamping error; cutting error; error compensation

Citation: Li, Z.; Dai, Y.; Guan, C.; Yong, J.; Sun, Z.; Du, C. High-Precision Machining Method of Weak-Stiffness Mirror Based on Fast Tool Servo Error Compensation Strategy. *Micromachines* **2021**, *12*, 607. https://doi.org/10.3390/mi12060607

Academic Editors: Benny C. F. Cheung and Jiang Guo

Received: 27 April 2021
Accepted: 19 May 2021
Published: 24 May 2021

Publisher's Note: MDPI stays neutral with regard to jurisdictional claims in published maps and institutional affiliations.

1. Introduction

In the fields of aerospace equipment, optical imaging systems are usually built with weak-stiffness mirrors in order to meet the need for light weight. A specially shaped aluminum mirror is used in the infrared horizon system to realize the imaging function of the optical system [1]. The shape accuracy of the mirror surface determines the performance of the instrument, and the higher the shape accuracy, the higher the imaging resolution of the optical system. In order to meet the requirements of visible light imaging, the peak to valley (PV) value of the surface shape accuracy should be better than two λ ($\lambda = 632.8$ nm) [2]. Compared with an ordinary optical mirror, weak-stiffness mirrors are difficult to clamp and have low stiffness, which makes them very prone to clamping deformation and cutting deformation [3]. General weak-stiffness mirrors will produce a clamping deformation greater than 5 μm after clamping, which obviously cannot meet the demand of visible light imaging [4]. How to effectively solve the issues of clamping deformation and cutting deformation is the key to achieving the high-precision machining of such mirrors.

One of the main machining difficulties of weak-stiffness mirrors is that after disassembling the mirror, the mirror surface will spring back and destroy the original machined surface. The magnitude and distribution of the springback is uncertain because it is difficult to accurately control the magnitude of the clamping force. The common solution to this problem is the optimization of the clamping method for controlling the clamping deformation within an acceptable range [5]. Pan et al. analyzed the effect of clamping force on the thin-walled part deformation at different clamping positions and obtained the nonlinear relationship between contact force, contact deformation, and contact area, which is beneficial for reducing clamping deformation [6]. Miao et al. chose a low rigidity uhull part ao a reoearch target, the machining process was designed and analyzed, different clamping schemes were designed for this part, and the comparison and analysis were finished by using the finite element method [7]. However, the clamping method is different for various mirrors, and it is difficult to control the clamping error within two λ. Second, the machining process of weak-stiffness mirrors will produce large cutting errors due to cutting deformation, machine motion errors, servo system tracking errors, and other factors [8,9]. Many models have been established to describe the cutting errors during the cutting process [10,11]. Liu et al. established a model of machining errors in three-axis ultra-precision lathes based on the theory of multi-body systems [12]. Zhou et al. established an orthogonal cutting force analysis model based on unequal parallel-edge shear bands and proposed a cutting force prediction algorithm [13]. However, various errors are coupled with each other in the cutting process, and it is difficult to establish an accurate model to describe these cutting errors.

Both clamping error and cutting error are randomly distributed non-rotational symmetric errors. Ordinary two-axis turning cannot realize the compensation of non-rotational symmetry errors, and slow tool servo cutting can realize the compensation of non-rotational symmetry errors, but its working frequency is low, generally only 10–20 Hz [14]. To realize the compensation of clamping errors and cutting errors, a higher working frequency is required. A fast tool servo (FTS) system is a high-frequency, high-precision machining method that is widely used in ultra-precision machining processes [15,16]. Compared to the slow tool servo, the FTS working frequency is greater than 100 Hz and has nanometer-level resolution, which can realize the cutting of micron-level complex error morphology [17]. FTS systems have been utilized for machine tool error compensation and the cutting of complex morphology. Kim et al. used an FTS system with a servo bandwidth of 100 Hz for the real-time compensation of machine tool axial errors, and completed the machining of a large aspheric off-axis aluminum mirror with a diameter of 620 mm [18]. Yu et al. used an FTS system to compensate for contour errors and to improve the machining accuracy of microarrays [19]. Few studies have been conducted on the use of fast servo tools for the cutting process of weak-stiffness mirrors, and this paper investigates the feasibility of using an FTS system to compensate for the clamping and cutting errors of weak-stiffness mirrors.

In order to improve the machining accuracy of weak-stiffness mirrors, based on the fast tool servo system, this paper proposes a compensation strategy for clamping errors and cutting errors during the turning of a weak-stiffness mirror. The structure of this paper is as follows: the second section introduces the compensation strategies for clamping errors and cutting errors. The third section introduces the simulation and analysis methods, including the FTS machining system, clamping deformation simulation, and the Z-axis working frequency analysis. The fourth section presents the experimental results and discussion. Finally, the conclusion is given in the fifth section.

2. Error Compensation Strategy

As shown in Figure 1, the machining errors of a weak-stiffness mirror are divided into clamping errors [20] and cutting errors [21]. Clamping errors includes elastic deformation errors and plastic deformation errors, while cutting errors can be divided into system errors and random errors.

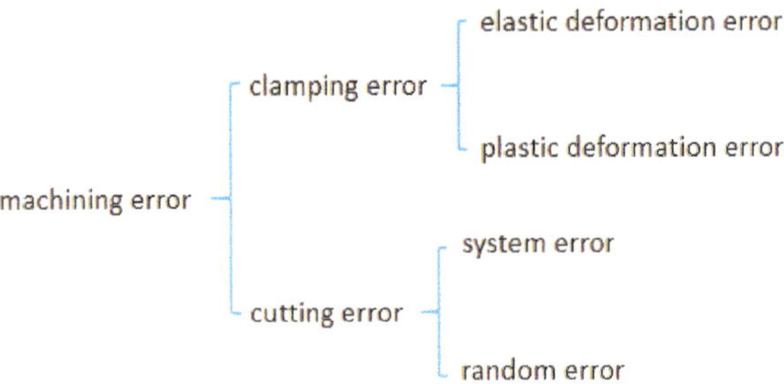

Figure 1. Schematic diagram of machining error distribution.

2.1. Clamping Error Compensation Strategy

As the contact surface between the mirror and the fixture is not an ideal plane, there is a contact area and a non-contact area, as shown in Figure 2a. Therefore, a large clamping force is generated during the clamping process, which will lead to clamping deformation on the machined surface [22]. Figure 2b shows the clamping deformation of a weak-stiffness mirror obtained through experimental measurement. Different colors represent different heights. The maximum height was about 2.4 μm, the minimum height was about −2.8 μm, and the PV value was about 5.2 μm. Clamping deformation includes elastic deformation and plastic deformation. Elastic deformation will recover after the removal of the fixture, while plastic deformation is a permanent deformation that cannot be recovered. As shown in Figure 3a, h_0 is the mirror surface before clamping. After clamping, the elastic deformation δ_1 and the plastic deformation δ_2 are produced. h_1 is the mirror surface after clamping. As in Figure 3b, after removing the fixture, the elastic deformation δ_1 recovers, while the plastic deformation δ_2 is still present. h_2 is the mirror surface after springback.

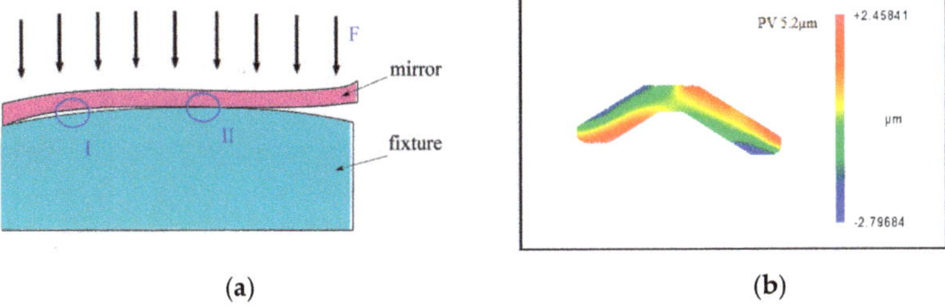

(a)　　　　　　　(b)

Figure 2. (**a**) Contact surface of the mirror and the fixture (I is the non-contact area, II is the contact area) and (**b**) the clamping deformation of a weak-stiffness mirror.

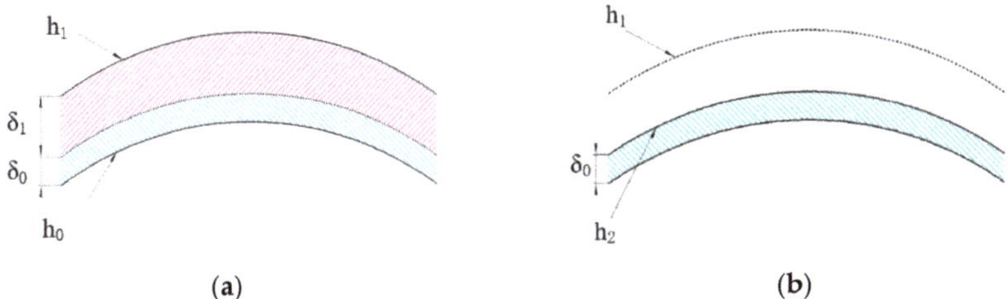

(a)　　　　　　　(b)

Figure 3. Changes in the mirror surface after clamping and removing. (**a**) The mirror surface after clamping and (**b**) the mirror surface after removing.

Based on the deformation process in Figure 3, this paper proposes a clamping deformation compensation strategy. Table 1 lists the relevant symbols. The compensation strategy is mainly for accurately obtaining the elastic deformation of the mirror surface, and then correcting the desired machining trajectory z_0 according to the elastic deformation. First, in order to accurately obtain the elastic deformation, the clamping method is optimized to a certain extent, and then the clamping deformation is controlled in the elastic region by finite element simulation. Therefore, there is no plastic deformation δ_2. The specific process of the simulation is described in Section 3.2. Then, the elastic deformation δ_1 can be obtained by measuring the mirror surface before and after clamping, as in Equation (1):

$$\delta_1 = h_1 - h_0 \tag{1}$$

Table 1. Some nomenclature.

Symbol	Meaning
z_0	the desired machining trajectory
z_1	the first machining trajectory
$z_{1'}$	the actual first machining trajectory
z_2	the second machining trajectory
$z_{2'}$	the actual second machining trajectory
z_3	the final equivalent machining trajectory
h_0	the mirror surface before clamping
h_1	the mirror surface after clamping

Second, the obtained elastic deformation δ_1 is used to correct the desired machining trajectory z_0, and then the first machining trajectory z_1 is obtained as Equation (2). The mirror is always clamped during the whole machining process, which means that there is always an elastic deformation δ_1 on the mirror surface. After the fixture is removed, the elastic deformation of the mirror surface will spring back, which will offset the corrected elastic deformation δ_1, and the equivalent machining trajectory z_3 after springback is consistent with the ideal machining trajectory z_0, so as to achieve clamping error compensation. The final equivalent machining trajectory z_3 after springback is given by Equation (3):

$$z_1 = z_0 + \delta_1 \tag{2}$$

$$z_3 = z_1 - \delta_1 = z_0 \tag{3}$$

2.2. The Cutting Error Compensation Strategy

Section 2.1 introduces the compensation for the clamping error, but a weak-stiffness mirror will also produce a larger cutting error, mainly due to cutting deformation in the actual machining process. As shown in Figure 4, according to the description in Section 2.1, the machining trajectory for the first machining is z_1, but the actual machining trajectory is z_{1}' under the influence of the cutting error δ_2.

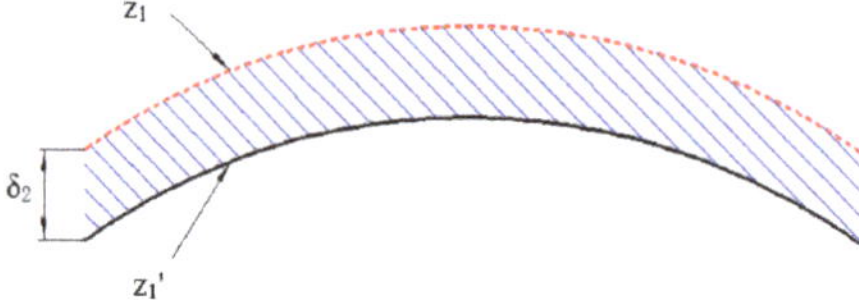

Figure 4. Cutting error.

There are many factors that can cause cutting errors. Some studies have shown that cutting errors can be divided into repeatable system errors and random errors caused by random factors such as machine tool status changes or tool wear under similar operating conditions [23]. Additionally, when the machine tool is in good condition and the operation is accurate, cutting errors are mainly repetitive system errors.

In order to verify that the cutting error was mainly a repetitive system error, repeated cutting experiments were performed on the same mirror under the same experimental conditions and machining trajectory. The experiment was carried out with an ordinary ultra-precision machine tool, and the machining result was measured with a vertical interferometer. The experiment speed was 300 r/min and the cutting depth was 1.4 μm. The cutting error produced by each cutting experiment was obtained by subtracting the machining trajectory from the actual measurement results. Figure 5 shows the distribution of cutting errors produced by each cutting experiment. Table 2 summarizes the PV and root mean square (RMS) values of the cutting error in each experiment. It can be seen from the experimental results that the magnitude and distribution of the cutting error generated by each cutting experiment were almost the same. The cutting errors were repeatable.

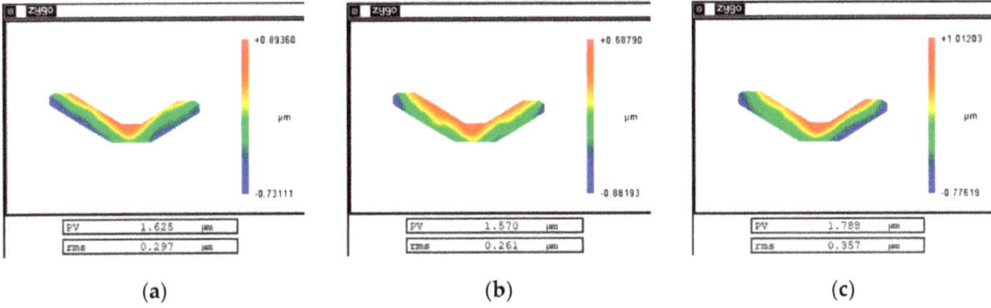

(a)   (b)   (c)

Figure 5. Morphological distribution of cutting errors at (**a**) the first cutting, (**b**) the second cutting, and (**c**) the third cutting.

Table 2. Cutting error repeatability test results.

Experimental Result	First Experiment	Second Experiment	Third Experiment
PV value	1.625 μm	1.570 μm	1.788 μm
RMS value	0.297 μm	0.261 μm	0.357 μm

Based on the above analysis, the compensation strategy for cutting errors is proposed. Firstly, the first machining is performed according to the description in Section 2.1, and the first machining trajectory is z_1. Then, the mirror is kept in the clamped state and the actual machining trajectory z_1' is obtained by measuring the machining results. The cutting error δ_2 generated during this machining process is obtained by Equation (4). Then, the obtained cutting error is used to correct the first machining trajectory and obtain the second machining trajectory z_2, as shown in Equation (5). Ignoring the influence of random error, it is believed that the second machining will produce a cutting error of the same magnitude and distribution as the first machining. The actual second machining trajectory z_2' will be equal to the first machining trajectory z_1, thereby realizing the compensation of the cutting error, as shown in Equation (6).

$$\delta_2 = z_1 - z_{1'} \tag{4}$$

$$z_2 = z_1 + \delta_2 \tag{5}$$

$$z_{2'} = z_2 - \delta_2 = z_1 \tag{6}$$

In summary, the compensation of clamping errors and cutting errors is achieved by two machining operations, respectively. The first machining compensates for the clamping errors and the second machining compensates for the cutting errors produced in the first machining. After considering the cutting error, the final equivalent trajectory z_3 is still equal to the ideal machining trajectory z_0.

$$z_3 = z_{2'} - \delta_1 = z_1 - \delta_1 = z_0 \tag{7}$$

3. Simulation and Analysis

3.1. FTS Machining System

Different from the ordinary machine tool error compensation, the clamping errors and cutting errors in this paper are randomly distributed errors on the order of microns. To realize this error compensation, the machining Z-axis needs to have C, X, and Z-axis linkage capability and a high working frequency.

In order to meet the basic requirements for error compensation, an ultra-precision cutting system based on FTS technology is established, as shown in Figure 6. The ultra-precision cutting system consists of the Z-axis, X-axis, and C-axis of the ultra-precision machine and FTS. During processing, the workpiece rotates with the C-axis, and the X-axis feeds along the radial direction of the workpiece. The FTS calculates the feed amount z required for the current machining position according to the current spindle angle θ and the X-axis position (i.e., the radial position of the workpiece) ρ, and then drives the diamond tool to perform a quantitative feed motion along the Z-axis to cut the workpiece. Since the feed amount z of the fast tool servo system is a function of the spindle angle and the position of the X-axis $z = f(\rho,\theta)$, it can realize machining with arbitrarily distributed errors. The working frequency of the Z-axis of the system is greater than 500 Hz, and the working stroke is greater than 100 μm.

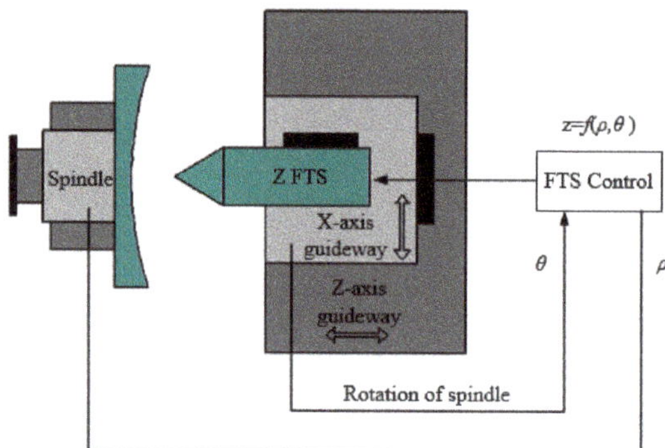

Figure 6. FTS machining system.

The basic process of the error compensation experiment is shown in Figure 7. It mainly includes the following steps: (a) real-time monitoring of the clamping process to extract the elastic deformation. First, measure the mirror surface before clamping. The mirror surface will change after clamping. After stabilization, measure the mirror surface after clamping. (b) Generate the first machining trajectory based on the extracted elastic deformation. (c) Compare the result after the first compensation machining with the desired machining result to obtain the cutting error. (d) Superimpose the cutting error on the basis of the first compensation machining and generate the second machining trajectory. (e) Disassemble the mirror and obtain the final result.

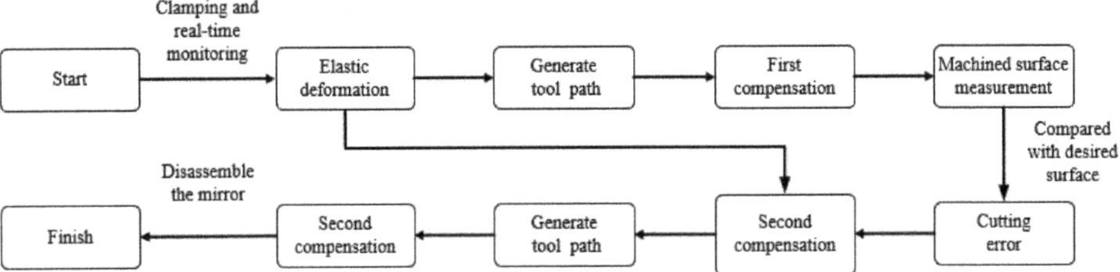

Figure 7. Flow chart of the error compensation experiment.

3.2. Clamping Deformation Simulation

According to the description in Section 2.1, the clamping deformation error needs to be controlled in the elastic region to accurately extract the elastic deformation error. In this section, we ensure that the clamping deformation is only elastic deformation through finite element simulation when the magnitude and distribution of the clamping deformation are known.

As shown in Figure 8, when the clamping force acts on the surface of the aluminum alloy, with the yield limit σ_s as the boundary, the work done by the clamping gradually accumulates in the material as elastic deformation energy and then produces elastic deformation. With the gradual increase in the degree of deformation, the accumulation of plastic deformation energy starts to occur after passing the yield point b, and then plastic deformation occurs [24]. According to the Mises yield criterion, Mises considers yield as an invariant state of the material, independent of the stress coordinate system. When the equivalent stress reaches the yield limit σ_s, the material begins to produce plastic deformation. The equivalent stress can be obtained by Equation (8):

$$\overline{\sigma} = \frac{1}{\sqrt{2}}\sqrt{(\sigma_1 - \sigma_2)^2 + (\sigma_2 - \sigma_3)^2 + (\sigma_3 - \sigma_1)^2} \tag{8}$$

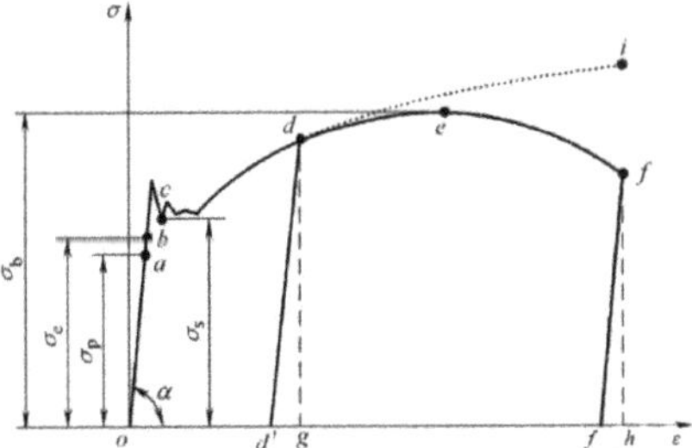

Figure 8. Stress–strain curve.

The experimental object is an infrared horizon mirror, as shown in Figure 9. The mirror is 295 mm long, 95 mm wide, 2.5 mm thick, and made of aluminum alloy with a yield limit σ_s 275 Mpa. In order to reduce the clamping force, the mirror is clamped by bonding, using low-stress extreme ultraviolet glue (uv-50). The glue will harden only after being irradiated by ultraviolet light. The bonding area is shown in Figure 10a. Figure 10b shows the magnitude and distribution of the clamping deformation obtained by measuring the mirror surface before and after clamping. Due to the complexity of the stresses generated by bonding, forced displacements are applied to the machined mirror surface based on the measured clamping deformation. Then, the stress distribution on the machined surface is obtained by the Mises yield criterion, as shown in Equation (8) [25]. The finite element simulation software was ansys workbench 19.0, the element was solid186, and the number of elements was 642485. Figure 10c is a displacement distribution diagram of the machining surface and Figure 10d is a stress distribution diagram of the machining surface.

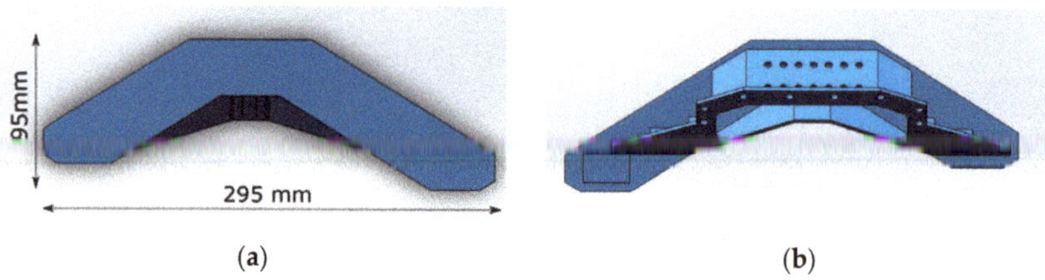

(a)   **(b)**

Figure 9. Infrared horizon mirror 3D model: (**a**) mirror machining surface and (**b**) mirror clamping surface.

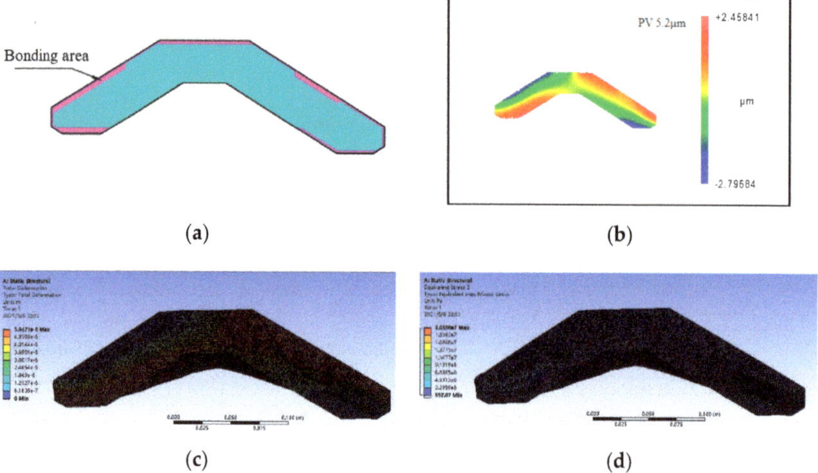

(a)   **(b)**

(c)   **(d)**

Figure 10. Clamping deformation simulation of a weak-stiffness mirror: (**a**) bonding area; (**b**) measurement results of clamping deformation; (**c**) machining surface displacement distribution map; and (**d**) machining surface stress distribution map.

From the experimental and simulation results, it can be seen that the clamping force of the extreme ultraviolet glue produced a clamping deformation with PV value of 5 μm on the machined surface. As shown in Figure 10d, it can be seen that the maximum stress was 20 MPa, which is far less than the yield limit σ_s of 275 MPa of the aluminum alloy material. According to the analysis in Figure 8, we can reach the conclusion that the mirror surface only has elastic deformation.

3.3. The Analysis of Z-Axis Working Frequency

3.3.1. Spatial Curve Extension Algorithm

To make the Z-axis work more smoothly and to reduce the Z-axis working frequency, the spatial curve extension algorithm is used to extend the Z-axis machining trajectory before machining. Since the infrared horizon mirror is not a regular circle, there must be a part of the machining trajectory outside the mirror, and the tool position outside the mirror needs to be processed. In addition, the mirror has beveled edges and is large. Due to the limited resolution of the interferometer measurement, the edge will have a jagged shape, as shown in Figure 11.

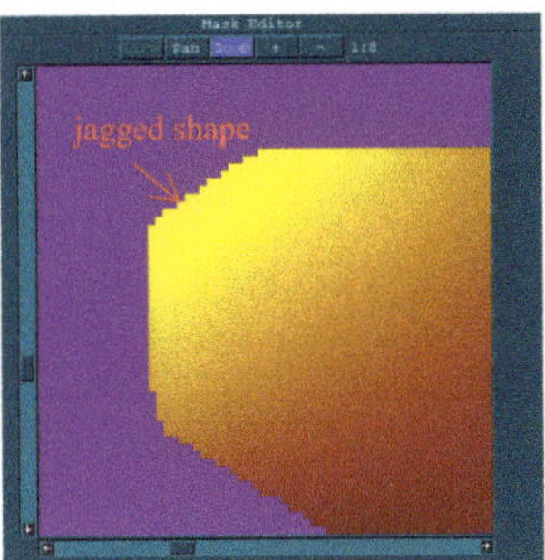

Figure 11. 24 inch interferometer measurement results (resolution 0.00066 m).

The space curve extension algorithm is divided into two steps. First, the edge data are extended by 3 pixel points (about 2 mm) along the vertical and horizontal directions, as shown in Figure 12a. Then, the data are extended along the spiral trajectory based on the cubic spline interpolation algorithm. Figure 12b shows the projection of the machining trajectory on the plane of the cylindrical coordinate system. The boundary point entering the mirror is $A(A\rho,A\theta,Az)$, and the boundary point leaving the mirror is $B(B\rho,B\theta,Bz)$. The tool moves from the point $B_i(B\rho_i,B\theta_i,Bz_i)$ to the point $A_{i+1}(A\rho_{i+1},A\theta_{i+1},Az_{i+1})$. $P_i^j(P\rho_i^j,P\theta_i^j,Pz_i^j)$ is a certain interpolation point of the curve B_iA_{i+1}, which is calculated by Equation (9):

$$Pz_i^j = \frac{(L+2l_i^j)(L-l_i^j)^2}{L^3}Bz_i + \frac{(3L-2l_i^j)(l_i^j)^2}{L^3}Az_{i+1} \\ + \frac{(l_i^j)(L-l_i^j)^2}{L^2}Bk_i + \frac{(l_i^j)^2(l_i^j-L)}{L^2}Ak_{i+1} \tag{9}$$

where L is the arc length of the projection of the space curve B_iA_{i+1} on the plane of the column coordinate system, l_i^j is the arc length of the projection of the curve $B_iP_i^j$ on the plane of the column coordinate system, Bk_i is the slope of the boundary point B_i, and Ak_{i+1} is the slope of the boundary point A_{i+1}. Since the highest item does not exceed 3 times in Equation (9), it is guaranteed that the Z value of the interpolated curve segment will not rise and fall more than 2 times.

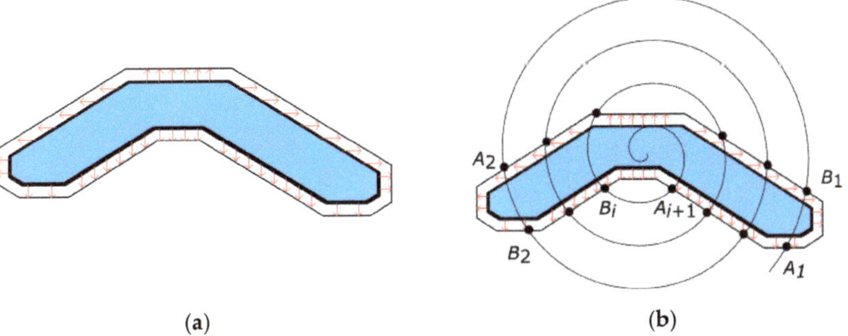

Figure 12. Spatial curve extension algorithm: (**a**) edge data extension and (**b**) data extension outside the boundary.

3.3.2. The Analysis of Z-Axis Working Frequency

Figure 13a is the result of the mirror surface data after extension. Figure 13b shows the spectrum of the machining trajectory after extension, which is obtained by spectrum analysis. N is the number of undulations in a rotation cycle of the Z-axis. According to Equation (10), when the spindle speed n = 300 r/min and the number of fluctuations N = 15 in a rotation period, the Z-axis working frequency $f = 75$ Hz. The working frequency of the FTS system built in this paper is greater than 500 Hz, which meets the needs of use.

$$f = \frac{N \times n}{60}$$ (10)

where f is the Z-axis working frequency Hz, n is the spindle speed r/min, and N is the number of undulations of the Z-axis in one rotation cycle.

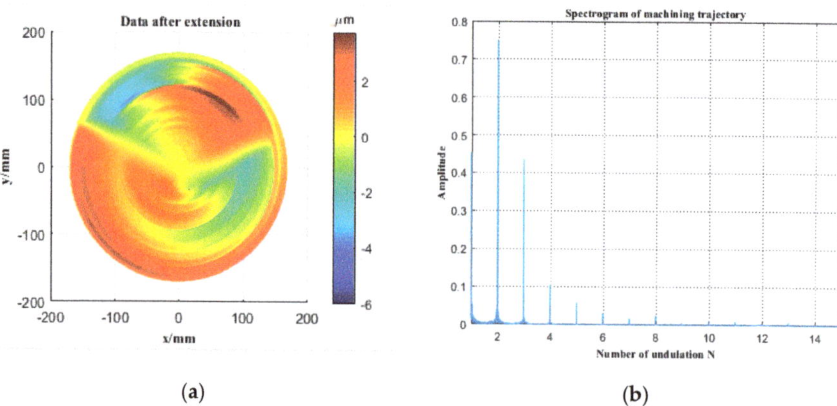

(a)

(b)

Figure 13. (**a**) Data after extension and (**b**) a spectrogram of the machining trajectory.

4. Experimental Results and Discussion

4.1. Experimental Results

Figure 14a shows the FTS machining system used in this experiment, with a maximum working stroke of 0.5 mm and a working frequency of 500 Hz. Figure 14b shows the experimental workpiece and fixture, positioned by the positioning pin holes on both sides. The center of mass of the workpiece coincides with the center of rotation of the spindle. A diamond tool with a radius of 0.2 mm is used. The spindle speed is 300 r/min, the finishing X-axis feed rate is 1.2 mm/min, and the depth of cut is 1.4 μm.

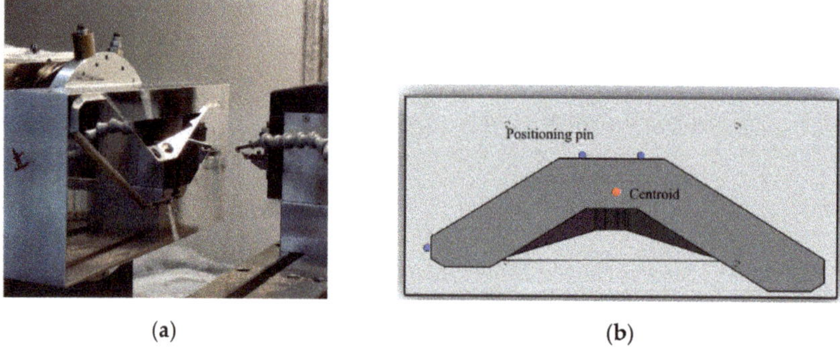

(a)

(b)

Figure 14. (**a**) FTS system and (**b**) a schematic diagram of the mirror and fixture.

Figure 15 shows the measurement results of the mirror surface during the experiment. As shown in Figure 15c,e, there was a large clamping error and cutting error generated in this experiment. The PV value of the clamping error was 5.2 μm and the PV value of the cutting error was 1.6 μm. The PV value of the final mirror surface was 0.7 μm after two compensations, as shown in Figure 15g. Additionally, the final surface roughness Ra was better than 10 nm, as shown in Figure 16.

(a)   (b)   (c)

(d)   (e)   (f)

(g)

Figure 15. Experimental results: (**a**) is the measurement result before clamping; (**b**) is the measurement result after clamping; (**c**) is the elastic deformation error; (**d**) is the measurement result after the first machining; (**e**) is the cutting error; (**f**) is the measurement result after the second machining; and (**g**) is the final measurement result after disassembling the mirror.

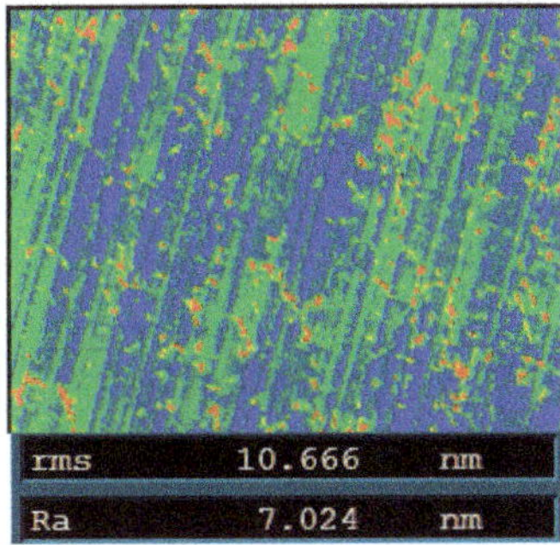

Figure 16. Surface roughness of the mirror.

4.2. Discussion

The experimental results show that the machining method based on the FTS error compensation strategy proposed in this paper can effectively overcome the clamping errors and cutting errors of a weak-stiffness mirror and significantly improve the machining accuracy. This machining method has strong applicability and can solve the machining problems of most mirrors with low stiffness and easy deformation. Based on the experimental results and the compensation strategy, there are a few points to discuss.

(1) In order to extract the elastic deformation of the mirror accurately, the compensation strategy proposed in this paper needs to control the clamping deformation in the elastic region. Through real-time monitoring and theoretical simulation, it ensures that the clamping deformation generated by each clamping is elastic. This method can be applied to most weak-stiffness mirrors. However, it may not be possible to control the clamping deformation in the elastic region for some mirrors that are very difficult to clamp or are very prone to plastic deformation. In this case, some adjustments can be made to the compensation strategy. First, the relationship between the elastic deformation δ_1 and the plastic deformation δ_0 needs to be obtained, as in Equation (11), and K is the scale factor. Then, based on Equation (11), the elastic deformation of the workpiece can still be obtained based on the measurement of the surface before clamping h_0 and the surface after clamping h_1, as in Equation (12) and Equation (13). The subsequent compensation strategy remains the same. To accurately obtain the scale factor K, we need to further investigate the elastic–plastic deformation theory.

$$\delta_0 = K\delta_1 \tag{11}$$

$$h_1 - h_0 = \delta_0 + \delta_1 = (1 + K)\delta_1 \tag{12}$$

$$\delta_1 = (h_1 - h)/K \tag{13}$$

(2) In this paper, cutting errors are compensated by iterative machining. It was verified by repeatability experiments that the cutting errors were dominated by system errors under the same experimental conditions. Figure 17 shows the cutting error after compensation. The PV value of the cutting error was reduced from 1.6 μm to 0.4 μm. In order to maintain a good surface roughness on the machined surface, each machining time was about 2.5 h. During the second compensated machining, the machine condition was changed due to the long machining time and the change of the ambient temperature, which led to a large random error. It can be assumed that the residual error of the mirror surface was mainly the random error generated by the two machining processes. By reducing the machining time and controlling the ambient temperature to maintain the machine in a similar state during the two machining processes, the machining accuracy of the weak-stiffness mirror can be further improved.

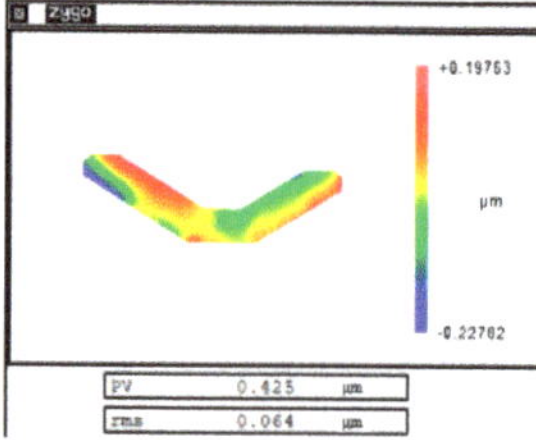

Figure 17. Cutting error after compensation.

(3) This paper gives the calculation of the Z-axis working frequency at different spindle speeds, which is instructive. Different weak-stiffness mirrors produce different magnitudes and distributions of errors and require different Z-axis working frequencies. The researcher can select a suitable FTS system based on the calculation result of Equation (10).

5. Conclusions

This paper proposes a high-precision machining method for weak-stiffness mirrors based on the FTS error compensation strategy. The elastic deformation of the mirror is accurately extracted by real-time monitoring and clamping deformation simulation. The compensation of the clamping error is realized by correcting the machining trajectory. Additionally, the cutting error is significantly reduced by iterative machining. The Z-axis working frequency required for realizing the error compensation was quantitatively calculated by a spatial curve extension algorithm and spectrum analysis. Based on the FTS system, a weak-stiffness mirror was machined twice. The final PV value of the machined surface was 0.7 μm. It effectively solved the clamping error with a PV value of 5.2 μm and a cutting error with a PV value of 1.6μm generated during the machining of the weak-stiffness mirror. The experimental results show that the machining method proposed in this paper effectively solves the clamping errors of weak-stiffness mirrors and significantly reduces the cutting errors during the machining process. The machining method proposed in this paper is instructive and can be widely applied to the machining of weak-stiffness mirrors.

Author Contributions: Conceptualization, Y.D.; Data curation, J.Y.; Formal analysis, Z.L. and C.G. Methodology, Z.L. and C.G.; Supervision, C.D.; Validation, Z.S.; Writing—original draft, Z.L.; Writing—review & editing, Z.L. All authors have read and agreed to the published version of the manuscript.

Funding: This research was funded by Science Challenge Project of China (Grant No.TZ2018006-0102-04).

Conflicts of Interest: The authors declare no conflict of interest.

References

1. Wang, H.; Wang, Z.; Wang, B.; Jin, Z.; Crassidis, J. Infrared Earth sensor with a large field of view for low-Earth-orbiting micro-satellites. *Front. Inf. Technol. Electron. Eng.* **2020**, *22*, 262–271. [CrossRef]
2. Huang, Y.; Zhao, F.; He, M.; He, X. Research on the control technology of machining marks in double-side polishing of large aperture optical element. In Proceedings of the 9th International Symposium on Advanced Optical Manufacturing and Testing Technologies: Advanced Optical Manufacturing Technologies, Chengdu, China, 16 January 2018; p. 108381O. [CrossRef]
3. Toubhans, B.; Fromentin, G.; Viprey, F.; Karaouni, H.; Dorlin, T. Machinability of inconel 718 during turning: Cutting force model considering tool wear, influence on surface integrity. *J. Mater. Process. Technol.* **2020**, *285*, 116809. [CrossRef]
4. Ju, K.; Duan, C.; Kong, J.; Chen, Y.; Sun, Y.; Wu, S. Prediction of clamping deformation in vacuum fixtureworkpiece system for low-rigidity thin-walled precision parts using finite element method. *Int. J. Adv. Manuf. Technol.* **2020**, *109*, 1895–1916. [CrossRef]
5. Chen, W.; Ni, L.; Xue, J. Deformation control through fixture layout design and clamping force optimization. *Int. J. Adv. Manuf. Technol.* **2008**, *38*, 860–867. [CrossRef]
6. Pan, M.; Tang, W.; Xing, Y.; Ni, J. The clamping position optimization and deformation analysis for an antenna thin wall parts assembly with ASA, MIGA and PSO algorithm. *Int. J. Precis. Eng. Manuf.* **2017**, *18*, 345–357. [CrossRef]
7. Miao, L.; Yu, H.; Cui, W.; Jiang, H.; Li, Y.; Wang, R.; Gao, X. Analysis and research of structure and process characteristic based on low rigidity shell part. In Proceedings of the the 7th International Conference on Mechanical and Electronics Engineering, Dalian, China, 26–27 September 2015; p. 03002. [CrossRef]
8. Kong, L.B.; Cheung, C.F.; Kwok, T.C. Theoretical and experimental analysis of the effect of error motions on surface generation in fast tool servo machining. *Precis. Eng. J. Int. Soc. Precis. Eng. Nanotechnol.* **2014**, *38*, 428–438. [CrossRef]
9. Penghao, R.; Aimin, W.; Long, W.; Dongxia, L. Simulation analysis of turning deformation of rotational thin-walled parts based on cutting force model. In Proceedings of the 2018 IEEE 9th International Conference on Mechanical and Intelligent Manufacturing Technologies, Cape Town, South Africa, 10–13 February 2018; pp. 21–25. [CrossRef]
10. Laghari, R.; Li, J.; Mia, M. Effects of Turning Parameters and Parametric Optimization of the Cutting Forces in Machining SiCp/Al 45 wt% Composite. *Metals* **2020**, *10*, 840. [CrossRef]
11. Zhang, L.; Sha, X.; Liu, M.; Wang, L.; Pang, Y. Cutting Force Prediction Models by FEA and RSM When Machining X56 Steel with Single Diamond Grit. *Micromachines* **2021**, *12*, 326. [CrossRef]
12. Liu, X.; Zhang, X.; Fang, F.; Liu, S. Identification and compensation of main machining errors on surface form accuracy in ultra-precision diamond turning. *Int. J. Mach. Tools Manuf.* **2016**, *105*, 45–57. [CrossRef]
13. Zhou, J.; Ren, J. Predicting cutting force with unequal division parallel-sided shear zone model for orthogonal cutting. *Int. J. Adv. Manuf. Technol.* **2020**, *107*, 4201–4211. [CrossRef]
14. Li, D.; Qiao, Z.; Walton, K.; Liu, Y.; Xue, J.; Wang, B.; Jiang, X. Theoretical and Experimental Investigation of Surface Topography Generation in Slow Tool Servo Ultra-Precision Machining of Freeform Surfaces. *Materials* **2018**, *11*, 2566. [CrossRef] [PubMed]

15. Liu, Y.; Zheng, Y.; Gu, Y.; Lin, J.; Lu, M.; Xu, Z.; Fu, B. Development of Piezo-Actuated Two-Degree-of-Freedom Fast Tool Servo System. *Micromachines* **2019**, *10*, 337. [CrossRef] [PubMed]
16. Zhu, L.; Li, Z.; Fang, F.; Huang, S.; Zhang, X. Review on fast tool servo machining of optical freeform surfaces. *Int. J. Adv. Manuf. Technol.* **2018**, *95*, 2071–2092. [CrossRef]
17. Kim, H.S.; Kim, E.J. Feed-forward control of fast tool servo for real-time correction of spindle error in diamond turning of flat surfaces. *Int. J. Mach. Tools Manuf.* **2003**, *43*, 1177–1183. [CrossRef]
18. Kim, H.S.; Kim, E.J.; Song, B.S. Diamond turning of large off-axis aspheric mirrors using a fast tool servo with on-machine measurement. *J. Mater. Process. Technol.* **2004**, *146*, 349–355. [CrossRef]
19. Yu, D.P.; Hong, G.S.; Wong, Y.S. Profile error compensation in fast tool servo diamond turning of micro-structured surfaces. *Int. J. Mach. Tools Manuf.* **2012**, *52*, 13–23. [CrossRef]
20. Zhang, C.; Jiao, S.; Wang, L. Clamping deformation analysis and machining parameter optimization of weak stiffness ring parts. In Proceedings of the the 2nd International Conference on Frontiers of Materials Synthesis and Processing, Sanya City, China, 10–11 November 2019; p. 012028. [CrossRef]
21. Li, Y.; Zhang, Y.; Lin, J.; Yi, A.; Zhou, X. Effects of Machining Errors on Optical Performance of Optical Aspheric Components in Ultra-Precision Diamond Turning. *Micromachines* **2020**, *11*, 331. [CrossRef]
22. Demeter, E.C. Restraint Analysis of Fixtures Which Rely on Surface Contact. *J. Eng. Ind.* **1994**, *116*, 207–215. [CrossRef]
23. Chen, Y.; Gao, J.; Deng, H.; Zheng, D.; Chen, X.; Kelly, R. Spatial statistical analysis and compensation of machining errors for complex surfaces. *Precis. Eng.* **2013**, *37*, 203–212. [CrossRef]
24. Wagoner, R.H.; Lim, H.; Lee, M.G. Advanced Issues in springback. *Int. J. Plast.* **2013**, *45*, 3–20. [CrossRef]
25. Wang, J.; Geng, S.M.; Zhang, L.Y.; Lu, Y.S. Finite element analysis and control of clamping deformation mechanism of thin-wall shell workpiece. *Acta Armamentarii* **2011**, *32*, 1008–1013. [CrossRef]

 *micromachines*

MDPI

Article

An Investigation of the High-Frequency Ultrasonic Vibration-Assisted Cutting of Steel Optical Moulds

Canbin Zhang [1,*], Chifai Cheung [1], Benjamin Bulla [2] and Chenyang Zhao [1,3]

[1] State Key Laboratory of Ultra-Precision Machining Technology, Department of Industrial and Systems Engineering, The Hong Kong Polytechnic University, Hung Hom, Kowloon, Hong Kong, China; benny.cheung@polyu.edu.hk (C.C.); zhaochenyang@hit.edu.cn (C.Z.)

[2] Son-x Gmbh, 52078 Aachen, Germany; benjamin.bulla@son-x.com

[3] School of Mechanical Engineering and Automation, Harbin Institute of Technology, Shenzhen 518055, China

* Correspondence: canbin.zhang@connect.polyu.hk; Tel.: +852-5623-2030

Abstract: Ultrasonic vibration-assisted cutting (UVAC) has been regarded as a promising technology to machine difficult-to-machine materials such as tungsten carbide, optical glass, and hardened steel in order to achieve superfinished surfaces. To increase vibration stability to achieve optical surface quality of a workpiece, a high-frequency ultrasonic vibration-assisted cutting system with a vibration frequency of about 104 kHz is used to machine spherical optical steel moulds. A series of experiments are conducted to investigate the effect of machining parameters on the surface roughness of the workpiece including nominal cutting speed, feed rate, tool nose radius, vibration amplitude, and cutting geometry. This research takes into account the effects of the constantly changing contact point on the tool edge with the workpiece induced by the cutting geometry when machining a spherical steel mould. The surface morphology and surface roughness at different regions on the machined mould, with slope degrees (SDs) of 0°, 5°, 10°, and 15°, were measured and analysed. The experimental results show that the arithmetic roughness S_a of the workpiece increases gradually with increasing slope degree. By using optimised cutting parameters, a constant surface roughness S_a of 3 nm to 4 nm at different slope degrees was achieved by the applied high-frequency UVAC technique. This study provides guidance for ultra-precision machining of steel moulds with great variation in slope degree in the pursuit of optical quality on the whole surface.

Keywords: high frequency; ultrasonic-assisted vibration cutting; difficult-to-machine material; spherical steel mould; ultra-precision machining

Citation: Canbin, Z.; Chifai, C.; Benjamin, B.; Chenyang, C. An Investigation of the High-Frequency Ultrasonic Vibration-Assisted Cutting of Steel Optical Moulds. *Micromachines* **2021**, *12*, 460. https://doi.org/10.3390/mi12040460

Academic Editor: Xichun Luo

Received: 31 March 2021
Accepted: 16 April 2021
Published: 19 April 2021

Publisher's Note: MDPI stays neutral with regard to jurisdictional claims in published maps and institutional affiliations.

1. Introduction

Steel is a fascinating die material in the precision mould industry for injection moulding of plastic optical lenses and low-Tg glass. To machine steel materials, coated carbide tools [1,2] and cubic boron nitride (CBN) tools [3] are widely applied. Diamond tools possess a nanometric edge radius, form reproducibility, and excellent wear resistance, so they are extensively used in ultra-precision machining technology, such as fly cutting and single-point diamond turning (SPDT), and are capable of producing components with submicrometric form accuracy and surface roughness in the nanometre range on plastic and non-ferrous materials such as copper, aluminium, brass, etc. [4]. However, diamond machining of ferrous materials such as steel is not amenable due to excessive tool wear caused by the chemical affinity of carbon to iron [5]. Guo et al. [6] conducted a critical review of the chemical wear and wear-suppressing methods of diamond tools in the diamond cutting of ferrous materials. To reduce the chemical reaction and wear on the diamond tool and to hence improve the surface quality of the workpiece, researchers have been putting great efforts into the development of a variety of strategies. These strategies include creating a cryogenic [7] or carbon-saturated inert gas atmosphere [8], applying vibration-assisted machining [9], depositing a protective coating on the cutting tool [10], and using a ceramic tool [11].

Taking into consideration ease and economical machine setup as well as great machining stability and reliability, vibration-assisted machining has been considered a promising technique with widespread use in the industry. Vibration-assisted machining enables superior cutting performance, such as smaller cutting forces, better surface finishing, higher cutting stability, and longer tool life [12,13]. Moriwaki and Shamoto [9] first used ultrasonic vibration-assisted cutting (UVAC, Conventional Ultrasonic Vibration Cutting, CUAC, or 1D UVC) for ultra-precision diamond turning of stainless steel. A mirror-like quality with surface roughness of 26 nm was achieved, and this demonstrated the technical feasibility of ultra-precision diamond machining of steel. Subsequently, the technique of ultrasonic elliptical vibration cutting (UEVC) was firstly proposed by Shamoto and Moriwaki [12] and was used to machine hardened steel [13]. The experimental results showed that smaller cutting forces and longer tool life were achieved by the UEVC method in comparison to those in both conventional cutting and CUVC.

The optical surface quality for which the required surface roughness can be roughly determined as less than 10 nm according to various optical functions [14,15] is vital for ensuring the functional performance of the workpiece for optical application. Zhang et al. [16] conducted face-turning experiments on hardened steel by using the UEVC method with poly-crystalline diamond (PCD) tools under various nominal cutting speeds, depths of cut, and feed rates. Mirror-like surfaces were obtained, and the surface roughness of 10 nm was achieved under an optimal combination of cutting parameters. However, a constant surface roughness of less than 10 nm is still challenging in UVAC or UEVC of steel with a common vibration frequency of less than 40 kHz.

To study the influence of vibration frequency on surface roughness, Klocke et al. [17] conducted ultrasonic vibration-assisted turning under various parameters on steel with hardnesses of both 35 HRC and 50 HRC at vibration frequencies of both 40 kHz and 60 kHz. By comparing the obtained surface roughness, they found that the surface roughness at a vibration frequency of 60 kHz was more stable and smaller (most are below 10 nm). This may be due to the fact that a smaller amplitude is required for the same contact ratio between the tool and workpiece during the cutting process by increasing the vibration frequency from 40 kHz to 60 kHz, leading to a stable cutting process and thus higher surface quality. More importantly, the stable cutting process is due to the smaller cutting force resulting from a smaller upfeed stroke in each vibration cutting cycle by increasing the vibration frequency.

Based on this, ultrasonic tooling systems with operating frequencies of 80 kHz [18] and 104 kHz [19] were developed by the German company Son-x GmbH and have been used in diamond turning of spherical and aspherical surfaces on steel with sub-micrometre form accuracy and nanometric surface roughness. Unlike the conventional ultrasonic tooling system in which the sonotrode vibrates in the longitudinal mode or the first-order bending mode, the sonotrode in the 104 kHz vibration system works in a multi-order bending mode. This mode produces a primary transversal vibration in the cutting direction as well as a secondary longitudinal vibration in the thrust direction, both of which are fused into a new vibration trajectory [19].

Although sub-micrometre form accuracy and nanometric surface roughness can be achieved with the newly developed 104 kHz ultrasonic tooling system, there are few studies that systematically investigate the effect of cutting parameters on surface roughness with this vibration setup so the setup's machining stability and reliability need to be further examined. Moreover, little research on the surface finish in ultrasonic vibration-assisted cutting of steel considers the effects of the machining geometry of the workpiece. Therefore, this study aimed to investigate the effect of cutting parameters such as nominal cutting speed, feed rate, tool nose radius, and vibration amplitude on the surface roughness of the workpiece by making use of the Taguchi orthogonal experiment. In addition, the surface roughness of spherical moulds with various slope degrees were measured and analysed to explore the effect of the machined geometry on surface quality in the ultrasonic vibration-assisted cutting of spherical optical steel moulds.

2. Working Principle and Cutting Mechanics Of UVAC

2.1. Cutting Mechanics

In ultrasonic vibration-assisted cutting, the cutting tool is equipped with vibration moving up and down at an ultrasonic frequency, parallel to the cutting direction. The vibration displacement and the vibration speed can be determined by Equations (1) and (2), respectively [20]:

$$x = A sin(2\pi f t) \tag{1}$$

$$x' = 2\pi f A cos(2\pi f t) \tag{2}$$

where A and f are the vibration amplitude and vibration frequency, respectively. If the maximal speed of vibration $(2\pi f A)$ is larger than the nominal cutting speed v_c, the separation between the rake face of the cutting tool and the cutting zone of the workpiece occurs. This separation is a characteristic of ultrasonic vibration-assisted cutting that can enhance lubrication and cooling conditions and thus reduce the cutting forces and tool wear, thereby improving the cutting stability and the surface finish [12,13].

To better understand the cutting process of UVAC, Figure 1a is a schematic diagram of the relative position between the tool tip and the workpiece in the UVAC process. Several important time moments at *t*1, *t*2, *t*3, *t*4, and *t*1′ in one cutting cycle are discussed and their corresponding dynamics are shown in Figure 1b. The three curves represent the vibration displacement, vibration speed, and nominal cutting speed relative to the static coordinate system. At the moments of *t*1 and *t*2, the vibration speed and the nominal cutting speed are equal. *t*1 and *t*2 can be determined by solving Equation (3) as follows [20]:

$$2\pi f A cos(2\pi f t) = -v_c \tag{3}$$

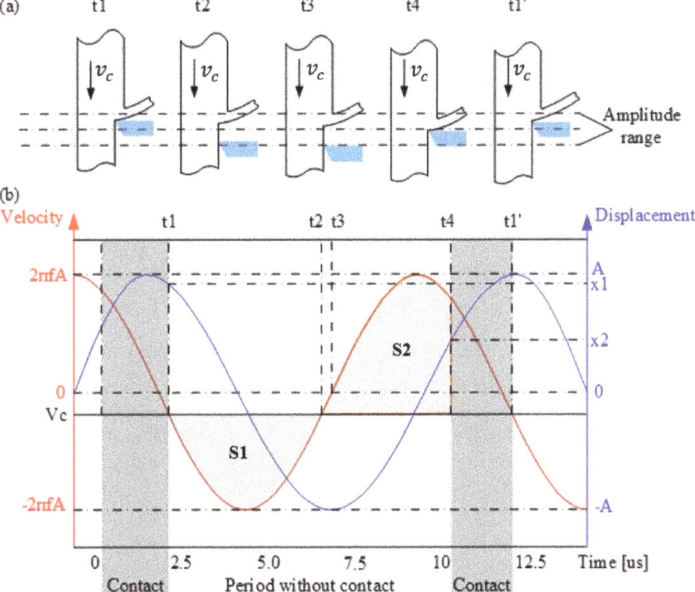

Figure 1. (**a**) Schematic diagram of the position relationship between the tool and the workpiece in UVAC. (**b**) The kinematics of UVAC.

At $t1$, the cutting tool starts to separate from the cutting zone of the workpiece, and the corresponding vibration displacement is $x_2 = Asin(2\pi ft1)$. At $t2$, the distance between the tool and the workpiece reaches the maximum value, which is the area of S1, as shown in Figure 1b. After t2, the distance between the tool and the workprice starts to decrease. At $t3$, the vibration motion reaches the lowest point, and then starts to move upward. The area of S2 as shown in Figure 1b shows a decreasing distance between the tool and the cutting zone of the workpiece after $t2$. When the areas of these two parts are equal (b1 = b2), at $t4$, the tool contacts the cutting zone of the workpiece again for cutting until $t1'$, and then one cutting cycle T is completed. After this, the tool and the cutting zone of the workpiece separate again and enter the next cutting cycle. The vibration position at $t4$ is $x_2 = Asin(2\pi ft4)$.

From $t1$ to $t4$, the displacement of vibration is Δx:

$$\Delta x = x_1 - x_2 = Asin((2\pi ft1) - Asin(2\pi ft4) \tag{4}$$

The displacement of the workpiece is $s = v_c(t4 - t1)$. The condition for the tool contacting the cutting zone of the workpiece again is $\Delta x = s$:

$$\Delta x = s \rightarrow Asin(2\pi ft1) - Asin(2\pi ft4) = v_c(t4 - t1) \tag{5}$$

The time point $t4$ can be solved numerically. To be specific, it can be solved by gradually increasing the number of time units dt, namely n, starting from $t1$. When $Asin(2\pi ft1) - Asin(2\pi f(t1 + ndt)) = v_c ndt$ is satisfied within the error tolerance, it is assumed that $n = N$. Therefore, $t4 = t1 + Ndt$.

The contact ratio of the tool tip with the cutting zone of the workpiece, namely 'Duty Cycle (DC)', is expressed as follows [20]:

$$DC = \frac{T - (t4 - t1)}{T} = (T - Ndt)f \tag{6}$$

Given that the vibration frequency f is constant, the speed ratio of nominal cutting speed v_c to the maximal vibration speed $2\pi f A$ can be varied by changing the value of v_c or A. When the speed ratio is ascertained, the corresponding duty cycle can be determined by solving Equations (3)–(6). In this way, the duty cycle for various speed ratios can be calculated. Figure 2 shows the duty cycle as a function of the speed ratio of the nominal cutting speed to the maximal vibration speed. The duty cycle decreases with decreasing speed ratio. It is believed that a smaller DC means better lubrication and cooling, which leads to a smaller cutting force and better surface quality [21]. In UVAC, under constant nominal cutting speed v_c, a higher vibration frequency leads to a smaller speed ratio and thus a smaller DC at a constant vibration amplitude or a smaller vibration amplitude at a constant DC. Both result in stabilised vibration and better surface quality.

2.2. The Applied High-Frequency UVAC

The vibration equipment Son-X UTS2 was used in this research. This ultrasonic tooling system was designed and described in the previous literature [19]. The transducer and sonotrode are actuated at the resonant frequency of 104 kHz, and the sonotrode works in the multi-order bending mode. During the cutting stage, the tool tip has both transversal vibration (blue line) and longitudinal vibration (red line), with a phase difference of 180°, as shown in Figure 3a. The amplitude ratio of the transversal and longitudinal vibration is about 5:1 (i.e., the longitudinal vibration amplitude is 200 nm when the transversal vibration amplitude is 1 μm). The differences in phase and amplitude between transversal and longitudinal vibration are finally fused into a slight-incline vibration track relative to the transversal direction, namely the cutting direction, and this vibration leads to a saw-like trajectory of the tool tip relative to the workpiece under a nominal cutting speed v_c, as shown in Figure 3b. Compared with UVAC, it is interesting to note that the tool tip includes an additional component motion in the thrust direction. This enables the tool tip to completely separate from the finished surface when the tool is withdrawn from the workpiece. This characteristic can facilitate contact between the coolant and lubricant, and the flank face of the cutting tool, or contact between oxygen in the air and the freshly cutting surface when cutting ferrous materials [22,23], thereby reducing the tool wear rate.

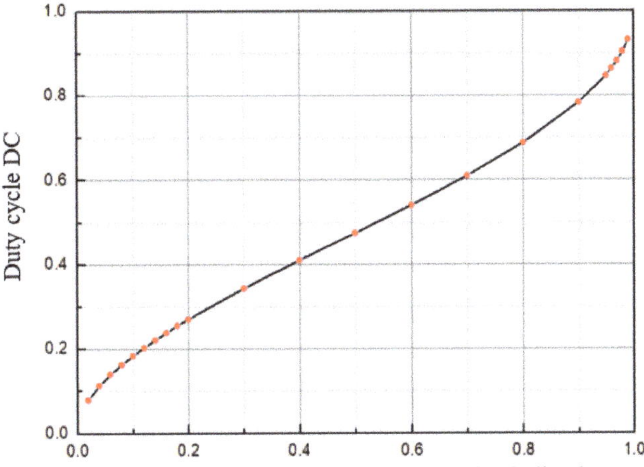

Figure 2. Relationship between the duty cycle and the speed ratio.

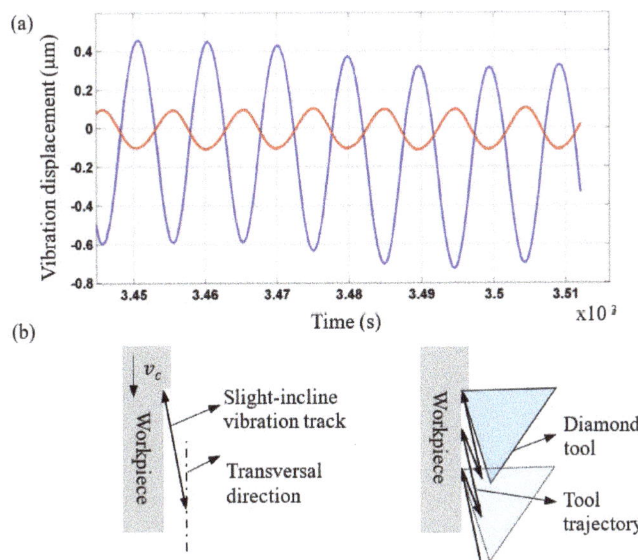

Figure 3. (**a**) Diamond tool tip transversal vibration displacement (blue line) and longitudinal vibration displacement (red line) during the cutting stage [19]. (**b**) Schematic of the tool tip movement of the slight-incline UVAC.

3. Experiment

3.1. Machining Dimension and Material of the Moulds

Figure 4a,b show diagrams of a concave and convex spherical mould, respectively. Both have a spherical surface with a radius of 15 mm. The concave mould has an aperture of 12 mm and a maximum slope degree of 23.6°. The convex mould has an aperture of 15 mm and a maximum slope degree of 30°.

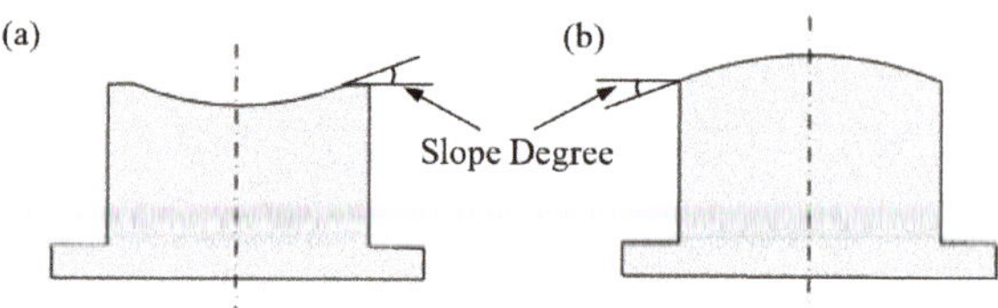

Figure 4. Diagram of the spherical moulds: (**a**) concave and (**b**) convex.

The geometry of the spherical surface could have an effect on the cutting force components as well as the surface roughness of the machined workpiece since it leads to variation in the cutting position of the tool edge. Taking the cutting process of the concave mould as shown in Figure 5, during the cutting process, the cutting edge of the diamond tool moves inwards to the center of the workpiece along the spherical surface. In this process, the point on the workpiece in contact with the cutting tool transfers from E to D, C, and B, and then to A, which has a slope degree of 20°, 15°, 10°, 5°, and 0°, respectively. This process also results in variation in the cutting point on the tool and their cutting angles θ, namely the angle between the symmetric line of the tool rake face and the normal line of the tool edge at these points, which are correspondingly 20°, 15°, 10°, 5°, and 0°. The variation in the cutting position and the cutting angle could possibly induce a difference in the cutting force and surface quality, which is worth studying.

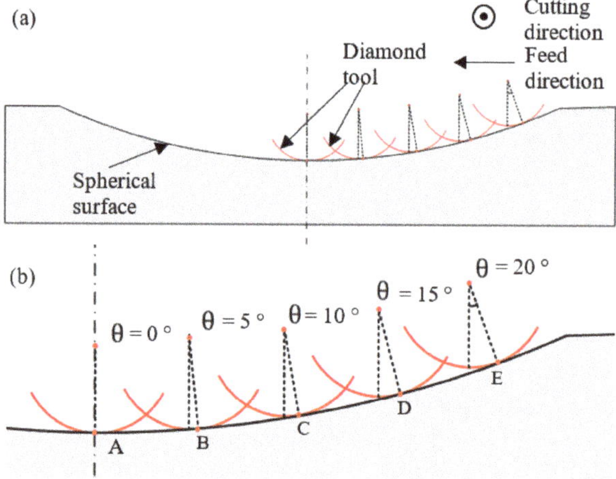

Figure 5. (**a**) Schematic diagram of cutting a concave spherical mould. (**b**) A partial enlarged view of (**a**).

The material used in this study was Mirrax 40 steel. Mirrax 40 is a remelted stainless steel prehardened to 40 HRC. It has excellent machinability, polishability, ductility, and toughness, which make it appropriate for injection moulds, extrusion dies, and flow moulds. The chemical composition of Mirrax 40 steel includes (wt.%): C: 0.21, Si: 0.9, Mn: 0.45, Cr: 13.5, Mo: 0.2, Ni: 0.6, and V: 0.25.

3.2. Procedures and Setup

The spherical surface turning experiments were conducted on an ultra-precision machine (Nanotech 450 from Nanotechnology Inc., Los Angeles, CA, USA) equipped with an ultrasonic vibration-assisted diamond machining system (i.e., Son-X UTS2 from Son-X GmbH Germany), which was installed on the Z-axis guide, as shown in Figure 6. In each experiment, the constant surface speed (CSS) mode was utilised. The spindle rotational speed increased steadily to keep the nominal cutting speed constant at different areas on the spherical surface with various diameters when the cutting tool moved inwards. Since the spindle rotational speed could not increase infinitely with the cutting tool moving close to the centre of the workpiece, it was kept constant until the cut was finished after the spindle speed reached 300 rpm. A rough cut of 10 μm per rev, 5 μm depth-of-cut, and 4 m/min cutting speed were firstly conducted to eliminate the dimensional error of the workpiece, followed by a finishing cut under specific parameters. After the finishing cut, the surface quality of the machined mould was evaluated and the surface roughness was measured using the New Zygo NexView Optical Profiler (USA). To directly and accurately measure the surface roughness of the areas at different slope degrees on the mould surface, an angle inclinometer was used for rotation of the workpiece. The arithmetic roughness S_a of the areas with the slope degrees of 0°, 5°, 10°, and 15° were measured and analysed.

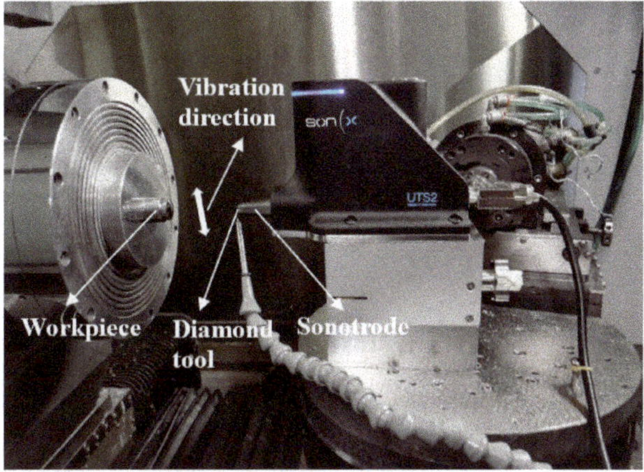

Figure 6. Photo of the machining setup.

3.3. Experimental Design Using the Taguchi Method

The Taguchi method [2,3,24] is an engineering approach used to investigate the effects of process parameters on the target performance for optimisation of these parameters. Designing an orthogonal array test can greatly reduce the experimental trials and then quickly identify key parameters. The possible key parameters in ultrasonic vibration-assisted cutting that may have significant effects on surface roughness are the tool radius (TR), cutting speed (CS, namely nominal cutting speed), vibration amplitude, feed rate (FR), and surface form. For the Son-X UTS2, the vibration amplitude of the vibrator is associated with the input current (IC). The larger the input current, the larger the vibration amplitude. As a result, the five parameters including TR, CS, IC, FR, and surface form were identified as five experimental factors for ultrasonic vibration-assisted cutting of the spherical moulds in this study. Based on the previous study, three levels for each factor (except the surface form with two levels of concave and convex) were selected to cover the research interest, as shown in Table 1. The rake angle and clearance angle of the cutting tool were 0° and 10°. The peak-to-peak vibration amplitudes were approximately 1.2, 1.8, and 2.4 μm in response to the input currents of 20, 30, and 40 mA, respectively.

Table 1. The experimental factors and their levels.

Factor	Level 1	Level 2	Level 3
A. Tool Radius TR (mm)	0.3	0.5	1.0
B. Cutting Speed CS (m/min)	2	3	4
C. Input Current IC (mA)	20	30	40
D. Feed Rate FR (μm/rev)	2.5	5	10
E. Surface form	Concave	Convex	

3.4. Data Analysis for the Taguchi Method

The target optimisation can be categorised into nominal-the-best type, the-smaller-the-better type, the-larger-the-better type, etc. Generally, the signal-to-noise ratio (S/N) is used as the objective function for optimising the process parameters. It should be noted that the equation for S/N could vary from one optimisation type to another. As the surface roughness of the spherical surface cutting process is a the-smaller-the-better type problem, S/N is defined by Equation (7) [24] as follows:

$$S/N = -10log_{10}[\frac{1}{n}\sum_{i=1}^{n} y_i^2] \tag{7}$$

where y_i is the value of the target characteristic under different noise conditions and n is the number of noise conditions.

In this experiment, y_i is the arithmetic roughness S_a of different areas on the mould surface, which are characterised by the slope degree, and n is the number of the measured slope degrees. To optimise the parameters for spherical surface cutting, levels that maximise S/N should be selected for the factors that have a significant effect on surface roughness.

4. Results and Discussions

Table 2 summarises the measured arithmetic roughness S_a at the four slope degrees on the machined spherical moulds, S/N, and $\overline{S_a}$ (the mean of S_a) for each L_{18} orthogonal array experiment. By comparing these values, the effect of the levels for each factor on surface roughness can be identified and the optimal level for each factor can be determined.

Table 2. Surface roughness of the machined spherical moulds.

Test No.	TR	CS	IC	FR	Surface Form	S_a(nm) SD = 0°	S_a(nm) SD = 5°	S_a(nm) SD = 10°	S_a(nm) SD = 15°	S/N	$\overline{S_a}$ (nm)
1	0.3	2	20	2.5	Convex	5.3	5.3	10.9	5.4	−17.1	6.7
2	0.3	4	30	10	Convex	12.5	11.0	12.1	11.5	−21.4	11.8
3	0.3	3	40	5	Convex	12.4	5.4	5.5	5.5	−17.9	7.2
4	0.3	2	20	2.5	Concave	6.1	6.4	9.5	10.9	−18.6	8.2
5	0.3	4	30	10	Concave	13.3	11.6	12.9	12.5	−22.0	12.6
6	0.3	3	40	5	Concave	8.1	5.4	6.1	5.8	−16.2	6.4
7	0.5	4	40	2.5	Concave	7.3	5.9	7.3	7.7	−17.0	7.0
8	0.5	2	30	5	Concave	5.7	5.9	6.5	7.4	−16.1	6.3
9	0.5	3	20	10	Concave	8.0	8.4	9.7	10.5	−19.3	9.1
10	0.5	4	40	2.5	Convex	6.8	5.9	9.8	11.6	−18.9	8.5
11	0.5	2	30	5	Convex	6.9	6.0	9.3	10.7	−18.5	8.2
12	0.5	3	20	10	Convex	13.9	8.6	9.5	9.7	−20.5	10.4
13	1	2	40	10	Convex	6.5	6.1	6.9	7.1	−16.5	6.7
14	1	3	30	2.5	Convex	5.2	6.0	7.2	7.5	−16.3	6.5
15	1	4	20	5	Convex	5.6	6.0	7.9	7.5	−16.7	6.8
16	1	2	40	10	Concave	8.0	8.3	8.4	8.5	−18.4	8.3
17	1	3	30	2.5	Concave	5.8	7.2	9.8	12.2	−19.2	8.8
18	1	4	20	5	Concave	5.7	6.9	9.4	10.3	−18.3	8.1

4.1. Surface Roughness Analysis

4.1.1. Average Arithmetic Roughness $\overline{S_a}$ for Various Levels of Each Factor

Figure 7 shows the average arithmetic roughness $\overline{S_a}$ for various levels of each factor. For the level of 0.3 mm of the tool radius, its value is determined as the mean value of $\overline{S_a}$ with the same tool radius. (i.e., (6.7 + 11.8 + 7.2 + 8.2 + 12.6 + 6.4) nm/6 = 8.8 nm). It is found that the feed rate has a great effect on the arithmetic roughness, while surface form has the least, which could be negligible compared with other factors. Tool radius, cutting speed, and input current have almost the same difference in arithmetic roughness under the selected levels. In view of the effect of the level for each factor, cutting speed and feed rate display upward and down-then-upward trends, respectively. This finding is consistent with previous study [16]. A larger cutting speed causes less overlapping cutting cycles and thus a larger upfeed stroke in each vibration cycle, requiring a relatively higher material load to be removed. Moreover, a larger cutting speed leads to a higher ratio of the cutting speed to the maximal vibration speed, increasing the contact time between the tool and the workpiece, which reduces the lubrication performance and heat transfer during the cutting process. Both can induce a larger cutting force and hence poorer surface quality. Regarding the feed rate, it appears that a high feed rate results in a high material removal load and cutting force, causing instability and thus poor surface quality in the cutting process. By contrast, a too low feed rate may cause the uncut chip thickness to become comparative to the radius of the tool edge, which can induce significant ploughing and rubbing and thus lead to poor surface quality [16].

The tool radius exhibits a peculiar and interesting downward trend of arithmetic roughness with increasing level value. It is noted that the thrust force increases with increasing tool radius under the same cutting conditions. Nath et al. [25] found that a 0.6-mm tool radius achieved better surface quality than a smaller or a larger one when machining tungsten carbide using ultrasonic elliptical vibration cutting with a vibration frequency of 40 kHz. It was found that worse surface quality for a larger tool radius (0.8 mm) was caused by the increasing cutting force, leading to a break in the regenerative chatter suppression dynamics [26] and thus instability of the cutting process. This means that the tool radius can be improved without causing cutting instability to achieve smaller theoretical surface roughness if the vibration-assisted cutting technique can produce a smaller average force under the same cutting conditions. As shown in Figure 7, the optimal tool radius for minimum surface roughness in the applied 104 kHz UVAC technique increases to equal to or maybe larger than 1 mm. This appears to indicate that this technique may have a better cutting performance for surface finishes in the machining of steel. Moreover, the results show that all of the arithmetic roughness values under various levels of each factor are below 10 nm, which could be regarded as another justification for the superiority of the surface quality in this high-frequency ultrasonic vibration-assisted cutting technique.

Input currents exhibit an up-then-downward trend of arithmetic roughness with increasing level value. It is understandable that the surface quality can be improved by increasing the input current of the vibrator, which generates increasing vibration amplitude. Increasing the vibration amplitude leads to an increase in the maximal vibration speed and thus a decrease in the speed ratio and cutting duty cycle. This results in better lubrication as well as heat transfer and hence better surface quality. However, it seems peculiar that the surface roughness with a current of 20 mA is lower than that for 30 mA. One possible reason is that a smaller vibration amplitude makes the vibrator have better stiffness and stability so it undergoes less deflection when subjected to certain loading.

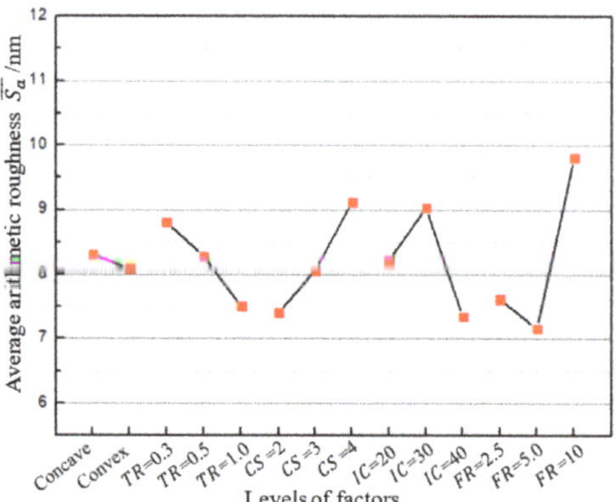

Figure 7. Average arithmetic roughness $\overline{S_a}$ of four slope degrees under various levels for each factor.

4.1.2. Arithmetic Roughness S_a for Each SD

To study the effect of various slope degrees on the surface roughness, Figure 8 summarises the arithmetic roughness S_a as a function of the slope degree of the workpiece. For the slope degrees 0°, 5°, 10°, and 15°, their values are determined as the mean value of S_a in all of the orthogonal array experiments at the same slope degrees of 0°, 5°, 10°, and 15°, respectively. It was found that the surface roughness exhibits an upward trend with increasing slope degree, except at the slope degree of 0°. The arithmetic roughness at 0° is higher than that for 5°. One possible reason for this is that the nominal cutting speed keeps changing with decreasing the cutting diameter in the central area of the workpiece because the rotational speed of the spindle remains constant after reaching 300 rpm in this area. This constant change in cutting speed could result in instability of the cutting process, reducing the surface quality. Another reason is that the surface morphology in this area is more likely to be scratched and influenced by the flow chips. Excluding this special area, the arithmetic roughness of other slope degrees exhibited an upward trend. This machining phenomenon is not preferable to produce a homogeneous surface finish on the surface with geometry of great difference in the slope degree so the underlying reasons for this need to be analysed for the purpose of suppression.

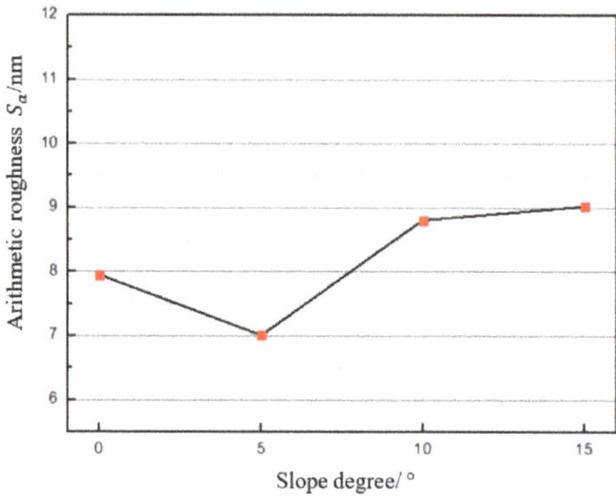

Figure 8. Arithmetic roughness S_a with respect to various slope degrees.

An attempt was made to determine the possible reasons for this phenomenon. One could be due to the variation in tool force under various tool cutting angles due to varying slope degrees of the workpiece. With the increase in slope degree of the workpiece during the cutting process, the cutting point on the tool edge changes and thus the cutting angle between the symmetric line of the tool rake face and the normal line of the tool edge at the cutting point increases as mentioned in Section 3.1. This leads to an increase in the lateral force loading on the cutting tool, which could probably cause cutting instability due to the relatively weak resistance to loading of the vibrator in the lateral direction, thus resulting in deterioration of the surface quality. To verify this speculation, tool force components during the cutting process at various cutting angles need to be measured and analysed.

To investigate the effect of cutting angle on the tool force during the cutting process, the tool force at the cutting angles of 0° and 15° was measured and compared. Figures 9 and 10 show a schematic and a photo of the experimental setup. The cutting tool with a nose radius of 1 mm and a rake angle of 0° was used. In the case of the cutting angle of 0°, the tool was installed with the symmetric line of the tool rake face parallel to the Z-axis of the machine. In contrast, in the case of the cutting angle of 15°, the cutting tool together with the force sensor were clockwise rotated along the Y-axis by 15° based on the condition of 0° by the B-axis of the machine (Top view). The desired nominal cutting speed, depth of cut, and feed rate of the practical cutting condition were 3 m/min, 5 μm, and 10 μm/rev, respectively. It is noted that it is impossible to acquire the actual transient tool force in the UVAC process as the vibration frequency is much larger than the natural frequency of the force sensor. As a result, the UVAC process was imitated by programming its tool trajectory as a conventional UVAC at a low frequency [27]. The imitated vibration frequency was set as 0.5 Hz with a peak-to-peak amplitude of 2 μm, and the actual nominal cutting speed was set as 14.4 μm/min to keep the upfeed stroke in each vibration cycle identical to that in the practical UVAC process.

In the experiment, face turning was conducted to flatten the workpiece. Following this, the cutting tool moved towards the workpiece at the given diameter by Z-axis until the desired depth of cut and started cutting for an rotation angle of 90° to make the cutting process stable. After this, a programmed grooving test was conducted in the C-axis mode and the tool force was acquired by the Kistler force senser 9256C1. According to Arcona's force model [28], the tool force is proportional to the material hardness of the workpiece. To avoid excessive tool wear in low-frequency vibration cutting of steel while achieving a tool force identical to that in the machining of steel, electroless plated nickel with a hardness close to that of steel was selected as the cutting material. The measured tool force at the cutting angles of 0° and 15° in the grooving tests with 3 vibration cycles is presented in Figure 11a,b, respectively.

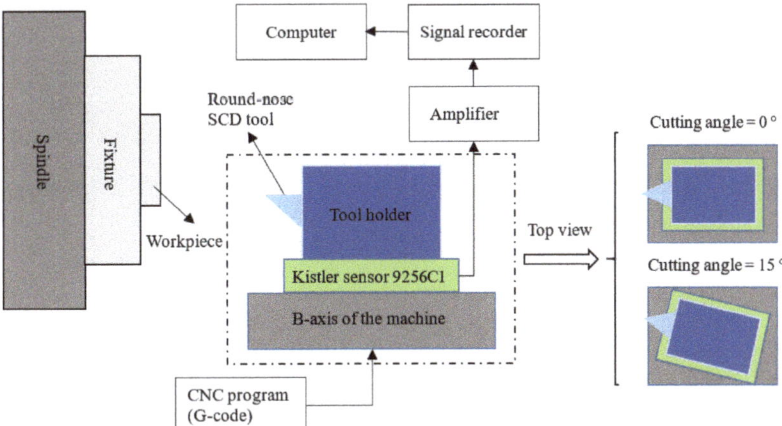

Figure 9. Schematic of the experimental setup for the low-frequency UVAC.

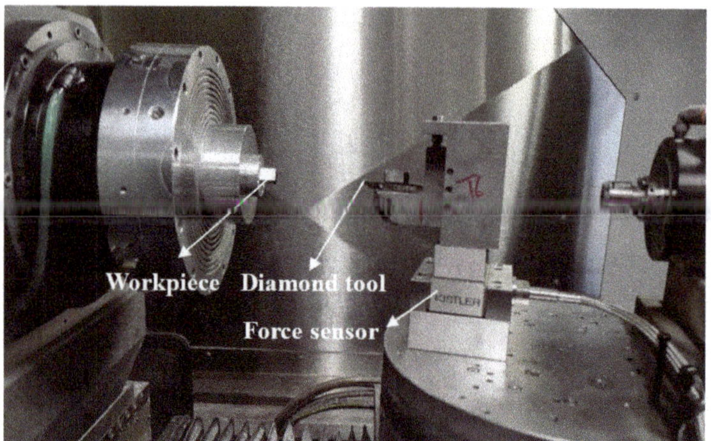

Figure 10. Photo of the machining setup for the low-frequency UVAC.

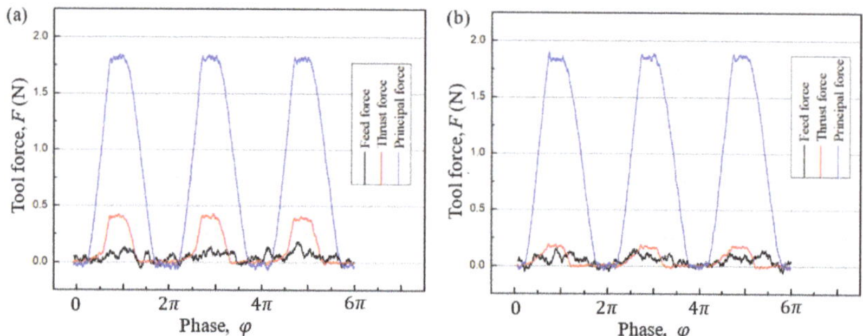

Figure 11. Tool force measured in the programmed grooving tests at various cutting angles: (**a**) 0° and (**b**) 15°, f = 0.5 Hz, A = 1 μm, v_c = 14.4 μm/min, depth of cut = 5 μm, and feed rate = 10 μm/rev.

With an increase in the cutting angle from 0° to 15°, the principal force remains constant, the thrust force decreases observably while the feed force increases slightly. Although this changing trend of tool force is consistent with our previous conjecturing, the thrust force and especially the feed force in the grooving test are within a small range (<0.5 N). The tool force in the turning machining process in which there is a smaller material removal rate is even smaller compared with the grooving test. As a result, the variation in tool force for the turning machining due to increasing cutting angle may have little influence on the cutting stability and therefore should not be regarded as the primary factor for declining surface quality at high cutting angles.

Another possible reason could be the variation in vibration amplitude along the tool edge. As the applied ultrasonic tooling system works in the multi-order bending mode, the sonotrode undergoes bending deflection during the vibrating process, as presented in Figure 12a. Consequently, the tool tip vibrates with the largest amplitude while the vibration amplitude decreases gradually from the centre of the tool edge to both of its sides, as shown in Figure 12b. The largest vibration amplitude at the central part of the tool edge allows for better machining conditions for lubrication and heat transfer to achieve a better surface finish. During UVAC of spherical steel moulds, the central part of the tool is used to machine the geometry with a small slope degree while the sideward part is used to machine the geometry with a large slope degree. As a result, the surface roughness of the machined spherical steel mould increases with increasing slope degree.

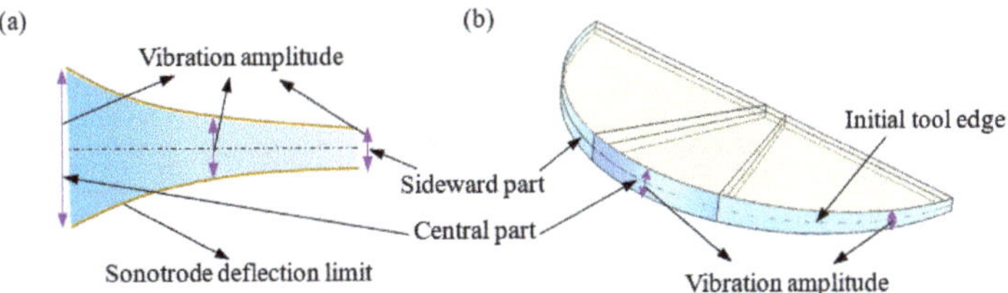

Figure 12. Schematic diagram of (**a**) the sonotrode deflection and (**b**) tool vibration amplitude distribution along the tool edge in the bending vibration mode.

4.2. Optimisation of the Level for Each Factor

To minimise the arithmetic roughness value of the machined spherical mould, the level for each factor that can maximise the S/N ratio should be selected. Figure 13 plots the S/N ratio of all levels for each factor. In practice, long-time use of UTS2 at the current of 40 mA possibly causes cutting instability, so the current of 20 mA is considered a better parameter value. As a result, the optimal combination of the level for each factor is determined to be 1 mm for tool radius, 2 m/min for cutting speed, 20 mA for input current, and 5 μm per rev for feed rate. Consequently, the convex mould was machined at the optimised parameters. Figure 14 shows the surface morphology and measured roughness of the machined convex mould at various slope degrees under optimised parameters. The arithmetic roughness S_a of 3–4 nm could be achieved for all slope degrees. Furthermore, the arithmetic roughness S_a is so small that the effect of the slope degree becomes insignificant. This means that the unevenness of the surface finish in machining the geometry with a great difference in slope degree using this ultrasonic tooling system can be alleviated by using an optimal combination of cutting parameters.

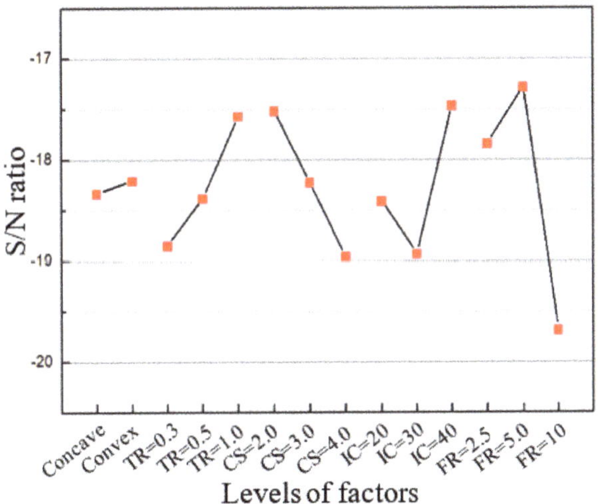

Figure 13. S/N ratio of levels for each factor.

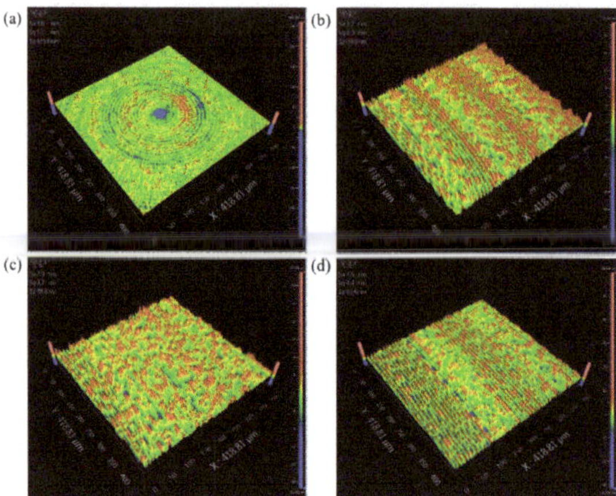

Figure 14. Surface morphology and measured roughness of the convex mould at different slope degrees under optimised parameters: (**a**) 0°, (**b**) 5°, (**c**) 10°, and (**d**) 15°.

5. Conclusions

This study investigated the effect of cutting parameters as well as the geometry of a workpiece on the surface roughness using high-frequency ultrasonic vibration-assisted cutting. The ultra-precision machined spherical moulds in this study were made of steel. Some findings are given as follows:

1. Feed rate has the greatest effect on the surface roughness of the workpiece: 5 um/res was able to achieve better surface roughness compared with 10 um/res and 2.5 um/res. Under the selected levels of the other factors, surface roughness decreased with increasing tool radius while it increased with increasing nominal cutting speed. The effect of input current on surface roughness remains unclear based on the obtained results.
2. To achieve a better surface finish, the optimal combination of the level for each cutting parameter is 1 mm for tool radius, 2 m/min for cutting speed, 20 mA for input current, and 5 μm per rev for feed rate.
3. Arithmetic roughness on the machined spherical steel mould exhibited a slight upward trend with an increase in slope degree from 5° to 15°, and the variation in vibration amplitude distribution along the tool edge arising in the bending vibration mode is speculated to be the possible reason.
4. Under optimal cutting conditions, the optical surface quality, with an arithmetic roughness S_a of 3–4 nm, can be achieved at various slope Degrees. Furthermore, the unevenness of surface finishes at various slope degrees was alleviated in this ultrasonic tooling system with a combination of optimal parameters.

The results demonstrate the technical merits of using high-frequency ultrasonic vibration-assisted cutting in the machining of steel material with a superior surface finish. Further research will be conducted to continuously examine and verify the reason for the unevenness of surface quality in the machining of workpieces with complex geometries where there is a great difference in the slope degree to identify a strategy to improve the vibration setup.

Author Contributions: C.Z. (Chenyang Zhao), B.B., and C.C. conceived and designed the methodology and experiments; C.C. and B.B. contributed to the materials and machining tools; C.Z. (Canbin Zhang) and C.Z. (Chenyang Zhao) performed the experiments; C.Z. (Canbin Zhang) analysed the measurement data and wrote the draft on the technical content of the manuscript; C.C. reviewed and revised the detailed content of the manuscript. All authors have read and agreed to the published version of the manuscript.

Funding: This research received no external funding.

Acknowledgments: The authors would like to express their sincere thanks for the financial support from the Research Office (project code: RK2Z) from The Hong Kong Polytechnic University. Special thanks are also due to the contract research project between the State Key Laboratory of Ultra-precision Machining Technology of The Hong Kong Polytechnic University and Son-X, Gmbh, Aachen, Germany.

Conflicts of Interest: The authors declare no conflicts of interest.

Nomenclature

The following abbreviations or physical quantities are used in this manuscript:

UVAC	Ultrasonic vibration-assisted cutting
SD	Slope degree
S_a	Arithmetic surface roughness
$\overline{S_a}$	Average arithmetic surface roughness
CBN	Cubic boron nitride
SPDT	Single point diamond turning
CUAC or 1D UVC	Conventional ultrasonic vibration cutting
UEVC	Ultrasonic elliptical vibration cutting
PCD	Poly-crystalline diamond
f	Vibration frequency
A	Vibration amplitude
v_c	Nominal cutting speed
DC	Duty cycle
TR	Tool radius
CS	Cutting speed, namely nominal cutting speed
IC	Input current
FR	Feed rate
S/N	Signal-to-noise ratio
CSS	Constant surface speed

References

1. Vopát, T.; Sahul, M.; Haršáni, M.; Vortel, O.; Zlámal, T. The tool life and coating-substrate adhesion of AlCrSiN-coated carbide cutting tools prepared by LARC with respect to the edge preparation and surface finishing. *Micromachines* **2020**, *11*, 166. [CrossRef]
2. Kuntoğlu, M.; Aslan, A.; Pimenov, D.Y.; Giasin, K.; Mikolajczyk, T.; Sharma, S. Modeling of cutting parameters and tool geometry for multi-criteria optimization of surface roughness and vibration via response surface methodology in turning of AISI 5140 steel. *Materials* **2020**, *13*, 4242. [CrossRef]
3. Kumar, P.; Chauhan, S.R.; Pruncu, C.I.; Gupta, M.K.; Pimenov, D.Y.; Mia, M.; Gill, H.S. Influence of different grades of CBN inserts on cutting force and surface roughness of AISI H13 die tool steel during hard turning operation. *Materials* **2019**, *12*, 177. [CrossRef] [PubMed]
4. Ikawa, N.; Donaldson, R.R.; Komanduri, R.; König, W.; Aachen, T.H.; McKeown, P.A.; Moriwaki, T.; Stowers, I.F. Ultraprecision metal cutting—The Past, the Present and the Future. *CIRP Ann. Manuf. Technol.* **1991**, *40*, 587–594. [CrossRef]
5. Paul, E.; Evans, C.J.; Mangamelli, A.; McGlauflin, M.L.; Polvani, R.S. Chemical aspects of tool wear in single point diamond turning. *Precis. Eng.* **1996**, *18*, 4–19. [CrossRef]
6. Guo, J.; Zhang, J.; Pan, Y.; Kang, R.; Namba, Y.; Shore, P.; Yue, X.; Wang, B.; Guo, D. A critical review on the chemical wear and wear suppression of diamond tools in diamond cutting of ferrous metals. *Int. J. Extrem. Manuf.* **2020**, *2*, 12001.
7. Evans, C.; Bryan, J.B. Cryogenic diamond turning of stainless steel. *CIRP Ann. Manuf. Technol.* **1991**, *40*, 571–575. [CrossRef]
8. Casstevens, J. Diamond turning of steel in carbon-saturated Diamond turning of steel in carbon-saturated atmospheres. *Precis. Eng.* **1983**, *5*, 9–15. [CrossRef]
9. Moriwaki, T.; Shamoto, E. Ultraprecision diamond turning of stainless steel by applying ultrasonic vibration. *CIRP Ann. Manuf. Technol.* **1991**, *40*, 559–562. [CrossRef]
10. Coelho, R.T.; Ng, E.-G.; Elbestawi, M.A. Tool wear when turning hardened AISI 4340 with coated PCBN tools using finishing cutting conditions. *Int. J. Mach. Tools Manuf.* **2007**, *47*, 263–272. [CrossRef]
11. Aslan, E.; Camuşcu, N.; Birgören, B. Design optimization of cutting parameters when turning hardened AISI 4140 steel (63 HRC) with Al2O3+TiCN mixed ceramic tool. *Mater. Des.* **2007**, *28*, 1618–1622. [CrossRef]
12. Shamoto, E.; Moriwaki, T. Study on Elliptical Vibration Cutting. *CIRP Ann-Manuf. Technol.* **1994**, *43*, 35–38. [CrossRef]

13. Shamoto, E.; Moriwaki, T. Ultaprecision diamond cutting of hardened steel by applying elliptical vibration cutting. *CIRP Ann. Manuf. Technol.* **1999**, *48*, 441–444. [CrossRef]

14. Li, L.; Yang, G.; Lee, W.B.; Ng, M.C.; Chan, K.L. Carbide-bonded graphene-based Joule heating for embossing fine microstructures on optical glass. *Appl. Surf. Sci.* **2020**, *500*, 144004. [CrossRef]

15. Mukaida, M.; Yan, J. Ductile machining of single-crystal silicon for microlens arrays by ultraprecision diamond turning using a slow tool servo. *Int. J. Mach. Tools Manuf.* **2017**, *115*, 2–14. [CrossRef]

16. Zhang, X.; Senthil Kumar, A.; Rahman, M.; Nath, C.; Liu, K. Experimental study on ultrasonic elliptical vibration cutting of hardened steel using PCD tools. *J. Mater. Process. Technol.* **2011**, *211*, 1701–1709. [CrossRef]

17. Klocke, F.; Dambon, O.; Bulla, B.; Heselhaus, M. Ultrasonic assisted turning of hardened steel with mono-crystalline ultrasonic assisted turning of hardened steel with mono-crystalline diamond. In Proceedings of the 10th International Euspen Conference, Zürich, Switzerland, 18–22 May 2008; pp. 165–169.

18. Klocke, F.; Dambon, O.; Bulla, B. Tooling system for diamond turning of hardened steel moulds with apsheric or non rotational symmetrical geometries. In Proceedings of the 11th EUSPEN International Conference, Como, Italy, 23–26 May 2011.

19. Gaidys, R.; Dambon, O.; Ostasevicius, V.; Dicke, C.; Narijauskaite, B. Ultrasonic tooling system design and development for single point diamond turning (SPDT) of ferrous metals. *Int. J. Adv. Manuf. Technol.* **2017**, *93*, 2841–2854. [CrossRef]

20. Brehl, D.E.; Dow, T.A. Review of vibration-assisted machining. *Precis. Eng.* **2008**, *32*, 153–172. [CrossRef]

21. Moriwaki, T.; Shamoto, E.; Inoue, K. Ultraprecision ductile cutting of glass by applying ultrasonic vibration. *CIRP Ann-Manuf. Technol.* **1992**, *41*, 141–144. [CrossRef]

22. Zhang, X.; Deng, H.; Liu, K. Oxygen-shielded ultrasonic vibration cutting to suppress the chemical wear of diamond tools. *CIRP Ann. Manuf. Technol.* **2019**, *68*, 69–72. [CrossRef]

23. Zhang, X.; Liu, K.; Kumar, A.S.; Rahman, M. A study of the diamond tool wear suppression mechanism in vibration-assisted machining of steel. *J. Mater. Process. Technol.* **2014**, *214*, 496–506. [CrossRef]

24. Shiou, F.J.; Cheng, C.H. Ultra-precision surface finish of NAK80 mould tool steel using sequential ball burnishing and ball polishing processes. *J. Mater. Process. Technol.* **2008**, *201*, 554–559. [CrossRef]

25. Nath, C.; Rahman, M.; Neo, K.S. A study on the effect of tool nose radius in ultrasonic elliptical vibration cutting of tungsten carbide. *J. Mater. Process. Technol.* **2009**, *209*, 5830–5836. [CrossRef]

26. Xiao, M.; Karube, S.; Soutome, T.; Sato, K. Analysis of chatter suppression in vibration cutting. *Int. J. Mach. Tools Manuf.* **2002**, *42*, 1677–1685. [CrossRef]

27. Arefin, S.; Zhang, X.; Anantharajan, S.K.; Liu, K.; Neo, D.W.K. An Analytical Model for Determining the Shear Angle in 1D Vibration-Assisted Micro Machining. *Nanomanuf. Metrol.* **2019**, *2*, 199–214. [CrossRef]

28. Arcona, C.; Dow, T.A. An empirical tool force model for precision machining. *J. Manuf. Sci. Eng. Trans. ASME* **1998**, *120*, 700–707. [CrossRef]

Article

Multi-Field Coupling Dynamics Modeling of Aerostatic Spindle

Guoda Chen [1,2,3,*] and Yijie Chen [2,3]

1 State Key Laboratory of Fluid Power and Mechatronic Systems, Zhejiang University, Hangzhou 310027, China
2 College of Mechanical Engineering, Zhejiang University of Technology, Hangzhou 310023, China; chenyj@zjut.edu.cn
3 Key Laboratory of Special Purpose Equipment and Advanced Processing Technology, Ministry of Education and Zhejiang Province, Zhejiang University of Technology, Hangzhou 310023, China
* Correspondence: gchen@zjut.edu.cn

Abstract: The aerostatic spindle in the ultra-precision machine tool shows the complex multi-field coupling dynamics behavior under working condition. The numerical investigation helps to better understand the dynamic characteristics of the aerostatic spindle and improve its structure and performance with low cost. A multi-field coupling 5-DOF dynamics model for the aerostatic spindle is proposed in this paper, which considers the interaction between the air film, spindle shaft and the motor. The restoring force method is employed to deal with the times varying air film force, the transient Reynolds equation of the aerostatic journal bearing and the aerostatic thrust bearing is solved using ADI method and Thomas method. The transient air film pressure of aerostatic bearings is obtained which clearly presents the influence induced by the tilt motion of the spindle shaft. The motion trajectory of the spindle shaft is obtained which shows different stability of the shaft under different external forces. The dynamics model shows good performance on simulating the multi-field coupling behavior of the aerostatic spindle under external force. which is quite meaningful and useful for the further research on the dynamic characteristics of the aerostatic spindle.

Keywords: dynamics modeling; aerostatic spindle; rotor trajectory; stability; Reynolds equation

Citation: Chen, G.; Chen, Y. Multi-Field Coupling Dynamics Modeling of Aerostatic Spindle. *Micromachines* **2021**, *12*, 251. https://doi.org/10.3390/mi12030251

Academic Editors: Benny C. F. Cheung and Jiang Guo

Received: 10 February 2021
Accepted: 25 February 2021
Published: 1 March 2021

Publisher's Note: MDPI stays neutral with regard to jurisdictional claims in published maps and institutional affiliations.

1. Introduction

The aerostatic spindle plays a key role in an ultra-precision machine tool in the nano-precision machining, which directly affects the machining quality [1,2]. Generally, the shaft of the aerostatic spindle is directly driven by a permanent magnet synchronous motor (PMSM) and supported by the aerostatic bearing, including the journal bearing and the thrust bearing [3].

Compared to the aerodynamic bearing, the aerostatic bearing provides higher capacity and steady force, while it contains more complex mechanism contributes to the dynamic behavior of aerostatic spindle. During the operation of the aerostatic spindle, it shows the multi-physics field coupling characteristics, in which the coupling effect of air bearing, the shaft and the motor should be considered for fully understanding its dynamics behavior [4]. However, the previous literatures commonly focused on single factor or two factors while investigating the dynamic characteristics of the aerostatic spindle [5–8]. It has been pointed out that the existence of unbalanced magnetic force (UMF) caused by rotor eccentricity at the motor can have a considerable impact on the machined surface [9–11], thus, the electromagnetic factor is not negligible in the dynamics model of aerostatic spindle.

Rotor dynamics modeling is the preferred method to study the dynamic characteristics of a rotor-bearing system, by which we can get detailed information of how the rotor acts under different condition. Among the previous literature, the rotor dynamics modeling method was widely used for modeling the aerodynamic bearing-rotor system with 2-DOF (Degree of Freedom) [12–16], the rotor trajectory and the stability problem were analyzed.

Since performance superiority of aerostatic spindle emerges in recent years, the analysis on aerostatic spindle came out frequently and have made great progress. Wang [17] implemented the rotor dynamic model of aerostatic bearing-rotor systems which applied a spherical aerostatic bearing, and the nonlinear dynamic behavior of the rotor bearing system is analyzed. Similar works were done by Zhang et al. [18], they proposed forecast orbit method to deal with the transient gas lubricated Reynold equation, and the 2D rotor trajectory under different rotor speed and imbalance mass was obtained. Two-dimensional analysis can reflect the characteristics of aerostatic bearing to some extent but not as precise as the result of the 5-DOF model. In a 5-DOF model, the tilt motion, the displacement in axial direction and the conical movement of the shaft can be acquired while the 2-DOF model cannot, and these forms of movement are also proved to have significant influence on the machined surface quality by Zhang et al. [19]. To achieve this, Li et al. [20,21] presented the 5-DOF model of an aerostatic spindle in a fly-cutting machine, in which the air pressure distribution of journal bearing and thrust bearing was calculated and the displacement of the shaft in different directions was obtained. Xu and Jiang [22,23] analyzed the 5-DOF rotor dynamics model of an aerostatic spindle, in which the stability, the unbalance response, and the forced response of the rotor-bearing system were investigated. Both Li and Xu adopted the dynamic coefficient method in their model, however it may lose the numerical accuracy when taking this method which indeed largely improved the efficiency. By contrast, the restoring force commonly used in 2-DOF models aims to obtain force directly by solving the transient Reynold equation [24,25], and it can also obtain the torque in a 5-DOF model. It can be optimized of the dynamic coefficient method, but if this measure is taken, massive extra work needs to be done. Besides, the available research on aerostatic spindle modeling fail to consider the electromagnetic factor, which has important effect on the motion error of the spindle.

In this paper, the electromagnetic factor is considered in the modeling of the aerostatic spindle, and the restoring force method is adopted. The finite difference method is adopted to discrete the transient Reynolds equation. The simulation for the motion trajectory of the spindle shaft is realized under the coupling of the restoring force of the aerostatic bearing, the UMF of PMSM and the external force on the shaft, where the shaft is regarded as a rigid rotor.

2. Mathematical Modeling of Aerostatic Spindle

The typical structure diagram of the aerostatic spindle is shown in Figure 1, the spindle shaft is directly driven by the motor, and the shaft is supported by an aerostatic journal bearing and two aerostatic thrust bearing.

Effectively modeling the dynamic behavior of the air film is quite necessary for the systematic and consistent air bearing design [26,27], thus, the modeling of the air film is of primary importance. The pressure distribution of the aerostatic bearing air film can be described by the Reynolds equation, and the Reynolds equation is obtained based on the Navier-Stokes Equations with following assumptions, the flow is isothermal, the gas viscosity is assumed to be constant, the pressure distribution in the vertical direction to the air film is assumed to be constant, the viscosity force is assumed to be much larger than inertia force, there is no velocity slip at the boundary, and the air is the ideal gas.

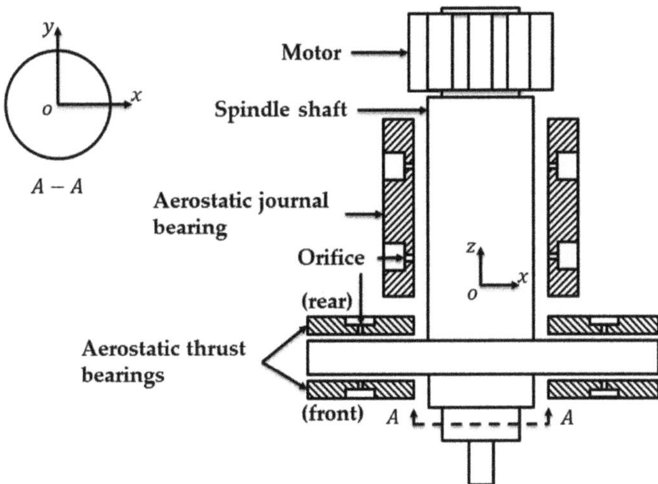

Figure 1. The structure diagram of the aerostatic spindle.

2.1. Modeling of Aerostatic Journal Bearing

The transient pressure distribution of aerostatic journal bearing can be modeled by the transient Reynolds equation in the Cartesian coordinate [28].

$$\frac{\partial}{\partial x}\left(ph^3\frac{\partial p}{\partial x}\right) + \frac{\partial}{\partial z}\left(ph^3\frac{\partial p}{\partial z}\right) + \frac{12\mu P_a}{\rho_a}\rho\tilde{v}\delta_k = 6\mu U_0\frac{\partial(ph)}{\partial x} + 12\mu\frac{\partial(ph)}{\partial t} \tag{1}$$

where x is the circumferential coordinates of journal bearing, z is the axial coordinate of journal bearing, h is the air film thickness, p is the air film pressure, μ is the dynamic viscosity of the air, U_0 is the surface velocity of the shaft, $\tilde{v}$ is the flow velocity of the supply air at the orifice, t represents the time, ρ is the air density, P_a and ρ_a are the pressure and density of the ambient air, $\delta_k = 1$ at the orifice and $\delta_k = 0$ at other position.

The dimensionless transient Reynolds equation is given by

$$\frac{\partial}{\partial\theta}\left(H^3\frac{\partial P^2}{\partial\theta}\right) + \frac{\partial}{\partial Z}\left(H^3\frac{\partial P^2}{\partial Z}\right) + Q\delta_k = \Lambda_J\frac{\partial(PH)}{\partial\theta} + 2\Lambda_J\frac{\partial(PH)}{\partial\tau} \tag{2}$$

The dimensionless parameters are defined as follows.

$$P = \frac{p}{P_s}, H = \frac{h}{C_r}, \theta = \frac{x}{L_j}, Z = \frac{z}{L_j}, Q = \frac{24\mu R^2 P_a}{C_r^3 P_s^2 \rho_a}\rho\tilde{v}, \tau = \omega t, \Lambda_J = \frac{12\mu\omega R L_j}{P_s C_r^2}$$

where L_j is the characteristic length of journal bearing, R is the radius of the journal bearing, C_r is the radial clearance between the journal bearing and the shaft, P_s is the supply pressure, Λ_J is the bearing number of the journal bearing, ω is the rotating speed of the shaft.

When the eccentricity or the tilt of the shaft happens as depicted in Figure 2, the thickness of the air film changes simultaneously, and the distribution of the air film thickness in the journal bearing can be expressed as

$$\begin{cases} h_{(i,j)} = C_r \cdot \left(1 + \varepsilon_{(i)} \cdot \cos\left(\theta_{(i,j)} - \alpha_{(i)}\right)\right) \\ H_{(i,j)} = 1 + \varepsilon_{(i)} \cdot \cos\left(\theta_{(i,j)} - \alpha_{(i)}\right) \end{cases} \tag{3}$$

$$\varepsilon_{(i)} = \varepsilon_{ib} + \frac{\sin(\beta) \cdot (i - i_b) \cdot L_s}{M_s} \tag{4}$$

$$\varepsilon_{ib} = \sqrt{(e_{ibx})^2 + \left(e_{iby}\right)^2} \tag{5}$$

$$\alpha_{(i)} = \begin{cases} arctg\frac{e_{iy}}{e_{ix}} & e_{ix} \geq 0 \\ arctg\frac{e_{iy}}{e_{ix}} + \pi & e_{ix} < 0 \end{cases} \tag{6}$$

where $\varepsilon_{(i)}$ is the eccentricity at the node i, $\theta_{(i,j)}$ is the angle of air film at node i and j, $\alpha_{(i)}$ is the eccentric angle at the node i, i_b is the node of the barycenter, L_s is the axial length of the shaft, M_s is the number of nodes that L_s divided into, ε_{ib} is the eccentricity at the node i_b, β is the tilt angle of the shaft, e_{ibx} and e_{iby} are the eccentricity of the node i_b in x direction and y direction respectively, e_{ix} and e_{iy} are the eccentricity of the node i in x direction and y direction respectively.

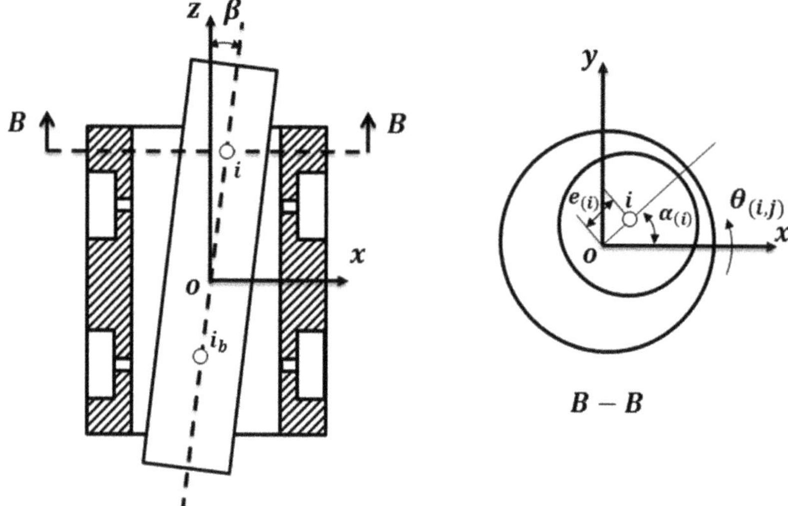

Figure 2. The tilt and the eccentricity of the shaft at the journal bearing.

The mass flow rate can be acquired based on the isothermal assumption given by Powell [29].

$$\dot{m} = \phi P_s A_{inlet} \sqrt{\frac{2\rho_a}{p_a}} \Psi \tag{7}$$

$$\Psi = \begin{cases} \left[\frac{k}{k-1}\left(\beta_i^{\frac{2}{k}} - \beta_i^{\frac{k+1}{k}} \right) \right]^{\frac{1}{2}} & \beta_i > \beta_k \\ \left[\frac{k}{2}\left(\frac{2}{k+1} \right)^{\frac{k+1}{k-1}} \right]^{\frac{1}{2}} & \beta_i \leq \beta_k \end{cases} \tag{8}$$

$$A_{inlet} = \begin{cases} \pi d C_r & \frac{d}{4} > C_r \\ \frac{\pi d^2}{4} & \frac{d}{4} \leq C_r \end{cases} \tag{9}$$

where ϕ is the coefficient of the mass flow rate, A_{inlet} is the minimum area of the flow channel, d is the diameter of the orifice, $\beta_i = p/P_s$ and $\beta_k = (2/(k+1))^{k/(k-1)}$.

As shown in Figure 3, the computational domain of the journal bearing film is meshed into $\theta(0 : N_j)$ and $Z(0 : M_j)$, the Periodic boundary condition, the Atmospheric boundary condition and the Mass flow boundary condition are also defined.

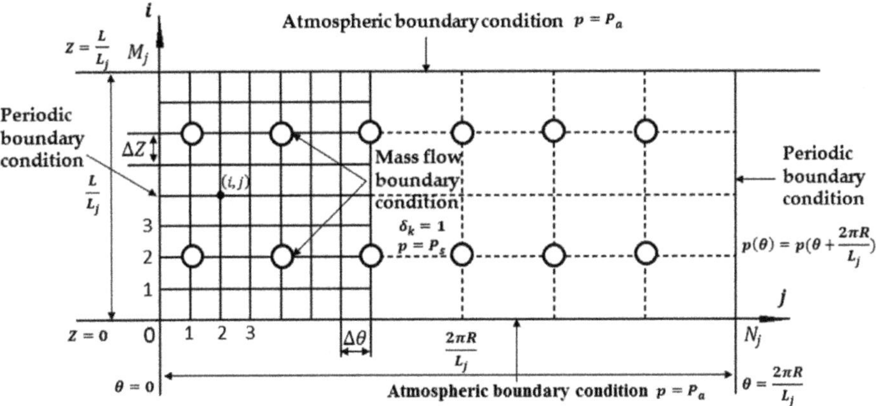

Figure 3. The computational domain of the journal bearing.

Submitting the expression $S = P^2$, the central difference method is adopted to discrete the variable in the θ and Z directions. The ADI method [18] is employed to simplify the implicit equation obtained after the discretization. The ADI format of the equation can be expressed as follows,

$$A_{(i,j)} S_{(i-1,j)}^{n+1} + B_{(i,j)} S_{(i,j)}^{n+1} + C_{(i,j)} S_{(i+1,j)}^{n+1} = D_{(i,j)} \tag{10}$$

$$A_{(i,j)} = -\frac{\left(H_{(i,j)}^n\right)^3}{(\Delta Z)^2} + 3\left(H_{(i,j)}^n\right)^2 \frac{H_{(i+1,j)}^n - H_{(i-1,j)}^n}{(2\Delta Z)^2} \tag{11}$$

$$B_{(i,j)} = \frac{2\left(H_{(i,j)}^n\right)^3}{(\Delta Z)^2} + \frac{\Lambda_J H_{(i,j)}^n}{P_{(i,j)}^n \Delta \tau} \tag{12}$$

$$C_{(i,j)} = -\frac{\left(H_{(i,j)}^n\right)^3}{(\Delta Z)^2} - 3\left(H_{(i,j)}^n\right)^2 \frac{H_{(i+1,j)}^n - H_{(i-1,j)}^n}{(2\Delta Z)^2} \tag{13}$$

$$\begin{aligned}
D_{(i,j)} - Q\delta_k + 3\left(H_{(i,j)}^n\right)^2 \frac{\left(H_{(i,j+1)}^n - H_{(i,j-1)}^n\right)\left(S_{(i,j+1)}^n - S_{(i,j-1)}^n\right)}{(2\Delta\theta)^2} + \\
\left(H_{(i,j)}^n\right)^3 \frac{\left(S_{(i,j+1)}^n - 2S_{(i,j)}^n + S_{(i,j-1)}^n\right)}{(\Delta\theta)^2} - \frac{\Lambda_J H_{(i,j)}^n}{2P_{(i,j)}^n} \frac{\left(S_{(i,j+1)}^n - S_{(i,j-1)}^n\right)}{2\Delta\theta} - \\
\Lambda_J P_{(i,j)}^n \frac{\left(H_{(i,j+1)}^n - H_{(i,j-1)}^n\right)}{2\Delta\theta} + \frac{\Lambda_J H_{(i,j)}^n S_{(i,j)}^n}{P_{(i,j)}^n \Delta\tau} - 2\Lambda_J P_{(i,j)}^n \frac{\left(H_{(i,j+1)}^{n+1} - H_{(i,j-1)}^n\right)}{\Delta\tau}
\end{aligned} \tag{14}$$

$$E_{(i,j)} S_{(i,j-1)}^{n+2} + F_{(i,j)} S_{(i,j)}^{n+2} + G_{(i,j)} S_{(i,j+1)}^{n+2} = I_{(i,j)} \tag{15}$$

$$E_{(i,j)} = -\frac{\left(H_{(i,j)}^{n+1}\right)^3}{(\Delta\theta)^2} - \frac{\Lambda_J H_{(i,j)}^{n+1}}{4P_{(i,j)}^{n+1}\Delta\theta} + 3\left(H_{(i,j)}^{n+1}\right)^2 \frac{H_{(i,j+1)}^{n+1} - H_{(i,j-1)}^{n+1}}{(2\Delta\theta)^2} \tag{16}$$

$$F_{(i,j)} = \frac{2\left(H_{(i,j)}^{n+1}\right)^3}{(\Delta\theta)^2} + \frac{\Lambda_J H_{(i,j)}^{n+1}}{P_{(i,j)}^{n+1}\Delta\tau} \tag{17}$$

$$G_{(i,j)} = -\frac{\left(H_{(i,j)}^{n+1}\right)^3}{(\Delta\theta)^2} + \frac{\Lambda_J H_{(i,j)}^{n+1}}{4P_{(i,j)}^{n+1}\Delta\theta} - 3\left(H_{(i,j)}^{n+1}\right)^2 \frac{H_{(i,j+1)}^{n+1} - H_{(i,j-1)}^{n+1}}{(2\Delta\theta)^2} \tag{18}$$

$$I_{(i,j)} = Q\delta_k + 3\left(H_{(i,j)}^{n+1}\right)^2 \frac{\left(H_{(i,j+1)}^{n+1} - H_{(i,j-1)}^{n+1}\right)\left(S_{(i,j+1)}^{n+1} - S_{(i,j-1)}^{n+1}\right)}{(2\Delta Z)^2} +$$

$$\left(H_{(i,j)}^{n+1}\right)^3 \frac{\left(S_{(i+1,j)}^{n+1} - 2S_{(i,j)}^{n+1} + S_{(i-1,j)}^{n+1}\right)}{(\Delta Z)^2} - \Lambda_J P_{(i,j)}^{n+1} \frac{H_{(i,j+1)}^{n+1} - H_{(i,j-1)}^{n+1}}{2\Delta\theta} - \tag{19}$$

$$2\Lambda_J P_{(i,j)}^{n+1} \frac{H_{(i,j)}^{n+2} - H_{(i,j)}^{n+1}}{\Delta\tau} + \frac{\Lambda_J H_{(i,j)}^{n+1} S_{(i,j)}^{n+1}}{P_{(i,j)}^{n+1}\Delta\tau}$$

The increment in the marching direction Z is carried out at the time step of $n + 1$, and the increment in the marching direction θ is carried out at the time step of $n + 2$. The Equation (10) and Equation (15) can be solved by the Thomas method, Then, the pressure distribution of the aerostatic journal bearing is obtained. Base on the load formulas given by Rowe [30], the air film force (i.e., the restoring force of the journal bearing) of the journal bearing in x and y direction is given by

$$f_{bjx} = P_s L_j^2 \int_0^{\frac{L}{L_j}} \int_0^{\frac{2\pi R}{L_j}} P(Z,\theta,\tau) \cos\left(\alpha_{(i)}\right) d\theta dZ \tag{20}$$

$$f_{bjy} = P_s L_j^2 \int_0^{\frac{L}{L_j}} \int_0^{\frac{2\pi R}{L_j}} P(Z,\theta,\tau) \sin\left(\alpha_{(i)}\right) d\theta dZ \tag{21}$$

And the torque of the air film force (i.e., the restoring torque of the journal bearing) on the barycenter of the spindle shaft respect to the x axis and y axis is given by

$$t_{bjx} = P_s L_j^2 L_s \frac{i - i_d}{M_s} \int_0^{\frac{L}{L_j}} \int_0^{\frac{2\pi R}{L_j}} P(Z,\theta,\tau) \sin\left(\alpha_{(i)}\right) d\theta dZ \tag{22}$$

$$t_{bjy} = P_s L_j^2 L_s \frac{i - i_d}{M_s} \int_0^{\frac{L}{L_j}} \int_0^{\frac{2\pi R}{L_j}} P(Z,\theta,\tau) \cos\left(\alpha_{(i)}\right) d\theta dZ \tag{23}$$

2.2. Modeling of Aerostatic Thrust Bearing

The transient pressure distribution of aerostatic thrust bearing can be modeled by the transient Reynolds equation in cylindrical coordinate.

$$\frac{1}{r}\frac{\partial}{\partial r}\left(rph^3\frac{\partial p}{\partial r}\right) + \frac{1}{r^2}\frac{\partial}{\partial\theta}\left(ph^3\frac{\partial p}{\partial\theta}\right) + \frac{12\mu P_a}{\rho_a}\rho\tilde{v}\delta_k = \frac{6}{r}\frac{\partial}{\partial\theta}(phV_0) + 12\mu\frac{\partial(ph)}{\partial t} \tag{24}$$

where θ is the circumferential coordinates of journal bearing, r is the radial coordinate of journal bearing, V_0 is the surface velocity of the shaft.

The dimensionless transient Reynolds equation is given by

$$\frac{\partial}{\partial\theta}\left(H^3\frac{\partial P^2}{\partial\theta}\right) + R_r\frac{\partial}{\partial R_r}\left(R_r H^3\frac{\partial P^2}{\partial R_r}\right) + Q\delta_k = \Lambda_T\frac{\partial(PH)}{\partial\theta} + 2\Lambda_T\frac{\partial(PH)}{\partial\tau} \tag{25}$$

The dimensionless parameters are defined as follows.

$$P = \frac{p}{P_s}, H = \frac{h}{C_r}, R_r = \frac{r}{L_t}, Q = \frac{24\mu r^2 P_a}{C_r^3 P_s^2 \rho_a}\rho\tilde{v}, \tau = \omega t, \Lambda_T = \frac{12\mu\omega r^2}{P_s C_r^2}$$

where L_t is the characteristic length of journal bearing, and Λ_T is the bearing number of the thrust bearing.

The tilt motion of the shaft is considered to have influence on the air film thickness of the thrust bearing while the influence of the eccentricity is neglected. The tilt motion at the thrust bearing is shown in Figure 4, and the distribution of the air film thickness at the thrust bearing can be expressed as

$$h_{(i,j)} = C_r - \sin(\beta) \sin\left(\theta_{(i,j)} - \gamma_{(i)} + \frac{\pi}{2}\right) \cdot \left(\frac{i}{M_t}(R_b - R_a) + R_a\right) + \Delta z \qquad (26)$$

where $\theta_{(i,j)}$ is the angle of air film at node i and j, $\gamma_{(i)}$ is the angle between the tile direction and the positive direction of x at the node i, i is the node number along the radial direction, R_a is the inner diameter of the thrust bearing, R_b is the out diameter of the thrust bearing, M_t is the number of thrust bearing nodes along the radial direction, Δz is the shaft displacement in the axial direction.

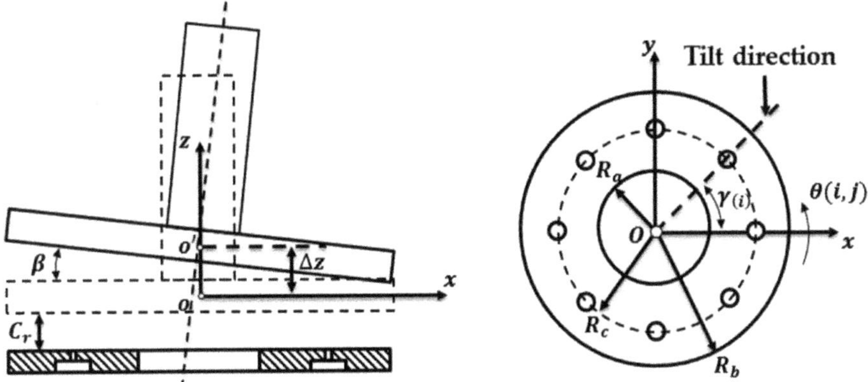

Figure 4. The tilt of the shaft at the thrust bearing.

As shown in Figure 5, the computational domain of the thrust bearing film is meshed into $\theta(0 : N_t)$ and $R_r(0 : M_t)$, the Periodic boundary condition, the Atmospheric boundary condition and the Mass flow boundary condition are also defined.

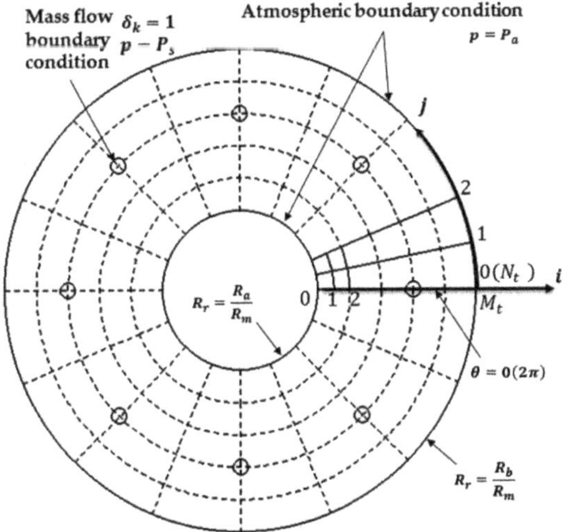

Figure 5. The computational domain of the thrust bearing.

Adopt the same treatment as the journal bearing, the ADI format of the equation obtained from thrust bearing can be expressed as

$$A_{(i,j)} S^{n+1}_{(i-1,j)} + B_{(i,j)} S^{n+1}_{(i,j)} + C_{(i,j)} S^{n+1}_{(i+1,j)} = D_{(i,j)} \tag{27}$$

$$A_{(i,j)} = -\frac{R_{r}{}^{2}_{(i,j)} \left(H^{n}_{(i,j)}\right)^{3}}{(\Delta R_{r})^{2}} + R_{r}{}^{2}_{(i,j)} 3\left(H^{n}_{(i,j)}\right)^{2} \frac{H^{n}_{(i+1,j)} - H^{n}_{(i-1,j)}}{(2\Delta R_{r})^{2}} + \frac{R_{r}{}_{(i,j)} \left(H^{n}_{(i,j)}\right)^{3}}{2\Delta R_{r}} \tag{28}$$

$$B_{(i,j)} = \frac{2R_{r}{}^{2}_{(i,j)} \left(H^{n}_{(i,j)}\right)^{3}}{(\Delta R_{r})^{2}} + \frac{\Lambda_{T} H^{n}_{(i,j)}}{P^{n}_{(i,j)} \Delta \tau} \tag{29}$$

$$C_{(i,j)} = -\frac{R_{r}{}^{2}_{(i,j)} \left(H^{n}_{(i,j)}\right)^{3}}{(\Delta R_{r})^{2}} - R_{r}{}^{2}_{(i,j)} 3\left(H^{n}_{(i,j)}\right)^{2} \frac{H^{n}_{(i+1,j)} - H^{n}_{(i-1,j)}}{(2\Delta R_{r})^{2}} - \frac{R_{r}{}_{(i,j)} \left(H^{n}_{(i,j)}\right)^{3}}{2\Delta R_{r}} \tag{30}$$

$$\begin{aligned} D_{(i,j)} = Q\delta_{k} &+ 3\left(H^{n}_{(i,j)}\right)^{2} \frac{\left(H^{n}_{(i,j+1)} - H^{n}_{(i,j-1)}\right)\left(S^{n}_{(i,j+1)} - S^{n}_{(i,j-1)}\right)}{(2\Delta\theta)^{2}} + \\ &\left(H^{n}_{(i,j)}\right)^{3} \frac{\left(S^{n}_{(i,j+1)} - 2S^{n}_{(i,j)} + S^{n}_{(i,j-1)}\right)}{(\Delta\theta)^{2}} - \frac{\Lambda_{T} H^{n}_{(i,j)}}{2P^{n}_{(i,j)}} \frac{\left(S^{n}_{(i,j+1)} - S^{n}_{(i,j-1)}\right)}{2\Delta\theta} \\ &-\Lambda_{T} P^{n}_{(i,j)} \frac{\left(H^{n}_{(i,j+1)} - H^{n}_{(i,j-1)}\right)}{2\Delta\theta} - \frac{\Lambda_{T} H^{n}_{(i,j)} S^{n}_{(i,j)}}{P^{n}_{(i,j)} \Delta\tau} - 2\Lambda_{T} P^{n}_{(i,j)} \frac{\left(H^{n+1}_{(i,j+1)} - H^{n}_{(i,j-1)}\right)}{\Delta\tau} \end{aligned} \tag{31}$$

$$E_{(i,j)} S^{n+2}_{(i,j-1)} + F_{(i,j)} S^{n+2}_{(i,j)} + G_{(i,j)} S^{n+2}_{(i,j+1)} = I_{(i,j)} \tag{32}$$

$$E_{(i,j)} = -\frac{\left(H^{n+1}_{(i,j)}\right)^{3}}{(\Delta\theta)^{2}} - \frac{\Lambda_{T} H^{n+1}_{(i,j)}}{4P^{n+1}_{(i,j)} \Delta\theta} + 3\left(H^{n+1}_{(i,j)}\right)^{2} \frac{H^{n+1}_{(i,j+1)} - H^{n+1}_{(i,j-1)}}{(2\Delta\theta)^{2}} \tag{33}$$

$$F_{(i,j)} = \frac{2\left(H^{n+1}_{(i,j)}\right)^{3}}{(\Delta\theta)^{2}} + \frac{\Lambda_{T} H^{n+1}_{(i,j)}}{P^{n+1}_{(i,j)} \Delta\tau} \tag{34}$$

$$G_{(i,j)} = -\frac{\left(H^{n+1}_{(i,j)}\right)^{3}}{(\Delta\theta)^{2}} + \frac{\Lambda_{T} H^{n+1}_{(i,j)}}{4P^{n+1}_{(i,j)} \Delta\theta} - \frac{R_{r}{}_{(i,j)} \left(H^{n+1}_{(i,j)}\right)^{3}}{2\Delta\theta} - 3\left(H^{n+1}_{(i,j)}\right)^{2} \frac{H^{n+1}_{(i,j+1)} - H^{n+1}_{(i,j-1)}}{(2\Delta\theta)^{2}} \tag{35}$$

$$\begin{aligned} I_{(i,j)} = Q\delta_{k} &+ 3R_{r}{}^{2}_{(i,j)} \left(H^{n+1}_{(i,j)}\right)^{2} \frac{\left(H^{n+1}_{(i,j+1)} - H^{n+1}_{(i,j-1)}\right)\left(S^{n+1}_{(i,j+1)} - S^{n+1}_{(i,j-1)}\right)}{(2\Delta R_{r})^{2}} + \\ R_{r}{}^{2}_{(i,j)} &\left(H^{n+1}_{(i,j)}\right)^{3} \frac{\left(S^{n+1}_{(i+1,j)} - 2S^{n+1}_{(i,j)} + S^{n+1}_{(i-1,j)}\right)}{(\Delta R_{r})^{2}} - \Lambda_{T} P^{n+1}_{(i,j)} \frac{H^{n+1}_{(i,j+1)} - H^{n+1}_{(i,j-1)}}{2\Delta\theta} - \\ 2\Lambda_{T} &P^{n+1}_{(i,j)} \frac{H^{n+2}_{(i,j)} - H^{n+1}_{(i,j)}}{\Delta\tau} + \frac{\Lambda_{T} H^{n+1}_{(i,j)} S^{n+1}_{(i,j)}}{P^{n+1}_{(i,j)} \Delta\tau} \end{aligned} \tag{36}$$

By solving the Equation (27) and (32) using Thomas method, the pressure distribution of the thrust bearing can be obtained. The air film force (i.e., the restoring force of the thrust bearing) of the thrust bearing in z direction is given by

$$f_{btz} = P_{s} L_{t}^{2} \int_{0}^{\frac{R_{b}-R_{a}}{L_{t}}} \int_{0}^{2\pi} P(R_{r}, \theta, \tau) \cos(\beta) R_{r} d\theta dR_{r} \tag{37}$$

And the torque of the air film force (i.e., the restoring torque of the thrust bearing) of the thrust bearing respect to the x axis and y axis is given by

$$t_{btx} = P_s L_t^2 \left(R_a + \frac{i(R_b - R_a)}{M_r} \right) \int_0^{\frac{R_b - R_a}{L_t}} \int_0^{2\pi} P(R_r, \theta, \tau) \cos(\beta) \sin\left(\alpha_{(i)}\right) R_r d\theta dR_r \qquad (38)$$

$$t_{bty} = P_s L_t^2 \left(R_a + \frac{i(R_b - R_a)}{M_r} \right) \int_0^{\frac{R_b - R_a}{L_t}} \int_0^{2\pi} P(R_r, \theta, \tau) \cos(\beta) \cos\left(\alpha_{(i)}\right) R_r d\theta dR_r \qquad (39)$$

2.3. Modeling of PMSM

Ideally, the magnetic force of PMSM is symmetric. However, the UMF will come out when the rotor eccentricity happens or the structure of the PMSM is not symmetric. Kawase et al. [31] used 3-D finite element method to analyze the UMF of the PMSM, and it is found that the axial component of the UMF is relatively small compared to other two components. Here, we assume the structure of the PMSM is symmetric and the axial component of the UMF is negligible. According to Maxwell stress tensor method, the 2D magnetic forces in the radial and tangential direction can be expressed as follows [32].

$$f_r = \int_0^{2\pi} \frac{1}{2\mu_0} \left(B_r^2 - B_\theta^2 \right) r d\theta \qquad (40)$$

$$f_\theta = \int_0^{2\pi} \frac{B_r B_\theta}{\mu_0} r d\theta \qquad (41)$$

where B_r is the radial flux density, B_θ is the tangential flux density, μ_0 is the permeability of the air.

Figure 6 shows the schematic structure of motor with eccentric rotor [4], and the 2-D magnetic force in cartesian coordinates can be expressed as

$$f_{mx(i)} = f_{r(i)} \cos\left(\alpha_{(i)}\right) + f_\theta \sin\left(\alpha_{(i)}\right) \qquad (42)$$

$$f_{mx(i)} = f_{r(i)} \sin\left(\alpha_{(i)}\right) + f_{\theta(i)} \cos\left(\alpha_{(i)}\right) \qquad (43)$$

To simulate the 3-D state of the PMSM, the multiple slice method is employed. As shown in Figure 7, the magnetic force of the 3-D PMSM can be expressed as

$$f_{mx} = \sum_{i=i_0}^{i=i_m} f_{mx(i)} \qquad (44)$$

$$f_{my} = \sum_{i=i_0}^{i=i_m} f_{my(i)} \qquad (45)$$

And the magnetic torque of the 3-D PMSM is given by

$$t_{mx} = \sum_{i=i_0}^{i=i_m} f_{my(i)} \frac{i - i_b}{M_s} L_s \qquad (46)$$

$$t_{my} = \sum_{i=i_0}^{i=i_m} f_{mx(i)} \frac{i - i_b}{M_s} L_s \qquad (47)$$

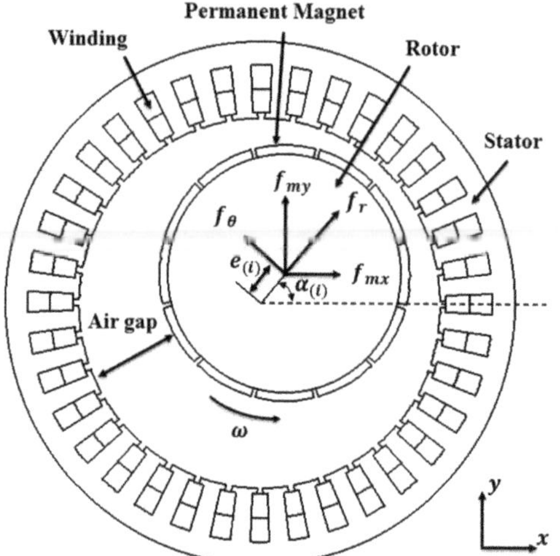

Figure 6. Schematic structure of the motor [4].

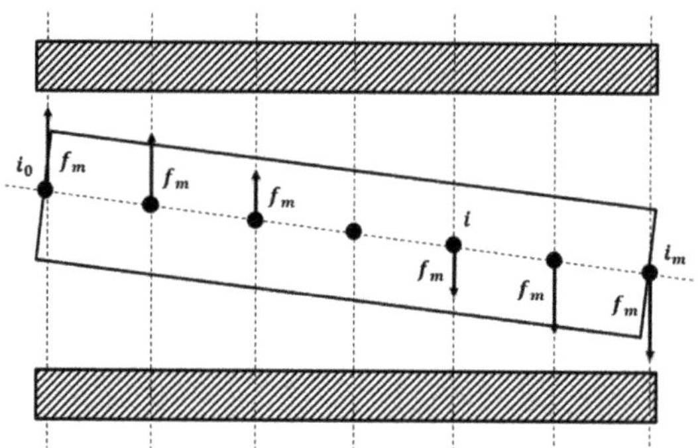

Figure 7. The simplified PMSM model using multiple slice method.

2.4. Dynamics Modeling of ABMS

The spindle shaft is regarded as a rigid rotor in the model, and the spindle shaft is considered to move with 5-DOF. The dimensionless acceleration, velocity and the attitude in the space can be calculated by

$$\left\{\ddot{\delta}\right\} = \frac{d^2\{\delta\}}{d\tau^2} = \{A\} \tag{48}$$

$$\left\{\dot{\delta}\right\} = \left\{\dot{\delta}_0\right\} + \left\{\ddot{\delta}\right\}\Delta\tau \tag{49}$$

$$\{\delta\} = \{\delta_0\} + \left\{\dot{\delta}\right\}\Delta\tau + \frac{1}{2}\left\{\ddot{\delta}\right\}(\Delta\tau)^2 \tag{50}$$

$$\{A\} = \left\{ \begin{array}{c} \frac{F_{eX}-F_{bX}+F_{mX}}{M} \\ \frac{F_{eY}-F_{bY}+F_{mY}}{M} \\ \frac{F_{eZ}-F_{bZ}+F_{mZ}}{M} \\ \frac{T_{eX}-T_{bX}+T_{mX}+(I_z-I_x)\dot{\Theta}_Y\dot{\Theta}_Z}{I_X} \\ \frac{T_{eY}-T_{bY}+T_{mY}+(I_y-I_z)\dot{\Theta}_Z\dot{\Theta}_X}{I_Y} \end{array} \right\} \tag{51}$$

$$\{\delta\} = \left\{ \begin{array}{c} X \\ Y \\ Z \\ \Theta_X \\ \Theta_Y \end{array} \right\} \tag{52}$$

where the dimensionless parameters are given as follows.

$$X = \frac{x}{C_r}, Y = \frac{y}{C_r}, Z = \frac{z}{C_r}, F_{ex} = \frac{f_{ex}}{P_s L_r^2}, F_{ey} = \frac{f_{ey}}{P_s L_r^2}, F_{ex} = \frac{f_{ez}}{P_s L_r^2}, F_{bx} = \frac{f_{bx}}{P_s L_r^2},$$
$$F_{by} = \frac{f_{by}}{P_s L_r^2}, F_{bz} = \frac{f_{bz}}{P_s L_r^2}, F_{mx} = \frac{f_{mx}}{P_s L_r^2}, F_{my} = \frac{f_{my}}{P_s L_r^2}, T_{ex} = \frac{t_{ex}}{P_s L_r^2 L_s}, T_{ey} = \frac{t_{ey}}{P_s L_r^2 L_s},$$
$$T_{bx} = \frac{t_{bjx} + t_{btx}}{P_s L_r^2 L_s}, T_{by} = \frac{t_{bjy} + t_{bty}}{P_s L_r^2 L_s}, T_{mx} = \frac{t_{mx}}{P_s L_r^2 L_s}, T_{my} = \frac{t_{my}}{P_s L_r^2 L_s},$$
$$M = \frac{mC_r \omega^2}{P_s L_r^2}, I_x = \frac{i_x \omega^2}{P_s L_r^2 L_s}, I_y = \frac{i_y \omega^2}{P_s L_r^2 L_s}, I_z = \frac{i_z \omega^2}{P_s L_r^2 L_s}$$

where F_{ex}, F_{ey}, F_{ez} are the dimensionless external force applied on the spindle shaft in x, y, z direction, F_{bx}, F_{by}, F_{bz} are the dimensionless air film force in x, y, z direction, F_{mx}, F_{my} are the dimensionless UMF in x, y direction, T_{ex}, T_{ey} are the dimensionless external torque with respect to x, y axis, T_{bx}, T_{by} are the dimensionless air film torque with respect to x, y axis, T_{mx}, T_{my} are the dimensionless magnetic torque with respect to x, y axis, M is the dimensionless mass of the spindle shaft, I_x, I_y, I_z are the rotational inertia of the spindle shaft with respect to x, y, z axis, X, Y, Z are the dimensionless displacement in x, y, z direction, Θ_X, Θ_Y are the rotation angle of the spindle shaft with respect to x, y axis.

The spindle shaft is still at the initial condition, thus $\left\{ \dot{\delta} \right\} = \{0\}$ at the initial time. The flow chat for calculating the 5-DOF dynamics model of the ABMS is shown in Figure 8.

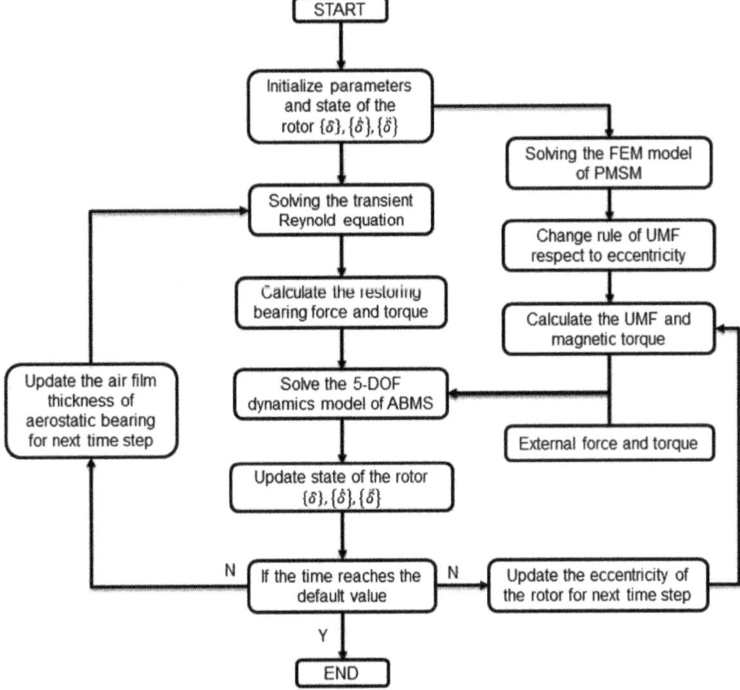

Figure 8. Flow chat for calculating the 5-DOF dynamics model of ABMS.

3. Numerical Simulation

3.1. Detailed Parameter

A case study is conducted to verify the effectiveness of the dynamics model of the ABMS. Table 1 lists the detailed parameters of the simulated model.

Table 1. The parameter of simulated ABMS.

Parameter (Unit)	Data
Length of shaft (m)	0.4
Diameter of journal bearing (m)	0.1
Length of journal bearing (m)	0.1
Inner diameter of thrust bearing (m)	0.1
Outer diameter of thrust bearing (m)	0.18
Bearing clearance (m)	1×10^{-5}
Diameter of orifice (m)	0.0002
Row number of Orifice on journal bearing	2
Row number of Orifice on thrust bearing	1
Orifice number of each row	8
Ambient pressure (Pa)	101325
Supply pressure (Pa)	405300
Density of air (kg/m^3)	1.204
Kinetic viscosity of air (Pa·s)	1.82×10^{-5}
Specific heat ratio of air	1.401
Flow coefficient	0.8
Outer Rotor diameter (m)	0.1
Outer stator diameter (m)	0.15
Inner rotor diameter (m)	0.106
Number of slots	36
Number of poles	12
Motor effective length (m)	0.08
Rotational speed (r/min)	3000
Winding form of motor	Three-phase of double layer winding
External force (N)	50 400

3.2. Numerical Result

The model selects the same motor parameter as that in reference [4], and the variation trend of UMF has been given as follows.

$$f_{mx} = \begin{cases} \left[\left(-4.4\varepsilon^2 + 7.4\varepsilon + 0.3\right) \cdot \sin\left(\frac{2\pi\omega}{60}t\right) + \left(9.4\varepsilon^2 - 3.1\varepsilon + 3.8\right)\right] \cdot \sin\theta_r & (\varepsilon \leq 0.5) \\ \left[\left(-4.4\varepsilon^2 + 7.4\varepsilon + 0.3\right) \cdot \sin\left(\frac{2\pi\omega}{60}t\right) + (7.0\varepsilon + 1.1)\right] \cdot \sin\theta_r & (\varepsilon > 0.5) \end{cases} \tag{53}$$

$$f_{my} = \begin{cases} \left[\left(-4.4\varepsilon^2 + 7.4\varepsilon + 0.3\right) \cdot \sin\left(\frac{2\pi\omega}{60}t\right) + \left(9.4\varepsilon^2 - 3.1\varepsilon + 3.8\right)\right] \cdot \cos\theta_r & (\varepsilon \leq 0.5) \\ \left[\left(-4.4\varepsilon^2 + 7.4\varepsilon + 0.3\right) \cdot \sin\left(\frac{2\pi\omega}{60}t\right) + (7.0\varepsilon + 1.1)\right] \cdot \cos\theta_r & (\varepsilon > 0.5) \end{cases} \tag{54}$$

$$\theta_r = \left(9.3\varepsilon^3 - 13.4\varepsilon^2 + 1.8\varepsilon + 2.8\right) \cdot \sin\left(\frac{2\pi\omega}{60}t\right) \tag{55}$$

Two cases are calculated, i.e., the dynamics model with 400 N and 50 N external force. In the first case, 400 N external force is applied on the shaft end in x direction, the default dimensionless time is set to be 100. Figure 9a shows the air pressure distribution of the aerostatic journal bearing, Figure 9b and c show the air pressure distribution of the front and rear aerostatic journal bearing respectively, And Figure 9d shows the integrated air pressure distribution vision of aerostatic bearing with 400 N external force at dimensionless time 100.

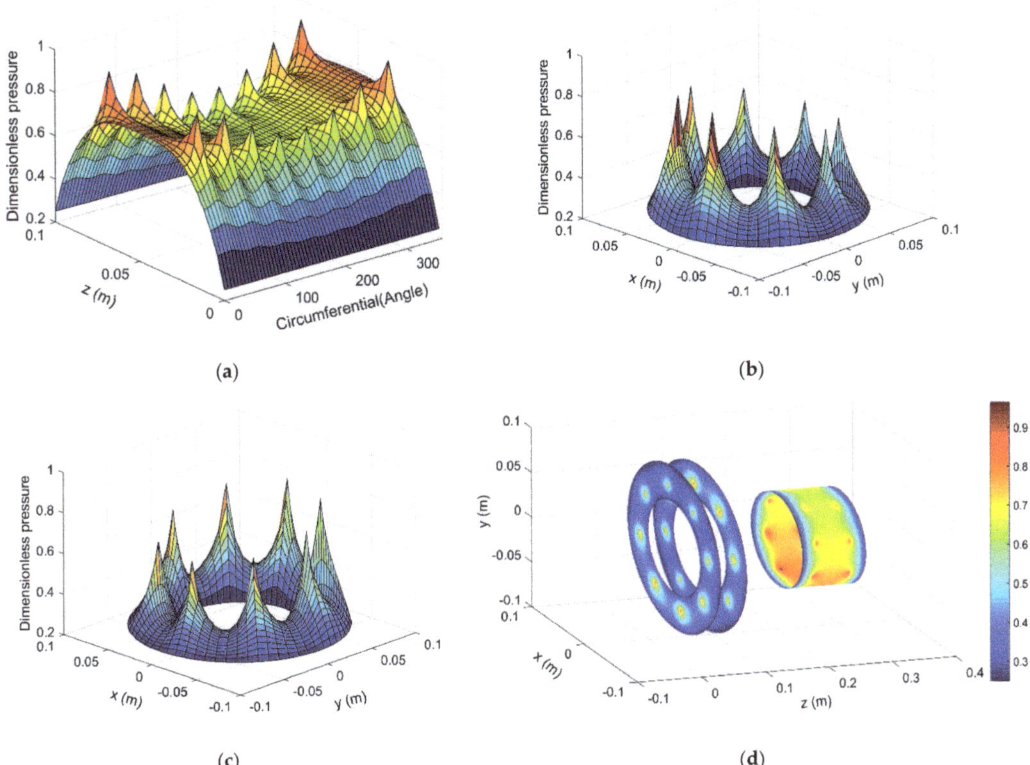

Figure 9. Pressure distribution of the aerostatic bearings with 400 N external force: (**a**) Journal bearing (**b**) Front thrust bearing (**c**) Rear thrust bearing (**d**) Integrated vision.

The result shows that when applies the external load on the shaft end, the spindle shaft can have a certain tilt angle and it affects the air pressure distribution of the aerostatic bearings obviously, which means the restoring force and torque of the air film applied on the spindle shaft are also changed. The tilt angle of the spindle shaft versus x axis and y axis is shown in Figure 10a,b, the tilt angle is converged and varies periodically with a certain amplitude, the extent of the tilt angle versus y axis is larger than the tilt angle versus x axis, this is mainly because the external force is applied in the x direction, thus the external torque is applied on y axis.

Figure 11a,b shows the motion trajectory of the shaft barycenter and the shaft end respectively. The result shows that the shaft acts stable with 400 N external force, its trajectory in x-y plane converged to a certain region. From a spatial perspective, the shaft end has left its initial position towards its balance position as shown in Figure 11c, and the displacement of the shaft end in z direction is shown in Figure 11d. The displacement in z direction is also converged and it varies periodically versus time.

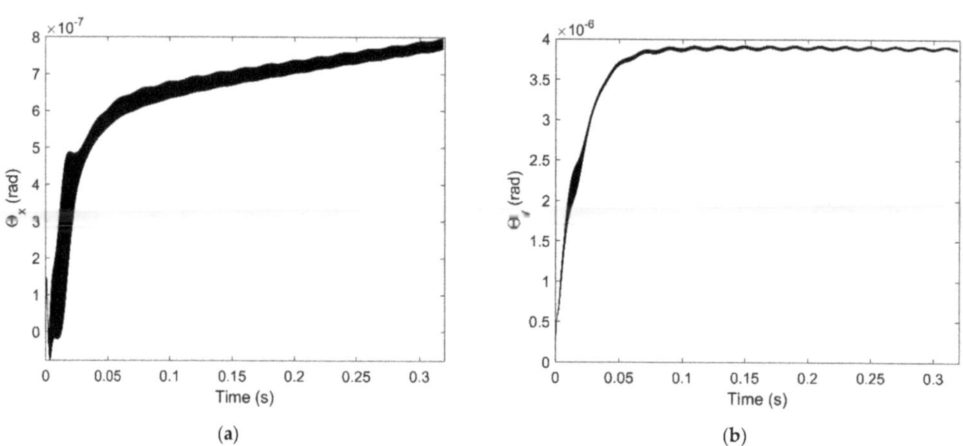

Figure 10. Tilt angle of the spindle shaft with 400 N external force: (**a**) Tilt angle versus x-axis (**b**) Tilt angle versus y-axis.

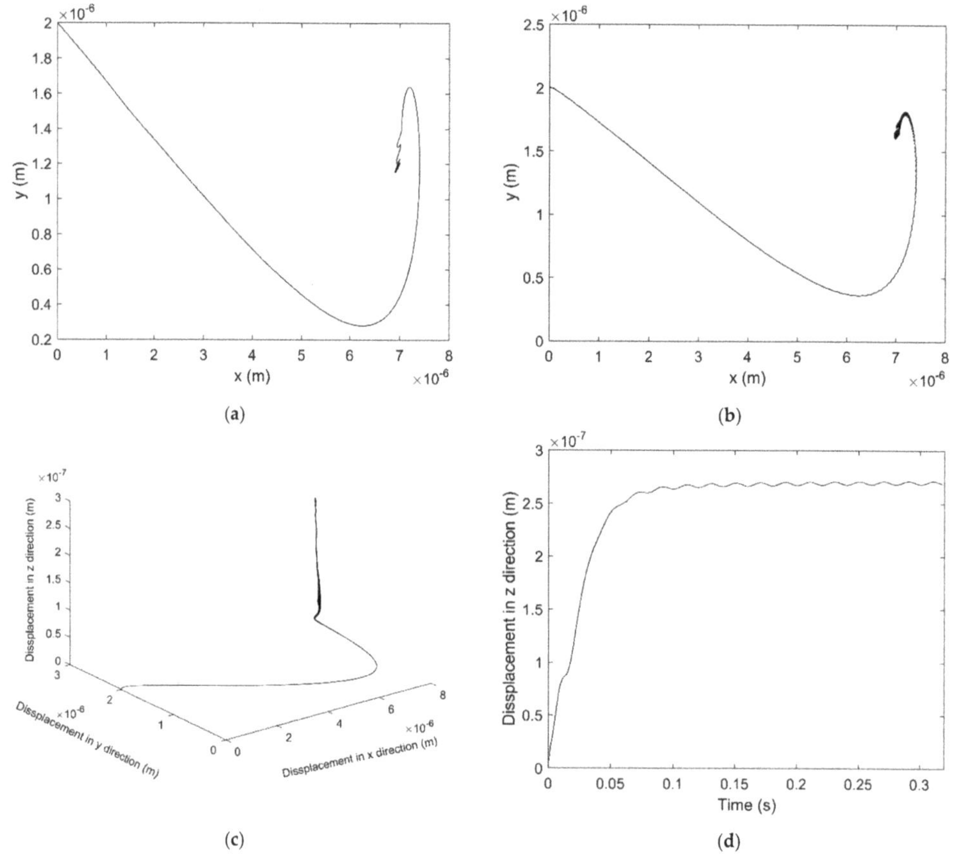

Figure 11. Motion trajectory of shaft with 400 N external force: (**a**) Position at the shaft barycenter in x-y plane (**b**) Position at the shaft end in x-y plane (**c**) Position at the shaft end in x-y-z coordinate (**d**) Position at the shaft end in z direction.

In the second case, 50 N external force is applied on the shaft end in x direction. The result is quite different with the first case, Figure 12a,b shows the tilt angle of the shaft versus x axis and y axis which are both diverged, which means the shaft is unstable under the current condition.

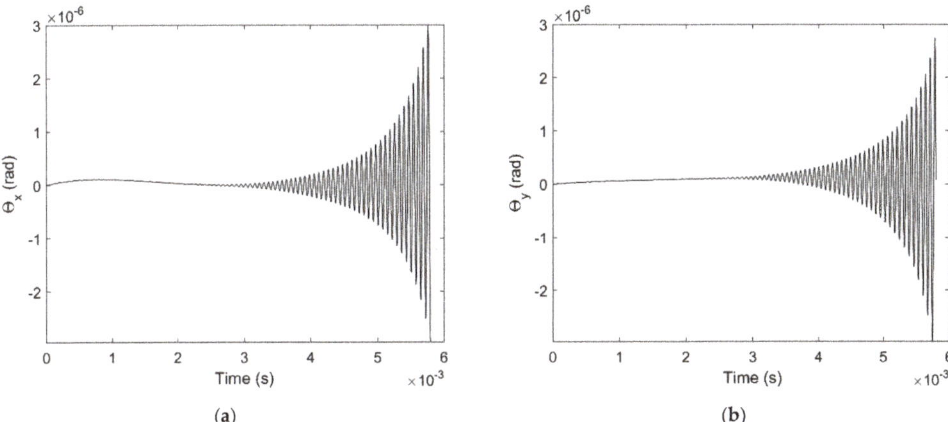

(a)

(b)

Figure 12. Tilt angle of the shaft with 50N external force: (**a**) Tilt angle versus x-axis (**b**) Tilt angle versus y-axis.

Figure 13a shows the corresponding air pressure distribution of the aerostatic journal bearing, Figure 13b,c show the air pressure distribution of the front and rear aerostatic journal bearing respectively, Figure 13d shows the integrated air film pressure distribution vision of aero-static bearing with 50 N external force at dimensionless time 100. According to the result, the air film pressure distribution of the aerostatic bearing becomes quite uneven, both the aerostatic journal bearing and the aerostatic thrust bearing show severe aerodynamic effect.

Figure 14a,b shows the motion trajectory of the shaft barycenter and the shaft end respectively. The result shows that the trajectory of shaft barycenter does not show the divergent trend while the trajectory of shaft end does, which means that the spindle shaft is at conical unstable state. It can be observed clearer in a spatial perspective as shown in Figure 14c, the trajectory of the shaft end moves as a spiral path. The displacement of the shaft end in the axial direction is also divergent as shown in Figure 14d.

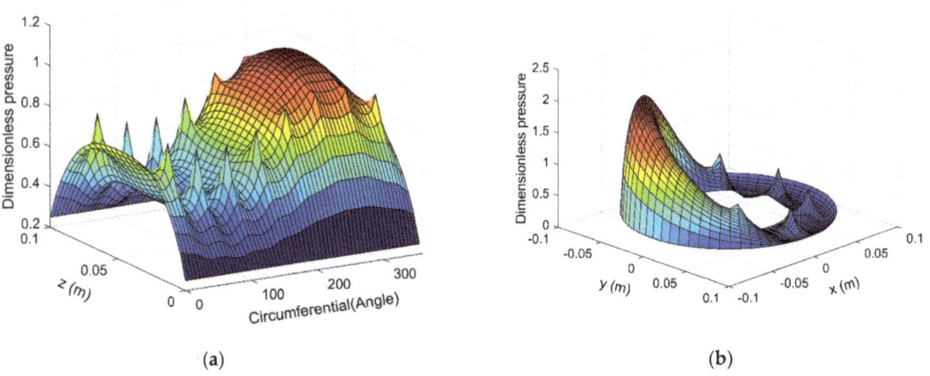

(a)

(b)

Figure 13. *Cont.*

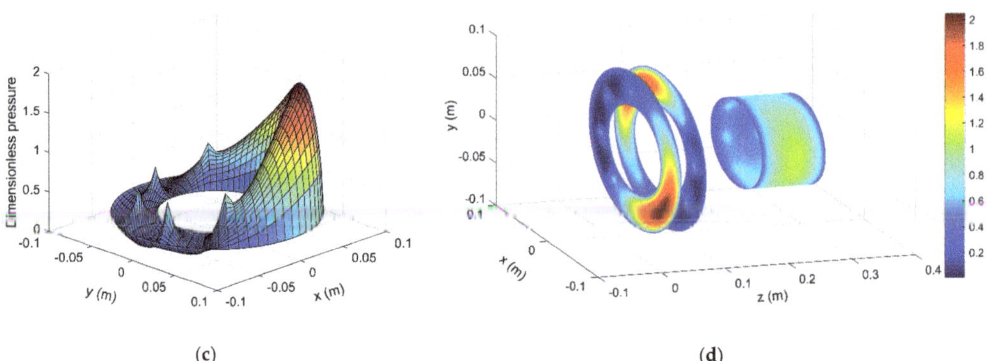

Figure 13. Pressure distribution of the aerostatic bearings with 50 N external force: (**a**) Journal bearing (**b**) Front thrust bearing (**c**) Rear thrust bearing (**d**) Integrated vision.

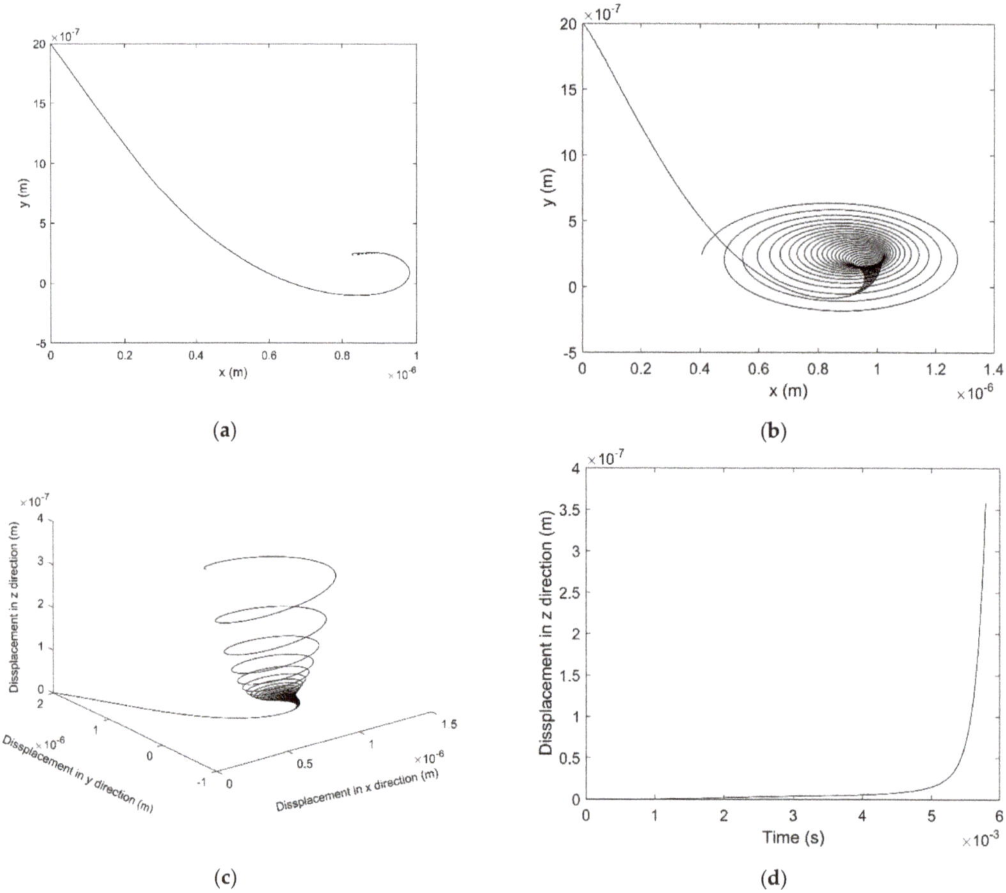

Figure 14. Motion trajectory of the aerostatic bearings with 50 N external force: (**a**) Journal bearing (**b**) Front thrust bearing (**c**) Rear thrust bearing (**d**) Integrated vision.

4. Discussion

In the motion process of the spindle shaft, the external force applied on the shaft end produce the torque on the shaft, resulting in the tilt motion of the shaft, then the distribution of the motor flux and the air film pressure is changed. Under the multi-field coupling effect of the external force, the air film force and the magnetic force, the shaft may finally stabilize in a certain spatial attitude or become unstable and even hit the bearing sleeve.

In this paper, a 5-DOF dynamics model of the aerostatic spindle was implemented using the restoring force method, while previous literatures mainly used the dynamic coefficient method [23,33]. The dynamic coefficient method improves the calculation speed compared to the restoring force method, but the dynamic behavior of the shaft can be obtained directly by employing the restoring force method. The proposed 5-DOF dynamic model of aerostatic spindle has considered the influence of the tilt motion and axial displacement of the shaft, which is neglected in the 2-DOF model. In the 2-DOF model of aerostatic spindle only the external force, the restoring force of the air film and the unbalanced magnetic force of the motor are considered, while the external torque, the restoring torque of the air film and the magnetic torque of the motor are not considered. However, they are all considered and integrated in the 5-DOF model. The motion trajectory of the shaft center in the 2D plane characterizes the dynamic behavior of the shaft, but it cannot show the conical motion of the shaft with the external force and external torque, which is quite important because the shaft has the spatial motion with 5-DOF.

In the simulation result, the dynamic behavior of the spindle shaft is stable with 400 N force and unstable with 50 N force. Similar phenomenon has been observed in the research of journal bearings, such as the research done by Khonsari et al. [34]. The stability boundary of the journal bearing varies with the change of the Sommerfeld Number, and the load on the bearing is one of the parameters that determine the value of the Sommerfeld Number. The stability boundary expands with the increase of the external load. To our knowledge, we think the reason behind dynamic behavior is probably the same as that of the journal bearings. That is, when the external load increase, the initial position locates in the unstable region under the low external load may locate in the stable region due to the expanding of the stability boundary. The Sommerfeld Number is also related to the rotation velocity [34]. Thus, the rotation velocity also affects the stability of the shaft, which is probably because that the change of the rotation velocity results in different aerodynamic effect and affects the stability. The critical value for distinguishing the stability of the shaft can be obtained in the 2-DOF model as the works done by Yang et al. [24]. It is sure that the critical value for the conical stability of the spindle shaft in the 5-DOF model also exists. It is meaningful to study the stability criterion of the shaft with 5-DOF. However, it will be much difficult to obtain the critical value in the 5-DOF model compared to that in the 2-DOF model. The 5-DOF model is much time costly to solve due to its high complexity and difficulty, thus, it will be further explored in the future research.

The proposed model can be improved in somehow. The shaft is regarded is a rigid body rather than the flexible body, and the structural deformation is neglected, however, it may have potential influence on the dynamic behavior of the shaft. Laha et al. [35] investigated the stability of the rotor supported by air journal bearing considering the rotor flexibility, in which the object was 2-DOF model. If the shaft is considered as a flexible rotor in a 5-DOF model, extra calculations need to be done, and the improvement of the solving code is also required. Due to the limitation of the current experimental condition, the validation of the theoretical results will be conducted in the future work.

5. Conclusions

In this study, a 5-DOF dynamics model of the aerostatic spindle is established. The modeling method and the numerical solution is given. Through the numerical model, the transient pressure distribution of the air film as well as the motion trajectory of the spindle shaft is obtained as expected. The result shows the significant meaning of considering the influence of the shaft tilt on the air film pressure distribution, which further determines the restoring force and torque of the aerostatic bearing. Besides, the different conical stability behavior of the spindle shaft is shown under different external load through the model.

Author Contributions: Conceptualization, G.C.; methodology, G.C. and Y.C.; Software, Y.C.; data analysis, Y.C.; original draft writing, Y.C.; review and editing G.C. and Y.C.; supervision, G.C. All authors have read and agreed to the published version of the manuscript.

Funding: This work was supported by the National Natural Science Foundation of China (Grant No. 51705462), the Fundamental Research Funds for the Provincial Universities of Zhejiang (Grant No. RF-A2020005) and the Talent Project of Zhejiang Association for Science and Technology (Grant No. 2018YCGC016).

Informed Consent Statement: Not applicable.

Data Availability Statement: No new data were created or analyzed in this study. Data sharing is not applicable to this article.

Conflicts of Interest: The authors declare no conflict of interest.

References

1. Chen, G.; Ju, B.; Fang, H.; Chen, Y.; Yu, N.; Wan, Y. Air bearing: Academic insights and trend analysis. *Int. J. Adv. Manuf. Technol.* **2019**, *106*, 1191–1202. [CrossRef]
2. Chen, G.; Sun, Y.; Zhang, F.; An, C.; Chen, W.; Su, H. Influence of ultra-precision flycutting spindle error on surface frequency domain error formation. *Int. J. Adv. Manuf. Technol.* **2017**, *88*, 3233–3241. [CrossRef]
3. Gao, Q.; Chen, W.; Lu, L.; Huo, D.; Cheng, K. Aerostatic bearings design and analysis with the application to precision engineering: State-of-the-art and future perspectives. *Tribol. Int.* **2019**, *135*, 1–17. [CrossRef]
4. Chen, G.; Chen, Y.; Lu, Q.; Wu, Q.; Wang, M. Multi-physics fields based nonlinear dynamic behavior analysis of air bearing motorized spindle. *Micromachines* **2020**, *11*, 723. [CrossRef]
5. Chen, Y.; Chiu, C.; Cheng, Y. Influences of operational conditions and geometric parameters on the stiffness of aerostatic journal bearings. *Precis. Eng. J. Int. Soc. Precis. Eng. Nanotechnol.* **2010**, *34*, 722–734. [CrossRef]
6. Miyatake, M.; Yoshimoto, S. Numerical investigation of static and dynamic characteristics of aerostatic thrust bearings with small feed holes. *Tribol. Int.* **2010**, *43*, 1353–1359. [CrossRef]
7. Yang, D.-W.; Chen, C.-H.; Kang, Y.; Hwang, R.-M.; Shyr, S.-S. Influence of orifices on stability of rotor-aerostatic bearing system. *Tribol. Int.* **2009**, *42*, 1206–1219. [CrossRef]
8. Sun, Y.; Wu, Q.; Chen, W.; Luo, X.; Chen, G. Influence of unbalanced electromagnetic force and air supply pressure fluctuation in air bearing spindles on machining surface topography. *Int. J. Precis. Eng. Manuf.* **2021**, *22*, 1–12. [CrossRef]
9. Zhang, S.; Yu, J.; To, S.; Xiong, Z. A theoretical and experimental study of spindle imbalance induced forced vibration and its effect on surface generation in diamond turning. *Int. J. Mach. Tools Manuf.* **2018**, *133*, 61–71. [CrossRef]
10. Wu, Q.; Sun, Y.; Chen, W.; Chen, G.; Bai, Q.; Zhang, Q. Effect of motor rotor eccentricity on aerostatic spindle vibration in machining processes. *Proc. Inst. Mech. Eng. Part C J. Mech. Eng. Sci.* **2018**, *232*, 1331–1342. [CrossRef]
11. Wu, Q.; Sun, Y.; Chen, W.; Wang, Q.; Chen, G. Theoretical prediction and experimental verification of the unbalanced magnetic force in air bearing motor spindles. *Proc. Inst. Mech. Eng. Part B J. Eng. Manuf.* **2019**, *233*, 2330–2344. [CrossRef]
12. Zhang, J. Analysis on motion stability of a high-speed rotor-bearing system. *Chin. J. Mech. Eng.* **2005**, *18*, 220–223. [CrossRef]
13. Wang, C.-C.; Lee, T.-E. Nonlinear dynamic analysis of bi-directional porous aero-thrust bearing systems. *Adv. Mech. Eng.* **2017**, *9*, 1–11. [CrossRef]
14. Wang, C.-C. Application of a hybrid numerical method to the nonlinear dynamic analysis of a micro gas bearing system. *Nonlinear Dyn.* **2009**, *59*, 695–710. [CrossRef]
15. Wang, C.-C. Theoretical and nonlinear behavior analysis of a flexible rotor supported by a relative short herringbone-grooved gas journal-bearing system. *Phys. D Nonlinear Phenom.* **2008**, *237*, 2282–2295. [CrossRef]
16. Hei, D.; Lu, Y.; Zhang, Y.; Liu, F.; Zhou, C.; Müller, N. Nonlinear dynamic behaviors of rod fastening rotor-hydrodynamic journal bearing system. *Arch. Appl. Mech.* **2015**, *85*, 855–875. [CrossRef]
17. Wang, C.-C. Application of a hybrid method to the nonlinear dynamic analysis of a flexible rotor supported by a spherical gas-lubricated bearing system. *Nonlinear Anal. Theory, Methods Appl.* **2009**, *70*, 2035–2053. [CrossRef]
18. Zhang, G.-H.; Sun, Y.; Liu, Z.-S.; Zhang, M.; Yan, J.-J. Dynamic characteristics of self-acting gas bearing–flexible rotor coupling system based on the forecasting orbit method. *Nonlinear Dyn.* **2011**, *69*, 341–355. [CrossRef]
19. Zhang, S.; To, S.; Wang, H. A theoretical and experimental investigation into five-DOF dynamic characteristics of an aerostatic bearing spindle in ultra-precision diamond turning. *Int. J. Mach. Tools Manuf.* **2013**, *71*, 1–10. [CrossRef]
20. Li, J.; Liu, P. Dynamic analysis of 5-DOFs aerostatic spindles considering tilting motion with varying stiffness and damping of thrust bearings. *J. Mech. Sci. Technol.* **2019**, *33*, 5199–5207. [CrossRef]
21. Li, J.; Huang, M.; Liu, P. Analysis and experimental verification of dynamic characteristics of air spindle considering varying stiffness and damping of radial bearings. *Int. J. Adv. Manuf. Technol.* **2019**, *104*, 2939–2950. [CrossRef]

22. Jiang, S.; Xu, C. Dynamics characteristics of a rotary table motorized spindle with externally pressurized air bearings. *J. Vibroengineering* **2017**, *19*, 801–811. [CrossRef]
23. Xu, C.; Jiang, S. Dynamic analysis of a motorized spindle with externally pressurized air bearings. *J. Vib. Acoust.* **2015**, *137*, 041001. [CrossRef]
24. Yang, P.; Zhu, K.-Q.; Wang, X.-L. On the non-linear stability of self-acting gas journal bearings. *Tribol. Int.* **2009**, *42*, 71–76. [CrossRef]
25. Wu, Y.; Feng, K.; Zhang, Y.; Liu, W.; Li, W. Nonlinear dynamic analysis of a rotor-bearing system with porous tilting pad bearing support. *Nonlinear Dyn.* **2018**, *94*, 1391–1408. [CrossRef]
26. Al-Bender, F. On the modelling of the dynamic characteristics of aerostatic bearing films: From stability analysis to active compensation. *Precis. Eng. J. Int. Soc. Precis. Eng. Nanotechnol.* **2009**, *33*, 117–126. [CrossRef]
27. Franssen, R.; Potze, W.; De Jong, P.; Fey, R.; Nijmeijer, H. Large amplitude dynamic behavior of thrust air bearings: Modeling and experiments. *Tribol. Int.* **2017**, *109*, 460–466. [CrossRef]
28. Lo, C.-Y.; Wang, C.-C.; Lee, Y.-H. Performance analysis of high-speed spindle aerostatic bearings. *Tribol. Int.* **2005**, *38*, 5–14. [CrossRef]
29. Powell, J.W. Theory of aerostatic lubrication. In *Design of Aerostatic Bearings*; The Machinery Publishing Co., Ltd.: Brighton, UK, 1970; pp. 35–66.
30. Rowe, W.B. Basic flow theory. In *Hydrostatic, Aerostatic and Hybrid Bearing Design*; Elsevier: Amsterdam, The Netherlands, 2012; pp. 25–48.
31. Kawase, Y.; Mimura, N.; Ida, K. 3-D electromagnetic force analysis of effects of off-center of rotor in interior permanent magnet synchronous motor. *IEEE Trans. Magn.* **2000**, *36*, 1858–1862. [CrossRef]
32. Meessen, K.K.; Paulides, J.J.; Lomonova, E.E. Force calculations in 3-D cylindrical structures using fourier analysis and the Maxwell Stress Tensor. *IEEE Trans. Magn.* **2012**, *49*, 536–545. [CrossRef]
33. Shi, J.; Cao, H.; Maroju, N.K.; Jin, X. Dynamic modeling of aerostatic spindle with shaft tilt deformation. *J. Manuf. Sci. Eng. Trans. ASME* **2019**, *142*, 1–43. [CrossRef]
34. Khonsari, M.M.; Chang, Y.J. Stability boundary of non-linear orbits within clearance circle of journal bearings. *J. Vib. Acoust.* **1993**, *115*, 303–307. [CrossRef]
35. Laha, S.K.; Banjare, H.; Kakoty, S.K. Stability analysis of a flexible rotor supported on finite hydrodynamic porous journal bearing using a non-linear transient method. *Proc. Inst. Mech. Eng. Part J J. Eng. Tribol.* **2008**, *222*, 963–973. [CrossRef]

MDPI AG
Grosspeteranlage 5
4052 Basel
Switzerland
Tel.: +41 61 683 77 34

Micromachines Editorial Office
E-mail: micromachines@mdpi.com
www.mdpi.com/journal/micromachines